Hans Pornschlegel (Editor)

Research and Development in Work and Technology

Proceedings of a European Workshop Dortmund, Germany, 23-25 October 1990

With a Preface by Dr. Ulrich Däunert,
Director of the Project Management Teams
Work, Environment and Health
part of the German Aerospace Research Establishment
(Deutsche Forschungsanstalt für Luft- und Raumfahrt, DLR)

With 69 Figures

Physica-Verlag
A Springer-Verlag Company

Professor Hans Pornschlegel
Sozialakademie Dortmund
Hohe Straße 141
D-4600 Dortmund 1, FRG

Published on behalf of the German Aerospace Research Establishment (Deutsche Forschungsanstalt für Luft- und Raumfahrt), Project Management "Work and Technology"

ISBN-13: 978-3-7908-0629-8 e-ISBN-13: 978-3-642-85278-7
DOI: 10.1007/978-3-642-85278-7

Die Deutsche Bibliothek - CIP-Einheitsaufnahme
Research and development in work and technology : proceedings of an European workshop, Dortmund, Germany, 23-25 October 1990 / Hans Pornschlegel (ed.). - Heidelberg : Physica-Verl., 1992
NE: Pornschlegel, Hans [Hrsg.]

Softcover reprint of the hardcover 1st edition 1992

7100/7130-543210 - Printed on acid-free paper

Preface by the Editor

After nearly 15 years of the German governmental Research and Development Programmes "Humanisation of Working Life", since 1989 "Work and Technology", the need to organize a "forum" for exchanges of experience between other European and national programmes has become urgent. Since the German programme spells out the need for mutual European and international information, coordination and cooperation in certain fields, a workshop was organized jointly by the Swedish Work Environment Fund (Arbetsmiljöfonden), the Federal Institute for Industrial Health and Safety (Bundesanstalt für Arbeitsschutz) and the Project Administration for Work and Technology (Projektträger Arbeit und Technik) on the initiative of the Sozialakademie Dortmund.

A European Workshop "Research and Development Strategies in the Field of Work and Technology" documented in this volume took place from 23rd to 25th October 1990 in Dortmund at the Federal Institute for Industrial Health and Safety with the participation of more than 100 representatives from 9 countries and from international as well as European institutions.

This volume presents the contributions given at the workshop, in some cases in a slightly expanded and revised form. Although the workshop was intended mainly as a forum to discuss research and development strategies, the manifold contributions show an impressive range of approaches, examples and results, and also of problems arising within this field. It also documents the implementation of approaches in different countries towards tackling the basic questions.

The editor, together with the organizers, expresses his hope and expectation that the present volume will contribute for some time to come to discussion within an interdisciplinary field of theoretical, empirical and practical work, which is becoming more and more European and international in its scope and perspectives.

The contributions remain fully within the respective authors' responsibility. The editor has tried to make the papers easier to read by marking their main points in italics.

Thanks is due to the Federal Minister for Technology and Research who supported the conference, to the co-organizers, the Swedish Work Environment Fund, Stockholm, the Federal Institute for Industrial Health and Safety, Dortmund, the European Foundation for the Improvement of Living and Working Conditions, Dublin, and the Project Administration for Work and Technology, Bonn, for their excellent organization of the occasion and of the follow-up; thanks also to all those concerned with the work on this volume, especially Mrs. G. Borgmann, Mr. R. Möhlmann, Mr. E.O. Nölle who cooperated amicably on the editorial and the technical side, Mr. Harrington as translator and Mr. S. Rissmann, responsible for layout.

Dortmund, April 1992

Preface by Dr. Ulrich Däunert

Director of the Project Management Teams Work, Environment, Health part of the German Aerospace Research Establishment (Deutsche Forschungsanstalt für Luft- und Raumfahrt, DLR)

At the beginning of the Seventies programmes for the improvement of working conditions were initiated in some European countries but although the liberalisation of the European market will become a reality as of 1993 and the creation of the European Social Area is on the agenda the cooperation of relevant national programmes is still insufficiently developed.

The subject of working conditions in Europe is extremely complex and difficult to generalize on the various members of the European Community; yet all European countries operate in the same economic environment with the only abundant and inexhaustible resource which Europe possesses, i.e. its labour force.

Under the pressure of international competition firms are finding it increasingly in their interest to make the best possible use of their staff's skill and training.

The experience and knowledge gained so far in various countries represents a potential which can only be made fully available by better international cooperation.

The more promising approach to achieve that might be to network scientists as human beings instead of linking programmes which are quite different in structure and funding sources. The "European Workshop on Research and Development Strategies in the Field of Work and Technology" should be regarded as an attempt to facilitate discussions amongst those who are addressed here.

It is reassuring to discover that something like a real European cooperative culture is evolving and that it is not based on bureaucracy but on mutual trust and on a growing awareness of a common European objective in transition of working life.

We are very much obliged to all those who have contributed to the success of this meeting, especially the "Dortmund Social Academy" and the "Swedish Work Environment Fund" as organizers, the "Federal Institute for Occupational Health and Safety" as host, the "European Foundation for the Improvement of Living and Working Conditions" as co-sponser and last but not least the speakers, who have spent a great deal of time and effort in helping to make this conference possible.

Bonn, April 1992

Contents

Contributions to Parallel Sessions

A: Economic Versus Human and Social Goals in Work and Technology Research

B: Integrated Concepts of Health Prevention, Industrial Safety and Ecological Requirements (including the European Dimension)

Introduction

European Problem Areas of Coordinating Research and Development Strategies in Work and Technology

Introduction

Hans Pornschlegel, Dortmund, Germany

1. Initiative and Organizers

During several meetings in Stockholm between the Swedish-German steering group of the Swedish Work Environment Fund (Arbetsmiljöfonden) and the Project Administration for Work and Technology (Projektträger Arbeit und Technik) of the DLR it was common opinion that the coordination of some programme areas and projects of both sides, and the cooperation within them, showed good progress and fruitful results. Contacts and cooperation between research institutions and researchers were also well underway. But there was never time *to discuss political, strategic and operational approaches* in the formulation, interpretation and implementation of *research and development (R&D) strategies* in the common fields of activities, labelled "quality of working life", "humanization", anthropocentric design concepts, work and technology, to mention the most common terms in English.

Last year the Sozialakademie Dortmund proposed to the Swedish and German parties to organize a workshop devoted to this cause. The idea was immediately taken up; the German side suggested that such a gathering should not only express German and Swedish voices but should be extended to a wider, European forum. The workshop could then better deal with the relations between the relevant national, European and possibly international programmes. It would allow deeper insights into the underlying political structures and mechanisms, the system of cooperation and conflict solving between publicly financed programmes, promoted institutions and expected results. The philosophies, and perhaps theories, behind R&D strategies and the risk of them becoming indistinct when implemented were also of concern.

At the same time, the Swedish colleagues had a keen interest in discussing *recent developments within the European Community*, with Swedish economic activities and investments playing an increasing and substantial role, and with the situation of the Single Market after 1992 in view.

The present workshop is the *result of a joint effort*, based on the initiative of the *Sozialakademie Dortmund* with the very close, intense and amicable cooperation of:

- the Swedish *Work Environment Fund (Arbetsmiljöfonden) (AMF)*, Stockholm, with major contributions from various Swedish quarters, and a very prominent and strong delegation;

- the *Federal Institute for Industrial Health and Safety (Bundesanstalt für Arbeitsschutz)*, Dortmund, with substantial conceptual, professional and organizational assistance, as illustrated by our presence, plus crucial contributions;
- the *Project Administration for Work and Technology (Projektträger Arbeit und Technik, DLR)*, Bonn, which stimulated the workshop with strong support as regards concept, contents and funding;
- the latter being provided ultimately by the *Federal Minister for Research and Technology*;
- the *Dublin-based European Foundation for the Improvement of Living and Working Conditions* who has kindly co-operated by presenting results and by taking care of the costs of simultaneous interpreting;
- the *EUROJOBS Working Group on Job Design and Work Organisation*, Brussels and Stockholm who participated actively with four members in the planning and running of the event.

2. Objectives and Target Groups

The European Workshop provides *a forum for analysing, comparing and assessing current policies and strategies in the field of work and technology in Europe* which are directed towards the improvement of working conditions and working life in a wider context.

The Workshop is addressed to representatives of the labour market parties, of the scientific community and of the institutions concerned, as well as to programme managers engaged in the design and development of relevant national and European programmes. The Workshop should provide *a "forum"* and an opportunity *to exchange experiences, discuss different approaches and facilitate the further development and coordination of national, bilateral, multilateral and European programmes and projects* in this field. Medium and long-term perspectives of policies and strategies in this field, selected central and exemplary issues as well as implications, barriers and problems concerning the programmes could be the focal point of these discussions. Insofar as *relevant major research activities of international organizations*, e.g. of the International Labour Organization (ILO), Geneva, and the Organization for Economic Cooperation and Development (OECD), Paris, include or even focus on national and European developments in the fields discussed, these should be part of the patterns considered.

3. Scope of the Questions Discussed in their Socio-Economic and Political Context

3.1 General Questions

- The workshop is *directed towards R&D strategies* in the fields of work and technology: It focuses therefore on *policy formulation by national, European or international publicly funded agencies*. I refer to A.C. Hubert's synopsis of national and European programmes.
- The workshop therefore *does not focus on findings of important R&D programmes and projects* as such, unless appear they relevant in their relation to the policies and strategies pursued.
- The workshop rather concentrates on the *planning processes, their implementation and evaluation*, on achievements, deficiencies and failures of the R&D policies pursued.
- Instead of undertaking a futile attempt to review the whole field of activities involved at national and European level, the organizers have *chosen some exemplary areas of R&D activities* which are *closely interrelated with current and future policies of the state, the labour market parties, companies and public organizations*.
- Since we are dealing with R&D, the *relations and interactions of the programme and policy-makers, managers and administrators with the scientific community* and the determining socio-political milieu have to be discussed. They are the factors which determine whether current and future R&D strategies should be pursued or not.
- It should be discussed to what extent *bilateral, multilateral, European or international cooperation* takes place in the areas selected, where there are deficiencies or even blank spots on the European and national R&D maps.
- It must be asked *what role the programmes discussed here* should and do in fact *play within the context of the national and European policies*:

  - aimed at *work and technology in general*, but also in their relationship, coordination and interaction with special programmes in related fields, e.g. information technology, computer assisted production, ecology, technologies and service concepts;
  - for the further development and adaptation of *social security systems and concepts*, by acknowledging underlying human and social needs under changing demographic conditions and values;
  - for *educational and training policies* by the state and other responsible institutions;
  - for *ecological goals and policies* to be pursued:
  - of *the labour market parties*, the companies and the manifold organizations and institutions involved in policy formulation and implementation (e.g. by chambers of commerce and industry and by professional bodies).

3.2 Questions Arising from the German Unification Process

Germany has to find out and assess the *consequences of German unification* - in many ways unforeseeable - which mean a fundamental and rather brutal transition from a planned into a socially oriented market economy, from centralized command to individual and collective initiatives, with new attitudes towards life and work. In the former GDR, research and development in these fields was organized by a number of relatively well-staffed - by our standards overstaffed - institutions. Even the university departments dealing with related problems had permanent staff, whereas West Germany draws largely on personnel with temporary contracts.

Two major questions then remain - and their urgency is demonstrated by the presence of some colleagues from the New Federal States: *How to utilize the existing capacities* and to allow at least some of the researchers and their institutions to play an adequate role on the German and European stage; the other big challenge is to promote the process of reconstruction in the former GDR by R&D in fields of work and technology and by an adequate transfer of know-how. Since new industrial, administrative and service structures will have to be built up at relatively high organizational and technological levels and under heavy restrictions in time and capacities, it seems crucial to work out the new patterns along innovative, as well as humanly and socially acceptable lines. In other words, the governmental programme "Work and Technology" from 1989 will have to play an important role, though it must be adapted to the conditions in the five new Federal States.

3.3 European Challenges

Within Europe, the impact of the *Single European Market 1993* is becoming more and more evident. In the fields under discussion, the following features appear to be paramount for the pursuit of adequate and future-oriented R&D policies and strategies:

- Art. 8 of the Single European Act requires the *free exchange of "goods, persons, services and capital"* under the conditions of the Act;
- Arts. 100 and 100a enforce, amongst others, the rapid development of *European ergonomic and safety standards* at advanced levels, thereby exercising a strong impact on the development of tools, machinery and equipment, including computer based systems, and their remaining national standardization.
- Art 118 urges *social harmonization* in Europe at relatively high levels with strong implications for future industrial safety and health policy and regulations which will become more European and be closely linked with ecological requirements.

The processes of *globalization of commerce, communications and technology* make even European developments appear as a minor element within international trends and processes. This has consequences not only for European and national policies in many major fields; they are having an immediate effect on the policies of companies and service organizations and even on the workplace situation (e.g. in the types, the range, the prices and the volume of products to be manufactured or assembled).

The formulation of R&D policies and strategies by publicly funded bodies - whether paid for out of tax-payers' money or from statutory levies - can hardly escape the need to make these developments integral parts of national programmes. This task has to be mastered without ceasing to pursue intensively concepts which serve national and often regional structures and policies, target groups and interests. These questions will have to be discussed within new systemic contexts.

An introduction to the *workshop* also requires a short indication of its intended *limitations*:

- There are many *closely related programmes, project complexes and individual projects* run by universities, by other institutions of higher learning and by other public and private research institutions *financed by other sources* at national or regional level; there are also very important, rather autonomous bodies (e.g. the Deutsche Forschungsgemeinschaft and the Max-Planck-Gesellschaft in Germany; the CNRS in France): These will not be dealt with unless they have a marked impact on or relationship with specific national programmes or European R&D strategies (which is difficult to ascertain).
- *More basic socio-economic or medical research*, for example, in the field is not dealt with specifically, though it is part of many programmes. The autonomously defined scope and contents of this kind of research usually go beyond the limits of the programmes discussed here. It is also promoted by other agencies.
- The *limitation of the problem areas to be discussed* is, as mentioned, the result of a rather pragmatic consensus among the organizing persons and institutions.

4. Problem Areas Chosen

It may be useful to make some indications aimed at highlighting some of the questions behind the following headings.

Parallel Session A

Economic versus Human and Social Objectives in Work and Technology Research.

This topic is nearly as old as ergonomic research itself, which often appears to contrast and conflict with objectives of technology and efficiency in theory and practice. German experience and research shows that the use of the traditional and classical *patterns of accounting, of cost and benefit analysis and of* valuation are mostly unable to reflect the medium and long-term costs and benefits of fairly costly human factors and of work organization in all its flexibility and social context.

The range of questions involved will also include individual and social costs arising from occupational risks and accidents. Modern technologies seem to create more diffuse and complex problems in this field than do most of the "classical" technologies where findings exist on a variety of interrelationships.

This problem area has so far had a high ranking in the German governmental programme "Work and Technology" of 1989 under the *leitmotiv* of integrated, innovative and preventive objectives and approaches. The question is whether the Europeanisation of businesses and organisations will put greater pressure on these issues.

Parallel Session B

Integrated Concepts of Health Protection, Industrial Safety and Ecological Requirements

The *pressure to foster such approaches* comes from various sides:

The *public* in highly developed industrial countries is becoming more and more aware of the interrelationship between its own health, general living conditions and working and safety conditions at the workplace, as well as ecological requirements.

At the same time, *the work place is becoming a kind of interface* where all these requirements meet through the human being. On the other hand, we have separate organisations and systems in our societies, dealing with the requirements in the fields of health, industrial safety and ecology. It seems to be a major challenge for present and future R&D strategies in our field to find integrated or at least closely coordinated ways and means of approaching these problems through extended basic research, applied research, experimentation and practice-oriented developments.

Since *EC policies* are placing more and more emphasis on health and ecological requirements, these dimensions must not be left out. Here we have a long-term field of research activities which will eventually go beyond EC borders and will have to be linked up with research on other continents, as well as in Eastern Europe, especially in view of the problems of working and ecological conditions in those regions.

Parallel Session C

Holistic and Innovative Concepts of Work and Technology

Since the Scandinavians brought their new forms of work organizations onto the international scene in the early seventies, Tayloristic concepts have been (and, it seems, still are) powerful rivals. Even now, the conflict between highly sophisticated, computer-assisted, neo-Tayloristic concepts and holistic, interactive concepts of work and technology appears undecided. But we can assume that the community of researchers and representatives from national, European and international organizations present at this workshop are supporters of holistic concepts of work and technology which we will discuss in one session here. Factors which either foster or impede such research and its implementation in actual practice should be borne in mind.

Parallel Session D
Strategies for Transferring and Disseminating Knowledge and Know-How in the Field of Work and Technology

It is a common impression within the German and the Swedish research community that much of the research findings and many of the developments and experiments have not found the echo in companies, trade unions, employers associations and professional organizations which they deserve and need. In many cases, research in these fields does not follow the classical lines of empirical sociological studies, as in the form of "action research", and is therefore difficult to transfer. This is especially so because many of the findings and results are tightly linked to their particular societal branch and company environment, which makes know-how transfer extremely difficult.

There is, however, no doubt in my mind that the efforts undertaken and the funds spent, e.g. in Germany, on the transfer and dissemination of knowledge appear rather insufficient to meet the potential demand. A well-coordinated, joint effort of all relevant agencies and actors is urgently needed to strengthen the infrastructures for qualification, motivation and flexibility required for future concepts of work and technology.

The need for adequate forms of dissemination and transfer of knowledge is underlined by the changing demographic patterns which will lead to an increasingly older working population (on average) which will have to carry out the work with new technologies under the changing value systems of the younger generations.

Due to the political constellations, expenditure in direct dissemination and transfer projects in Germany is restricted at present to 10 percent of the total "Work and Technology" programme.

5. Some Basic Assumptions

1. The existence, structure and volume of national and European R&D *programmes* in the field of work and technology are *indicators of the relative socio-political importance* placed on the particular fields of action by the legislators, governments and institutions concerned.

2. The structure and volume of these programmes *reflect the outcome of sometimes very long processes of institutional and personal cooperation*, developed after many external and internal conflicts, including at the legislative level. Insofar as the longevity of such programmes is not guaranteed by statutory levies and relative autonomy, the *actual emphasis* on target areas and funding will vary with:

   - political shifts in majorities, government members;
   - public discussions between the labour market parties;
   - the impact of the scientific community on policy approaches and concepts;
   - the roles of different, often invisible lobbies and forces, hidden interest-group cartels.

   Research programmes of such a nature have political ends, and they are also based on political processes with many known and unknown variables.
3. *Systemic structures of the problems to be tackled* are increasingly determining the scope, content, character and methods of R&D in work and technology. Complex, interdisciplinary approaches (e.g. "action research", the concept of "holistic approaches" and of "learning organisations") are therefore fostered while the role of "classical" scientific research seems to become somewhat secondary (e.g. the approaches to the complex projects "Labour Market *Flexibility and Work Organisation" of the OECD, "Work in Transition" of the ILO/AMF).*
4. *The genesis and further development of R&D programmes on work and technology are also part of systemic structures which are closely interdependent or correlated.*
5. *National and European programmes in the field show many areas of common interest. This reflects:*

   - *a common, though highly differentiated awareness of socio-political and economic problem areas requiring changing roles and activities among national, European and international public bodies concerned with R&D in one way or another;*
   - *similar structures and approaches to R&D strategies and policies;*
   - *but also diverging approaches at national level, dependent on the specific societal, cultural, ethnic, political and legal factors.*

6. *The developments under discussion have increasingly European and international dimensions (cf. above). It is therefore imperative to respect, take into account and possibly include these dimensions and aspects in all R&D strategies concerned, in their implementation and administration.*
7. *While a globalization and Europeanisation of R&D has taken place*, not only under world-wide cooperation of multinational companies but also between smaller units, between research institutions and researchers, these dimensions *are not reflected adequately in many national programmes* with the urgency, precision and positive orientation which they deserve.

8. Speaking for the German side in particular, I wish to express the hope that this workshop will have one positive side-effect: That our national R&D Work and Technology Programme should be developed and implemented more and more in this direction, also with a view to the well-understood national interest.
9. In spite of a *larger Germany*, there are the new *dangers of a rather provincial kind of research*, i.e. provincial in that:

   - it lacks the indispensable, future-oriented European and international dimension in its sub-programmes (*Arbeitsschwerpunkte* and *Bekanntmachungen*);
   - its findings are usually publicized only within the borders of German-speaking countries (or regions; language barriers; cf. den Hertog);
   - it is not sufficiently part of the European and international R&D community in the field;
   - it does sufficiently not render frequently useful and transferable results for the process of European developments, especially in the fields of harmonisation and standardisation.

10. These dangers are becoming aggravated under the political, economic and material *pressures to concentrate capacities and efforts, in this field also, on the reconstruction of the East German economy and of the social structures there*. In the worst case, seen overall, this could eventually lead to a "knowledge and technology gap". This risk will need to be systematically avoided by balanced R&D strategies in the field, which will boost the innovative and preventative approaches of the present programme.

Introduction

Carl A. Asklöf, Stockholm, Sweden

We have met here to discuss a very important topic, the development of working life. There is an urgent need to improve working conditions and to use human resources better. This is not only for the good of the people at the work place, but it is also necessary for efficiency and competitiveness.

Flexibility at the work place is needed to adjust to external and internal forces. Rapid changes have to be made readily, although individuals and working groups are parts of complex systems within the organization. The organization as a whole is dependent on many external conditions. To be flexible despite prevailing constraints is the challenge. And, of course, to make the necessary changes in the organizations and systems that exercise a restraining influence.

In Sweden some system changes are made for the purpose of solving the persisting problems of too many work injuries and too many days of absence due to sick leave. There are proposals that employers should be given more responsibility for rehabilitation and also have to pay for the first two weeks of sick leave. Moreover, it is proposed *to change the workers' compensation system* so that employers pay scaled fees in proportion to the cost of work injuries in the respective industries. The supervisory authorities, rather than going into details in their inspection, are adopting a more system-oriented approach and are developing methods to supervise the employer's management of the work environment.

Some of you may have heard about the newly created *Work Life Funds in Sweden for rehabilitation and work environment measures*. The aim of these activities is to obtain practical results at the work place and to reduce work related absenteeism and lift some of the burden off the workers' compensation system. These superfunds dispose of 15 billion SEK to be used over the next seven years. The Work Life Funds rely on industry and sectoral programmes developed by the labour market partners which set general priorities and objectives.

Whereas the Work Life Funds are temporary and aim at immediate practical improvements, the *Work Environment Fund* also has more long-ranging objectives. The programmes of the Work Environment Fund are not primarily sectoral, dealing with the problems of specific industries, but focus on specific problem orientated themes.

There are always *policy* implications in our programmes and more often than not, a *research component* aimed at building up a broad and deep understanding of the underlying causal factors and at developing instruments to tackle present and future problems. Some of these programmes will be presented later in this workshop.

In research on work and technology, on the methods and means to improve working life, it is of vital importance *to maintain a strong international connection.* National border-lines are vanishing as a result of deliberate long-term action, such as in the European community, or sweeping popular movement, as witnessed in the reunification of the two Germanies. We are developing not only a new world of work but also *a common world of work*, and we need to collaborate more in doing so.

This seminar on Research and Development Strategies for Work and Technology is an important *cornerstone of German-Swedish collaboration.* The Work Environment Fund and the Arbeit und Technik programme, with its predecessor HdA, have worked together since the beginning of the 1980s. It has been a good and mutually rewarding collaboration, especially as we have many things in common. It is my hope that this seminar will be a starting point for many more collaborative ventures between Germany and Sweden, and also within a broader European context.

I am confident that we will have three very rewarding and stimulating days here in Dortmund. It has been a privilege for us to work with Professor Pornschlegel in organizing this meeting, and I would like to end my speech by thanking Sozialakademie Dortmund and the Bundesanstalt für Arbeitsschutz for all the effort that you have put into organizing this meeting.

Plenary Session

Goals and Strategies
of the Labour Market Parties
and Institutions Involved

Trends and Perspectives in Work Life

Rolf Lindholm, Stockholm, Sweden

A European Workshop dealing with research and development in the field of work and technology cannot discuss its immediate topics and problems without placing them into *a perspective of developing business trends and policies*. These formulate and form essential terms of reference as well as the course of action to be pursued also in the field of work and technology. Let me do so by sketching *five fundamental statements* in this respect:

1. Business is becoming more and more *global.*
2. Business of the future *is people*, not jobs.
3. People need *more and new competences.*
4. Patterns of *management strategies* will have to change, especially in production.
5. We have to strive to create *learning organisations.*

1. Business Going Global

The *impacts of globalization* can be clearly demonstrated by a couple of figures: Between 1970 and 1990 world trade doubled in size in relation to production throughout the world, from 22 to 27 per cent. An even huger flow on financial markets is indicated by an increase to an estimated 50 trillion US$ from 3 trillion US$ over the same period.

Computer and information technology allow an abundant and fast flow of informations around the world as seen in the immediate information of Tokyo and London with changing data on the Wall Street Stock Exchange. It is more or less true also for all trade transactions and logistic functions. A movement of hundreds and thousands of business people, of technicians and experts takes place all the time. Business environments are improving world wide, overcoming distances, language and cultural differences.

We all know the effects of such a transformation in the *shift of labour-intensive industries* out of Europe and the United States *into south-east Asia*. Many business systems with low or medium demands for competence have been moving from developed industrial countries to those newer industrial countries (textiles and clothing, shipbuilding, mining, commercial steel).

This process forces business in the industrial countries since decades, with increasing intensity, to find *new combinations of innovative business ideas and systems* with changing combinations and concepts of labour and technology. Management is therefore forced to find new ways to *keep employment in the industrial countries*. Entering world wide competition in certain major fields may be imperative for some multinational, big companies and their networks. On the other hand, there is always scope for *"niche" strategies* in local, regional and even international markets; successful examples - so far

- have been fashion garments (Italy and West Germany) and cars (West Germany, Sweden). In these strategies direct labour costs become less important than value added in processes and successful marketing.

The development of transnational enterprises operating internationally is another major feature of the global business society. Some of the national subsidiaries are generally more at home in the global environment than in a particular national economy. This is a challenge to the structures and institutions of the nation state.

2. Business is People...

In the past we have thought of business in its internal structures of hierarchies, staff and work organisation, of a division of work as expressed in job descriptions and well defined jobs with clearly demarcated responsibilities. But the demands on business do not allow to rest on such traditional structures. The more complex tasks in industrial business and in services become, the more it is necessary no longer just *to think* in predetermined jobs but rather *in people who are able, willing and flexible* enough to make the tasks to be performed part of the business' success. The orientation towards customers in global markets (or complex national and regional markets) with varying, individualised demands needs people not just job incumbents.

High capital investments in industrialised countries put pressure on management systems, usually combined with high technological standards, *to create value added* to sophisticated products and services through people. That is where competition against countries under development still has chances. The quickly changing structures of demand and supply in many markets make human beings eventually the most essential asset for intelligent and swift adaptation of intelligently managed business systems.

3. People with More and New Competences

As mentioned already, business systems in mature branches of industry with lower or medium demands for competence tend to move into younger industrialising nations. It is therefore a major *characteristic of new business systems* developing in highly industrialised countries *to demand high competence on the part of everybody involved*. Competence is therefore a key asset and an essential precondition for the success of business based on people. In manufacturing the formula for success rests mostly on a combination of high level capital intensity with high levels of competence under conditions where direct about cost tends to be less important than in the past. Examples have been given already; bio-technology and materials technology might be mentioned as other fields.

A growing proportion of salaried personnel and indirect production *activities become more competence-intensive* as e. g. in research and development, management and sales, training and personnel development, but also in indirect production functions where the "product" is a service, not a good. There are also strong repercussions on the working practices and work organisation in production processes.

The general economic shift towards services shows also a competence oriented bias towards ever more sophisticated consulting, training, insurance, banking and finance, software industries as well as information industries and services. In industrialised countries the manufacturing sector employs only 25 to 35 % of the gainfully employed population by 1990. This gives roughly 15 % on the shop floor - but only a few of those are working on direct production, the remainder is busy in knowledge-intensive services like maintenance, repair and build-up of complex installations. In Sweden it is only one in four blue collar workers involved in direct production! Some of these aspects are taken up in the next paragraphs.

MASS PRODUCTION	LEAN PRODUCTION
Large Series	Batches
Integration	Specialists
Labour Exchangeable	People Fixed Asset
Own Design, Sub-Contractors	Partners, Involvement
Buffer Stocks	Kanban
Product Specialists	Design Teams
Independent Dealers Maximizing Profits	Dealers First Step In the Kanban System
Maximizing Selling Today	Life-Long Customer Relationship

4. Changing Patterns of Management Strategies and Systems, Especially in Production

As we move from business administration to global business development, *demands to create new business* increase often dramatically. There is a need to increase entrepreneurial approaches and thinking at all levels of enterprises, especially in larger ones. This frequently finds its expression in either creating new smaller companies out of existing larger ones, often under the umbrella of the old company. Alternatively, new forms of project orientation, of profit centres try to create artificially market and entrepreneurial orientation in smaller units run rather autonomously within big and bigger companies. The ability to react to and exploit sudden, unexpected changes in the external environment becomes more important than sticking to prefixed long time strategic goals and approaches.

One of the consequences is the buy-in of components, building parts and even complete products *reducing depth of own production and improving its flexibility*. The diffusion of technology is therefore quicker. Product life cycles are shortened, market prices of products tend to fall rapidly (e. g. computer hardware).
In this climate innovation, flexibility, closeness to the market and globalization become key features.

A traditional organisation of industrial production cannot survive under these new requirements. These find their expression in the highly costumized production of cars. There are some *basic changes in the industrial production system*:

- *Unbroken, complete flows of production* for an entire product or major componentsare spreading, making high flexibility a must.
- Pushing *mechanisation and automation* strongly in direction of flexibility, a high level of competence of the personnel operating such an equipment becomes crucialand indispensable.
- *New logistics concepts* such as Just-in-time depend on interaction between the links of the chains with perfection.
- *Zero defect strategies* and quality of the line personnel require highly skilled personnel in the production acting with great care and responsibility.

Regardless of the production strategy chosen in each case in industrialised countries, the *key elements for success* are three aspects of the human element: *skill, flexibility and motivation.*

These changes in the economic and cultural setting made *a shift in paradigm especially for engineers* necessary: The report "The Machine that Changed the World" (Macmillan New York 1990) by the Massachusetts Institute of Technology (MIT) covers the automobile industry in fourteen countries. It describes in a very well informed and masterly way two paradigm shifts in the process of industrial development. The first one took place under Henry Ford and Alfred Sloane around 1908 by creating new ways of thinking and new production methods which unfolded during the next two decades in industrial countries.

Since the fifties, even more fundamental changes have been initiated in which Toyota played a major role. The *new concepts* have been spreading since then. Production flows are increasingly characterized by flexible work organisations, no buffer-stocks, a minimum of raw materials stocks and of semi-manufactured parts, operator responsibility for the production process and quality of the product.

The *Japanese factories* have overtaken the American and European counterparts tremendously in labour productivity by nearly 2 : 1. At the same time they improved quality and reduced amounts of space and stocks. This is a crucial challenge to the European automobile and related industries to fight for competitiveness during the commencing decade and the beginning new century.

Workers' role is changing towards a better working life in the car industry. Individual workers are stimulated to recognise and understand their own roles in the overall process, to accept responsibility for the functioning of the production system and process, also for quality, without avoidable delays and extra work. This goes along with increasing automation of handling and assembly. It becomes clear from the MIT report that western

European companies do by far not meet the Japanese standards of productivity. They are less inclined to automate more in assembly.

The report also coined the expression of *lean management* covering new concepts of overall and especially production management. It covers the product design as well production and marketing. It intends to mobilise different competences and skills within and outside the company, also affecting relations with sub-contractors playing an active part in the product development.

Summary Of Assembly Plant Characteristcs 1983				
	Japanese In Japan	Japanese In North America	American In North America	All Europe
Performance				
Productivity (Hours/Veh.)	16,8	21,2	25,1	36,2
Quality (Assembly Defects /100 Vehicles	60,0	65,0	82,3	97,0
Layout				
Space (SQ,FT, / Vehicle/Year)	5,7	9,1	7,8	7,8
Size of Repair Area (As % of Assembly Spa-ce)	4,1	4,9	12,9	14,4
Inventories (Days, For 8 Sample Parts	0,2	1,6	2,9	2,0

MIT 1990

Summary of Assembly Plant Characteristcs 1989				
	Japanese In Japan	Japanese In North America	American In North America	All Europe
Workforce				
% of Workforce In Teams	96,3	71,3	17,3	0,6
Job Rotation 0 = None 4 = Frequent	3,0	2,7	0,9	1,9
Suggestions/ Employee	61,6	1,4	0,4	0,4
Number of Job Classes	11,9	8,7	67,1	14,8
Training of New Production Wor-kers (Hours)	380,3	370,3	46,4	173,3
Absenteeism	5,0	4,8	11,7	12,1

In this context, the new view means again that *people become more important*. A basic understanding of the technology applied as well as its systemic functioning is required from the worker. A strongly compartmentalized outlook on work is supplanted by a holistic one. It means a slow but steady turn-round from Tayloristic concepts. *New management strategies* therefore have the following contours:

- Managers see *above all the people* in the organisation as a complex living organism. Interaction and development of people matters.

- *People are involved in searching for the best approaches* and ideas for solutions and take responsibility for it within clearly defined business goals.
- *"Every man is a businessman"* denotes this new style of leadership in which new business, new production ideas are constantly developed by ever more people in the business.

Towards a New Generation of Leadership

TRADITIONAL	NEW
Machinery	People, Complex Living Systems
Pre-Determined	Not Pre-Determined
Routines	Objektives
Controll-System	Formation of Ideas, Communication
Sanctions	Rewards
Production Orientation	Whole Business System

Nevertheless, *these concepts cannot be uncritically applied* to all business situations and systems. Much depends on the stage of development of products, services and markets as well as on the culture historically developed in a company and its social environment.
The demands on *the manager* of today and tomorrow are *to be a leader as a person*, understanding himself and the complex requirements of his social environment better, more deeply and therefore to react and act more effectively. So yesterday's positions will always be subject to rigid questions.

5. The "Learning Organisation"

We need business systems which *ensure* that *the competence of all participants* for the solving of individually relevant problems within the business is renewed to a sufficient extent.

This requires to have *access to the necessary and business-generated information and know-how* by utilising and at the same time rationalizing its volume and depth as necessary. Networks within and between enterprises and with research institutes, commercial appliers and the like are needed as well as adequate information technology and training concepts.
Life-long learning is a task for each individual's development. It may be also a source of insecurity and instability for many untrained in permanent learning.
There will be a definite shift from traditional forms of learning by schoolroom study, courses, seminars, study circles and the like. It will be a question of *developing work organisations which stimulate learning at work*, of individuals learning by going through important development stages in their own fields and by changes to new activities. Learning should become an integrated part of the work organisation.

Short term types of organisation like project organisation with a corresponding reduction of more permanent organisational structures should lead the way towards more "pedagogic structures" in organisations.

There are many indications that *the rediscovered and resharpened role of competence and skills* will be the most crucial question for developing business in the nineties.

Trade Unions' Objectives and Demands for Research and Development in the Field of Work and Technology

Siegfried Bleicher, Frankfurt/Main, Germany

I would like to use the opportunity presented by this European workshop on the organisation of work and technology to discuss certain approaches and requirements from *the point of view of IG Metall.*

It is undisputed that *the place of work and the working environment constitute a major part of peoples' lives.*

Nevertheless, even in spite of a humanisation programme which has been running in the Federal Republic of Germany for 15 years now, there is *little evidence that this fact has been recognised* in the structuring of factories, whether "externally" (architecture, industrial and urban culture) or "internally" (technology, organisation and structuring of work).

Yet *the factory* of today and tomorrow *is a politico-cultural environment* which expresses itself in various corporate social forms.

Thus the *knowledge and experience of the workers* are inextricably linked not only with the technical organisation of work but also with its social and political organisation.

The *factory of today*, however, is still characterised not only by horizontal and vertical divisions of labour, by hierarchy and dependency and by power interests of capital over work but also by the sometimes contradictory interests of the individual staff groups, blue and white collar workers, men and women endeavouring to better themselves.

On the other hand, *the cultural dimension of work*, the participation of people in the organisation of their labour, is now to be found in many company guidelines relating to values. However, these will remain guidelines and, consequently, unused resources if demands for humanisation intended to benefit society, the environment and culture are not translated into action and do not determine the future form of work.

The starting point has to be *the inclusion of ergonomic and ethical parameters* into a company policy which will change the nature of work. They must become an integral part of corporate investment and innovation policy.

This would by no means contradict company interests as these are more reliant than ever on qualified and motivated employees.

Innovation in the spheres of technology and work organisation is required because flexible reactions are needed to meet the wishes of customers, suppliers and partners in cooperation, because customers are to be offered expert services and because production deadlines must be met. Now more than ever before, human qualities, specialist skills, creativity, willingness to accept responsiblity and the potential of the workers represent an instrument and an opportunity for individual employees.

As trade unionists, we want to take full advantage of this opportunity. Many of our officals, above all members of the works councils who represent the social movement in the workplace, are now trying to bring about just such a change in the everyday routine of work.

"The Future" is the name of the *project which IG Metall* is pursuing. It is a project of encouragement! At the invitation of IG Metall in 1988, scientists, artists, officials and politicians of completely differing creeds discussed, under the theme "How do we want to work and live tomorrow", the paths to take and the way forward towards unity of purpose and freedom.

In company practice, this means striving as far as possible *towards self-determined work processes* which use technology as a tool for the good of humanity. What is not wanted is a form of work in which people degenerate into mere appendants to technology.

Attempts to change inherited, traditional work structures and create production processes which are determined to a large extent by the people themselves already exist in numerous companies in the metal-working industry, for example, in the form of production islands. The forward-looking concept of teamwork is currently being implemented at VOLVO in Uddevalla, Sweden.

By means of *such organisational solutions* for jobs, work areas, production planning or designing, which are consistent with social and environmental concerns, ethical parameters are to be converted into work activities which promote the personality.

This policy is complemented by *demands for products* which are useful to society and by concepts for moving from military to civilian production which IG Metall in particular has been developing.

It aims at a transition from a defensive and protective stance against the negative effects of corporate rationalisation policies to *a preventive and effective commitment* to the humane organisation of work and the organisation of production and products which benefit society and the environment.

With *the growing destruction of the environment*, as rivers, lakes, and forests die and numerous toxic waste scandals come to light, and with high unemployment and alienated working conditions, technological progress has become the target of criticism.

The ideology of "technological progress" which was long equated with social progress - also by trades unionists - made us blind to the problems of reality. Our approach to technology and humanisation policies, which promote personality and preserve the environment, bids farewell to this naive belief in progress. It complements our protective policy with a united organisational policy which pushes products and production and thus work and life to the forefront.

I would now like to move on from my introduction by giving a *brief summary* of seven hypotheses for achieving unity of purpose in policies for the organisation of work and technology, policies which are value-oriented, promote the personality and benefit the environment.

Hypothesis 1

Industrialisation and socio-cultural development have always belonged together and will continue to do so.

Progressive industrialisation and computerisation will create a new rhythm of life, changing our temporal conception of day and night, mid-week and weekend.

If the resulting combination of work and leisure is not to be felt as an impingement by work on leisure, the work process must become freer, more humane and more democratic for all employees.

Hypothesis 2

On the one hand, the heteronomy to which the individual is subjected by work is diminishing as technology takes over work previously done by humans; on the other hand, this heteronomy is increasing because the purpose of systematic technological development is to increase the productivity and efficiency of production methods. The only way to resolve this contradiction is to move towards *greater self-determination and participation*.

As workers become more qualified and the type of qualifications change, the individual's expectations with regard to work will grow. There will be *a shift from predominantly material values* to immaterial values.

Production methods which are compatible with society and products which do not harm the environment, as well as the new value-related demands for self-determined work, make it necessary to develop the *corresponding industrial architecture and human-oriented production concepts*.

Hypothesis 3

The factory of the year 2000 will be a central point where communal areas will take on the aspect of public domains, where the compulsion to work will rub shoulders with the wish for self- development, where social and environmental solutions will have to be found through cooperation between companies, trade unions, the employees themselves and the outside political and cultural spheres. Both corporate and work-organisation culture will be put to the test.

Hypothesis 4

In the past, the company-oriented, trade-union collective approach has led to *a separation between working and living* and between company and social policies.

Future aims in the field of work organisation must not only strive towards the holism of work and be dedicated to the realisation of cooperative achievement. They must also take into account the interaction between the world of work and the world of living, between the creation of things and the creation of life, and integrate them into a self-contained yet flexible political concept.

Hypothesis 5

The organisation of work and of the future requires *the intelligence and autonomy of all specialists* inside and outside the company, in trade unions, in communities, in the Federal State Government and in the Federal Government, amalgamated in a continuous dialogue which extends across borders.

Specialists are, on the one hand, the theoreticians at colleges, institutes and universities; on the other hand, they are the skilled, semi-skilled and unskilled workers, the engineers, the technicians, the typists, the secretaries in the offices. Their specialisation is *potential based on experience, production knowledge and corporate identity*. Through cooperation and inter-disciplinary holism, work organisation will become their subject.

Demands for environment-friendly products and socio-cultural production methods can only be met effectively if the workers are included in the discussions on how to meet such demands, if the abilities of the workers and of the members of the works councils and trade unions can develop and be implemented. Humanisation then vests work with meaning.

Hypothesis 6

The main concern of governmental and private policies on technology is to maintain and *improve their competitive position* on the international market.

It is, therefore, not social demand which determines the structure and orientation of technology policy but *the economic interests of companies*. Although other aims such as the improvement of working and living conditions or the preservation and safeguarding of our natural environment are propagated, they are too seldom put into effect. If, within the framework of State research and technology policy, gigantic political markets for high technology, micro technology and so on are created, why, I ask myself, is this not happening more in *the spheres of environmental policies and health?*

The enduring success of *Japanese clean-air policies* shows us that it is possible to protect the environment to a greater extent without harming economic development. The Japanese example also shows that an adequate environmental policy is capable of eliminating not only environmental disasters but also technologically disadvantageous consequences.

In our own experience *humane work and technology* solutions have almost always proved to be the more competitive.

Hypothesis 7

State research and technology policies are characterised by a crass disparity between, on the one hand, the amounts spent on promoting the development and application of "technologies of the future" and, on the other, the study of their social and ecological consequences and the development of possible alternatives.

I agree with the Federal Government that research and technology require widespread approval in a democracy in order to maintain social progress. I am, however, convinced that in stark contrast to its declaration, *the Federal Government is not employing an ordered process of weighing up risks*, forming a considered opinion and making a decision in order to reduce the potential harm of technological progress to a reasonable level and to encourage solutions to the problems of social requirements.

State research and technology policy must reconcile varying interests through debate and guarantee that all those involved have their say, if socially viable solutions to present and future problems are to be found. It must open up alternative channels of organisation. Or, to express it even more positively, State research and technology policy must provide financial support to all protagonists and encourage their development if it wishes to find answers to the problems of the future and not just degenerate into a policy of mere industrial promotion and subsidisation.

The Role of Research in the Development of Work and Technology

Jacques T. Allegro, Amsterdam, Netherlands
Jaap Boonstra, Leiden, Netherlands

To start with I would like to present to you a short review of research results on work and technology in the Eighties. I shall then handle some questions which will be of interest to organizations when dealing with the role of research in the Nineties. I would like to finish with a number of specific themes, which in my view, need more effective handling in a European context.

Obviously the *review of the Eighties* will have to be restricted to a few main themes and even these will not be representative. The usual fundamental and applied research projects on the theme work and technology in organizations have provided *interesting results*:

1. there is *new knowledge on automation*: these results could contribute to an improvement in the quality of work, but in reality this is seldom the case;
2. there is *new knowledge on the influence of toxic substances* on the environment: there we are dealing with important unforeseen side-effects;
3. there is *new knowledge on the significance of new materials* including bio-technology;
4. there is *new knowledge on the significance of job-content* in relation to stress symptoms; and the last point - crucial for this paper:
5. there is *new knowledge on the background of the failure of a great deal of research to actually change organizations.*

Summarizing, it must be quite clear that in the Nineties there will be a *permanent need for the type of research on work and technology outlined above.* It can provide a great number of new insights and at the same time often help to eliminate mistaken ideas. However, increased knowledge is no guarantee in itself that the information will be applied to administer change, particularly when the matter at hand is the changing of an organization as a system. One researcher (Barton, 1981), after an in-depth study, goes as far as calling this phenomenon "Ode to an unused research report".

We shall need to take a closer look at why research findings fail to contribute to the organizational changes originally aimed at. To my mind this is important because the world in which we live, that is the *environment of organizations* which surround us and the actual organizations themselves, is *characterized by a spiral of ever increasing change and uncertainty*,, as Peters (1987) points out in his book "'Thriving on chaos" handbook for a management revolution'.

Flexibility is indispensable for any organization. In addition, *decentralized power and responsibility* - as well as creativity - will have to play an important part in the future. All in all, this involves a *great demands being made on the employees* and the organization. But there is no reason why these demands should not be made, because in organizations nowadays the employees are far better educated than in the past and, with the aid of fast-moving developments in automation, equipped to meet the challenge. The employees are furthermore highly interested in a good quality of working life. Therefore, integral innovation directed towards making a contribution to greater flexibility and efficiency and towards improving the quality of work should be a permanent dimension in organizations in the Nineties. For many organizations this may lead to adjustments in the organization's structure and an adjustment of the basic paradigms.

The above-mentioned developments will involve *a shift (in emphasis) when doing research*. This shift will, in my view, be more suited to the changes and challenges awaiting us in the field of work and technology.

What we are talking about is *essentially a more user-friendly form of research*. The most progressive form is to be found in *action research*, but important changes are possible in the more traditional kinds of research too. As far as the latter is concerned, I shall pay particular attention to research as instrument development and research as an integral part of the implementation process.

Another important theme in this lecture is that *the consequences of new technology for work and working conditions should be seen over a wide spectrum*. It is *not just a question of job-content*, we are also dealing *with possible health and safety risks*. The fact that the consequences of safety, health and welfare are often the responsibility of different ministries, research institutes or advisory boards, should not close our eyes to the fact that in practice these consequences are closely linked at (lower) management, employees and shop-floor levels.

1. Research as Action Research

In my introduction I pointed out that organizations can expect a great number of changes and that adequate strategies will have to be developed to meet these changes - strategies based on a different organizational-paradigm. It is *not changes in parts* we are concerned with here, *but* with *integral changes*, whereby it will be essential to have an understanding of the dynamics involved in the processes of change. Traditional, bureaucratically-inclined organizations are not equipped to do this, which means a new approach is called for; in my opinion, this already has taken shape in the open socio-technical systems approach.

If we compare both of these organizational systems judging them on the basis of their specific characteristics we get the following picture (Van Beinum, 1988):

Scientific Management	Open Socio-technical System
Old paradigm	**New paradigm**
- overcapacity of parts	- overcapacity of functions
- emphasis on external coordination and control	- emphasis on internal coordination and control
- technical system of primary importance	- joint optimalization of the social and technical systems
- people as extensions of machines and commodities	- people as complementary to machine and recource to be developed
- restricted skills	- skilled
- alienation	- involvement
- organizational-design based on total specification	- organizational-design based on minimal critical specification

The new paradigm also demands a *new approach to research*, namely an approach which attaches *importance to the processes of change* and the *role of those organization-members* involved; they themselves having direct shared-responsibility in the process of research. This different type of research, based on different theoretical, planned change and methodological paradigms is best described as *action-research*. Taking action research to *mean the following* (Rapport, 1970)

- a *collaboration between researchers and a client system* to contribute to the solution of a problem;
- making a *contribution to the aims of science*.

In my opinion a third dimension has to be added:

- a *focus on change processes.*

Gustavsen, closely involved with the Swedish LOM programme, states (1989) that action-research has frequently been criticized for its dialogue with the field of research. However, on further consideration, it is an adequate departure point for contributing to change in organizations by means of research. An essential principle is that the position adopted by many theorists, namely that *'non involvement in practical matters is essential to the concept of science' has to be abandoned.*

A great number of case studies have been carried out using this action-research approach, with the aim of developing jobs, departments and organizations which will contribute to the flexibility, efficiency and quality of work. These studies have produced a number of interesting practice-theoretical *views about how to set up change processes* (Boonstra, 1990). The emphasis, as far as the quality of the work is concerned has been on the creation of jobs that produce challenging work with *a high level of autonomy*. The results which can be read off from these (and other case studies) show a high level of job-satisfaction, good quality work and reasonable productivity levels. However, it is not possible to introduce this action research approach without running into problems, because this different approach to organizations and organizing - which is contradictory to the scientific management approach - demands substantially different attitudes and conduct from management and employees. Which problems have arisen, or do arise, and what guide-lines can be given for tackling these problems? In the process, some important views, which will be shortly indicated, have been developed with respect to:

the preparations of projects, in which broad-spectrum work-conferences involving representatives from all levels of the company play an important role; *the projects' implementation structures*, in which further knowledge has been gained on the significance of temporary project-structures, in particular the form and procedure of steering committees and working parties;

the *phasing of projects*, in which, in particular, attention is drawn to the cyclical character of diagnosis-implementation and evaluation with a strong emphasis on tailor-made programmes.

This *cycle,* which is common to projects *will have to be worked through a number of times* and this will certainly be the case if the project is moved from an experimental department to other departments. For the diagnoses it is essential that the researcher/adviser builds up a *reasonable knowledge of the organization*, both in terms of its structure and culture. This diagnosis can be carried out in a number of ways, by means of questionnaires, check-lists, participatory observation and content analyses. A *short recognizable report*, complete with the views of the different interested parties, complemented by a clear set of priorities is a good basis for the setting up of a plan. It is difficult to give guide-lines for the implementation, as this is so specific. It is possible, however to give a number of *suggestions about the adviser/researcher's method of work*. These would include:

- the necessary integration of the job-structural and group dynamics approach;
- the *integration from top-down and bottom-up*;
- the recognition of the importance of individual differences, historical circumstances, conflict of interests and their regulation;
- key figures/container concept;
- formalization as a minimal specification;
- a number of suggestions concerning the role behaviour of consultants.

The *evaluation of (parts) of projects* should not be seen as being solely or chiefly for scientific purposes, although more familiarity with great amount of work which is carried out in other organizations is extremely valuable. What is important is that the results of the evaluation *be used for the further stages of the process*.

I would also like to point out that there is an essential principle running through this development, namely that *management and employees work in close cooperation* despite the fact that they can have in other areas different interests to protect. Summing-up, I hope that I have made it quite clear that action-research, based on a different paradigm is able to make a contribution to the development of work and new technology, which is *complementary to the benefits of traditional research*. A high level of efficiency and high quality in the actual work are the central focus and often the result.

2. Research for the Purposes of Instrument Development

Traditional research could also contribute more direct to the field of work and technology in the future. Management and employees are expressing a *growing need for (new) instruments in the development of work and new technology*, to measure the change-process, the quality of the end product (the automated system) and the quality of the work itself. There is a trend, away from the overpowering expert towards taking matters into one's own hands. This development should be whole-heartedly supported since the introduction of new technology has too often been left to the experts.

In my view, research has, up to now, been *too little directed towards instrument development for users*. I would put forward the proposition that the development of work and technology can be greatly helped by developing these users instruments. A good example of this development can be seen in *the Netherlands* with the *operationalization of the concept 'welfare'*. This concept is laid down in the aims of the 'Arbowet' (law on working conditions) along with safety and health regulations. In the Act (section 3) the concept of welfare is described in the following way:

- adaptation of working conditions and
- *work to the qualities of the employee* including his/her professional skill
- room for the employee *to arrange their work in their own way*
- room for the employee *to keep in contact with other employees*
- *information on* the aims and results of the *work*.

For the everyday running of organizations this is a far too general description. How can you measure whether the welfare of employees has been effected by the introduction of new technology? How can welfare be improved? How can management and employees work on improving this? A great number of practical problems which demand practical solutions remains to be solved.

What follows is an example of such an instrument to measure the quality of work. The *WEBA-instrument* indicated here (Welfare at Work; WEBA group 1990) addresses the problem of describing and assessing the job-content and the organization of the work - the ultimate aim being to improve the quality of the work.

The instrument actually *measures* the possible *psychological strain* and *interference with the learning processes* in and through work. To that end, a number of welfare risks are assessed, namely:

- the number of organizational duties/tasks
- the difficulty of the job
- the level of autonomy in the work
- the opportunities for contact
- the amount of information given

Four *courses of action* can be singled out, namely adaptive measures, measures for improvement, innovatory measures, and other measures. The development of this instrument (commissioned by the DGA) is a long process. The instrument has certainly not been fully developed yet. One thing is certain, there is a great deal of interest for working with this instrument - different groups in organizations have already asked to use it.

1. It is used by experts in close collaboration with employees.
2. A special, simplified version, has been made for the unions in the form of a checklist which they are to publish themselves.
3. It has been incorporated in an inspection programme which measures the overall quality of the work.

This operationalization of the concept welfare is further part of special courses for personnel managers.

I hope that his part of my paper has made it clear that research for the development of instruments can directly or indirectly make a substantial contribution to integral innovation.

3. Research as an Integral Part of the Dissemination Process

A third dimension to research which I would like to underline is *research as an integral part of the dissemination process*. In concrete terms what is meant here is that in order to implement substantial changes in the field of work and new technology, research must be embedded in an implementation process which covers information-giving, training and consulting. Research results on their own are not enough to bring about these changes.

It is also imperative that the four components of the implementation process mentioned above be regarded as integral. In other words, not first embarking on research and then on information, but when setting up the research taking into account the implementation and therefore involving the future users.

This same stand-point is defended in the *German research and development programme 'Work and Technology'*: 'Often the translation of research results into practice was particularly successful when those using and implementing those results later on were involved not only in the definition of the objectives but also in the research process itself and its current evaluation and carried out their own implementation projects. This close and direct interrelation of research and its practical implementation was reached in particular by branch-oriented projects' (Research and Development Program 'Work and Technology', 1990). This approach can be particularly well-illustrated as cited above in the *field of branch-oriented projects*. The most important justification for a branch-oriented approach is based on the assumption that problems and difficulties in all kinds of areas will occur in more than one company in a certain sector. For example, by carrying out intensive case studies on the problems concerning the quality of work in connection with the introduction of new technology in a dozen companies, a number of recommendations can be developed on informationgiving, training and consulting for the whole branch. In concrete terms one could think of the following *recommendations:*

- setting up or improving companies' approaches to automation;
- the development of a better, ergonomically sound workplace;
- changing the organization of work and consequently the content of a number of jobs within the company;
- the setting up or improving of information services and training for employees and management;
- the development of joint initiatives at branch-level, for example through collective bargaining agreements.

The approach outlined above would appear to be particularly well-suited to small and medium-sized businesses, which in many cases do not have the means or skills to start for themselves. An essential characteristic of this *branch-oriented approach* is that employers and employees organizations are both closely involved in this important issue.

One of the important *pre-requisites for the successful implementation* of the integral approach outlined above, is that the research to be carried out and the various components of the implementation process namely information-giving, training and consulting should all *be in one institute*, and that researchers, trainers, advisers and information--givers *work together in a number of projects*, wherever possible. Only then would it seem to be possible to arrive at an integral approach. It is also important to work with a great number of disciplines, as the influence of new technology on work and working conditions makes itself felt in many areas.

Should this approach to research as part of the dissemination-process live up to expectations in terms of concrete changes in organizations, it will clearly make an important contribution to integral innovation.

In conclusion I would like to state that a different role for research in the developmentof work and technology, due to more attention being paid to the forms of research I have discussed, would make an important contribution to integral innovation, that is to say flexibility and efficiency on the one hand and quality of the work on the other. Cooperation between researchers, management and employees is a basic requirement. I should like to emphasize the fact that this could be extended to a *European context* and I would like to put forward the following *concrete suggestions*:

- The greater emphasis on *action research* which I have proposed implies a close emphasis on the contents of the jobs employees (and managers) do. It is striking that despite the interesting developments in Sweden, Germany and France in particular, the European Commission has paid almost no attention to these problems in their third action programme for safety and health. I would therefore like to make a case for *more emphasis* being placed *on action research* at *EC level* in the four-year action programmes.
- With respect to *research for instrument* development I would call attention for *closer European cooperation*. What I have in mind in particular is the founding of a *central data bank for toxic substances*, instead of the different international databanks we have now. At the moment the feeling is that a great deal of work is being duplicated, despite the fact that resources and expertise are scarce. The International Chemical Map book, a cooperative effort involving the World Health Organization and the EC is a step in the right direction.
- As far as research as an integral part of the implementation process is concerned, I would like to make a strong case for the setting up of an *integral branch-project on quality of working life in an underdeveloped branch of industry* in a number of European countries.

References

Barton, A. (1981), *Ode to an unused research report*. Knowledge: Creation, Diffusion, Utilization, 3 (2), 141-142.

Beinum, van, H.J.J. (1988), *New Technology and Organizational Choice*. The News Journal of the Ontario Quality of Working Life Centre.

Boonstra, J.J. (1990), *Integrale Organisatie-ontwikkeling*. Leiden, in preparation.

Federal Minister for Research and Technology (1990), *Research and Development Program 'Work and Technology'*. Bonn.

Gustavsen, B. (1989), *The LOM-programma as research policy, Problems, Considerations and Solutions*. Stockholm.

Peters, Th.J. (1987), *Thriving on Chaos, handbook for a management revolution*. London.

Rapoport, R.N. (1970), *Three dilemmas in action research;* Mimeo, given to the Social Science Research Council Conference, York.

Weba-group, (1990), *Outlines of the WEBA-instrument*; a conditional approach for the assessment of the quality of working life, WEBA, Leiden/Amsterdam.

European and National Programmes in the Field of Work and Technology: Achievements, Shortcomings, Perspectives

Tony Hubert, Brussels, Belgium

The purpose of this paper is to make a comparative examination of the achievements, shortcomings and perspectives of national R&D strategies of European countries in the field of work and technology. However, the variegated European scene gives me the feeling that I am constantly comparing apples with pears: there are more differences than similarities between countries' ethoses and approaches. Take *Sweden* as the pivotal country:

- its culture can be considered Germanic, with concentration on "teknik"; but the national funds it devotes to "working life" research and actions far exceed (in relative per capita GNP terms) those of Germany. And Germany, in turn, invests far more money in such research than any other Community country.
- it is a Scandinavian country, like ... Norway. But Norway's national "R&D" programmes (HABUT and SBA) receive no government funds, being financed, respectively, solely by a joint trade union-employers' association development fund and by membership contributions. The Swedes, on the other hand, apparently love paying taxes for working life research.
- or ... Denmark. Here the names are often virtually the same: Arbeitsmiliöfonds (my pan-Scandinavian spelling) but the resources available and even the activities are significantly different: the Danes are concerned more with getting messages out, the Swedes on structuring research into programmes, albeit with a large built-in information output.
- finally, like Belgium, Sweden is highly export orientated, with a higher proportion of large companies. Yet the productivity and value-added of its labour force are below those of Belgium and its absenteeism rates only rivalled by the Dutch. Belgium tried in the late Seventies with little apparent success to follow foreign examples of work humanisation funds, and when they were scrapped no-one cried "au secours".
- Yet all six countries' governments (and all others in Europe) are engaged in helping enterprises located on their territory (though predominantly national companies) to enhance their competitiveness in particular by the better use of their productive resources. And one of the most obvious ways of improving this use is to improve the management of technology ie. linking work and technology. Indeed, the "management of technology" is, with "internationalisation", one of the most rapidly growing areas of teaching and research in business schools in Europe.

- Despite the "apples and pears syndrome", the presentation of national programmes and discussions with some of the directors does enable a series of tentative suggestions to be made about the achievements so far, shortcomings and prospects. As with most such analyses, perceived shortcomings contain the germs of major opportunities for the future.

1. Achievements so Far

- The programmes and funds exist.
This is an achievement since research linking work to technology has not always been approved of by the technological fraternity nor various breeds of politicians. And some countries are strikingly absent from this survey (Italy and Spain) because there are no national programmes.
  Funds have not been without their critics. Some have in fact disappeared (the Dutch and Belgian (1978) work humanisation funds and the Danish IHS programme - though they were in fact all set up for specified periods of time, like ... the German HdA). Most programmes have gone through critical evaluations: the German HdA on three occasions, WRU in the mid Eighties, the Swedish AMFO and ALC (and the total situation) at the present time.
  The continuation of programmes is supposed to have been ensured by the strong backing by the trade unions; but behind the scenes, trade union leaders do not seem to have been very enthusiastic, even when unions overcame their hang-ups about "rationalisation being performed in the name of humanisation". Rather, it would seem to be "branch organisations" (trade and professional groupings) and their associated smaller firms which are the bigger supporters of national programmes.
- Funds have helped build up a *cadre* (to use a good gauchiste term) *of academic researchers* who have gained experience of working with individual enterprises and organisations. This is strongly underlined in Germany and Sweden where a feature of recent developments has been the commitment of national "work" funds to the "educational" domain; and AuT is now directly sponsored by the ministry of education. If the future of European enterprises is in high value-added goods and services, the higher learning sector must be more significantly involved in their design and production.
- There is a greater recognition now than a decade ago that technological advance does not necessarily lead to *innovation and change*: that "organisation" and "management" can make all the difference in technological success or failure.
  The activities of national research programmes have probably contributed to this growing awareness, though they are only one influence among many. There is a growing recognition in private enterprise that competitiveness depends to a significant extent on the quality and the competence of the human resources

available. "Value added" and "quality" are key words in all companies, and generally speaking this goes well beyond lip-service.

- Some high tech areas attach considerable importance to the "human" (menschengerecht, now jargoned by the European Commission as "anthropocentric") design and implementation of systems: *preventive approaches* are beginning to be recognized as being much more cost effective than curative approaches.

2. Shortcomings

- Does research lead to change? And, if so, under what circumstances?

  In Switzerland the chairman of the group of experts of national research programme NFP 15 on "Improving Working Life" freely admits that it was financially too expensive to assess, through formal evaluation, whether the research really did lead to change at the work unit level. The head of the British Work Research Unit, though having "monitoring research at home and abroad" as part of his remit, concedes that, because of pressure of action orientations, this is the Cinderella task. In (Flemish) Belgium there would appear to be consensus that, acting as *consultants*, staff of a research organisation (STV) have been instrumental in ensuring that the "human factor" has been fully involved in the design stage of a major governmental technological renewal programme (computerising public administration).

  So perhaps researchers need to develop consultancy skills if the purpose of national programmes is to foster change.

- Organisational change, often associated with technological change, increasingly requires the *involvement and commitment of the actors concerned*. "Local" decisions are needed to ensure rapid flexibility. Research at the local level, even "action research", is not necessarily conducive to action. Such thinking seems to be very much behind the thrust of developments in the Norwegian, Swedish and French (ANACT) programmes.

- There is still little real understanding of how *diffusion* takes place. Why, if the recurring ideas really are so good, does not more of industry (notably Japanese industry, those rapid implementers of best corporate practices) adopt more of them more rapidly?

  Moreover, there would seem to be a case for arguing that ideas which have been spread around most and incorporated into working life have not benefited from researchers' support. Indeed, researchers have expressed considerable amazement that "Taylorism" or "Fordism" have to a large extent been eliminated in manufacturing as a result of organisations' doing things differently under the pressure of competition rather than heeding the advice of social scientists. Work-and-technology researchers could well be behind the times, rather than at the frontiers of new knowledge.

On the other hand, the concepts of "transfer" and "transformation" of knowledge and know-how are now built into many major research projects of national programmes from the very beginning. So there is generally greater room for manoeuvre than in the past for communicating with a variety of target groups.

- *Inter-disciplinary approaches* still remain an issue, even in Sweden: it is difficult to get technology researchers and work researchers to cooperate smoothly. *Technologists* still do not appear to take "work scientists" too seriously. This is perhaps because "work scientists" do not understand enough of the technologies involved. Are the best "work scientists" those persons who have had their basic training in technologies and then converted? And should national programmes be more selective in the researchers they sponsor?

  However, the past decades have seen innovations in inter-disciplinary cooperation which is opening up new vistas: behaviouralists with information technologists, archeologists with forensic scientists. The time has perhaps come to try to take stock again of the failures, achievements and prospects of inter-disciplinary research.
- The *national research programmes remain minute* and marginal compared with national technological programmes; there are few instances where the two meet formally (or even informally); and there is no documentation on the respect (or otherwise) with which work-and-technology results are held by those in charge of "hard" technology programmes.

  One instructive exception is to be found in Sweden: the MDA programme examining the links between people, computers and work, which is jointly financed by the technological research agency, STU. A significant emerging bridge between work and technology is the concept of *learning*.
- Academic advance still remains to a significant extent focused on traditional approaches such as *"publish or perish"*. These do not necessarily concord with the needs of work-and-technology research.

  Yet, arguably, it is mainstream academics who make the most significant advances in bringing about understanding of the broad developments (if not knowledge) through university and higher level teaching. Hence the importance of direct support to the innovative university community.
- The focus of national research programmes has been on *private sector manufacturing* enterprises. However, the bulk of employment is now in services, particularly in public sector, non-market services. And what projects have been undertaken in public services have focused on health care, admittedly a growing consumer of technology and public funds, but not the only one.
- As the world has been *internationalising*, national programmes have remained almost completely national. There have been influences and "oil-stain effects": the grass often seems to grow greener over the garden wall. But only one joint project based on real needs across borders has been detected.

The contrast with pan-European technological programmes, pooling resources to avoid duplication and to gain time, is crass.

3. Perspectives

As indicated in the introduction, the shortcomings outlined above can be seen as major challenges and prospects. Over and above them, three matters should be kept in mind:

- Why is it that, two generations after the Longwall experiments and a blather of lip-service of the "no single technological solution", *technologists rarely or reluctantly design the "human factor" into the blueprints* for new machines, products and services? The explanations advanced appear to be somewhat simplistic, concerned with jargon and pecking orders of different disciplines. Actions towards the higher learning sector have started to be undertaken and, more embryonicly, towards schools. National programmes are likely to undertake more efforts in the future to enhance understanding in schools.
- *Preventive approaches are more significant than curative approaches*. The 80:20 rule appears to hold true as regards costs, effectiveness and efficiency when dealing with projects or even individual items of machinery: incorporating improvements at the beginning will be less costly and more effective than tinkering with existing technologies. So programmes are likely to become more targeted and selective in what they finance.
- There are a number of *common concerns*: the diffusion of know-how in general; influencing the manufacturers of capital goods to take an "anthropocentric" approach; improving the working conditions of specific branches of industry; .... As European level technological programmes have grown in stature, why should not "technology and work" programmes assume European dimensions? FAST's proposed "Human Work in Advanced Technical Environments" could be one starting point; the new *EUROFORM* initiative (requiring governmental proposals and hence particularly apt in the present context) in the human resources' field, another.

  But it should be remembered that the international dimension of dissemination is complex. On the one hand, even countries which are very similar in language, culture and temperament (such as Denmark and Norway) have had little success in using each others' material concerning the design of work; on the other, US (and perhaps to a small extent British) material still circulates internationally: Swedes go on sabbaticals to the Australia and the USA, rather than Austria and Liechtenstein.

Worker Training: Competing in the New International Economy

Wendell Fletcher, Washington, D.C., USA

I am most pleased to have this opportunity to address the European Workshop on Research and Development Strategies for Work and Technology. My remarks today focus on a report, *Worker Training: Competing in the New International Economy*, released in September 1990 by the Office of Technology Assessment of the United States Congress. I served as the project director for the report. The Office of Technology Assessment (OTA) is a nonpartisan analytical arm of the United States Congress. Its function is to help legislators anticipate and plan for the positive and negative impacts of technological changes.

The study involved 18 months of research on the way American firms train their employees. It focused on how firms use training to improve their competitiveness. To a lesser extent, the study examined training systems in other countries, especially Germany and Japan. At the outset, I would like to note that a key purpose of OTA studies, by their nature, is to identify areas of needed improvements. Thus, my remarks today focus on the kinds of changes U.S. industry and management need to make - rather than on the considerable strengths they possess.

Research and development on human resource policy has been given much more emphasis in European countries than in the United States, and I believe this has important implications for U.S. competitiveness. Traditionally, factory floor and office jobs in much of American industry have not required much training. These jobs were purposely designed to be simple, reflecting the needs of standardized mass production. Today, the demands of new technology often call for *alternative forms of work organization, approaches* that enlarge the job responsibilities of workers. Yet, so far, few American firms have attempted to make use of these alternatives. I believe that strong government support for a program of research, development and dissemination of best practices in the fields of technology, work organization and training would encourage more firms to act. Of course, other policies would be needed in addition to research. To place these R&D needs in context, I would like to briefly discuss some of the major findings from the Office of Technology Assessment report.

A major conclusion of our report is that *U.S. industry needs to deploy new technology more effectively*. Obviously, effective use of technology depends heavily on people. Too often, American firms seek to improve production by focusing primarily on hardware - equipment and physical plant - rather than on the people that will make the hardware perform. In many cases, the greatest need is *to reorganize work in ways that take advantage of the spread of decentralized computing*. Last year, computers and related information processing technologies accounted for nearly half of U.S. business spending on capital equipment.

Our analysis suggests that firms that *use new technology* most effectively do so *by adopting new approaches to management and work organization*. Moreover, they give their workers substantial training to help them learn new and different skills. But the firms that take this approach are a *very small minority*. This will have to change if firms are to take advantage of the flexibility conferred by decentralized computing. As decentralized computing becomes more common in auto plants, textile mills, banks and insurance offices, more employees will be expected to make decisions and take action on their own initiative. American firms far too frequently expect this to happen by instinct - automatically - and do not plan for the training of their workers that will maximize their abilities to take the initiative.

Very often, *it is the organization of production*, not just advanced technology use, *that distinguish the best performing firms*. In some cases, firms are redesigning their workplaces to improve employee morale and to create positive attitudes about quality and costumer service. Some accomplish this by introducing multiskilled work groups with self management responsibilities. Other techniques, such as just-in-time manufacturing and exposure of factory floor workers to statistical process control, are beginning to be more widely used by U.S. industry.

Of course, some U.S. companies employ new technology, work reorganization and workforce skills very effectively - and this has a direct effect on the quality of their products and often their productivity as well. For example, Motorola, a major U.S. semiconductor producer, more than doubled its production at one of its factories through reorganizing work and developing the skills of its employees through training. This was before adoption of new technology. Plumley Companies, a medium size supplier to auto-companies, credits employee education efforts for major advances in quality of its products.

American firms are finding that training their employees for new technologies and work roles *is a difficult and challenging task*. To function effectively in the new environment employees need adequate basic education. They also need to be effective in interpersonal relations and in communicating with co-workers. They need a broader range of task specific skills than industry used to require in decades past.

The United States is not a training intensive culture - and so it is in training where some of the greatest improvements are needed and the greatest challenges for research in work and technology will be faced. Overall, only 35 percent of U.S. workers reported any kind of skill upgrade training, formal or informal, in their current job in the last national survey. American companies do train technicians, professionals and managers. However, few companies systematically train factory floor workers or nonsupervisory clerical or support workers. It is these workers who are particularly unlikely to take training on their own.

And yet it *is these workers who are the focal point* of much of the current effort to reorganize work. Many of them are not comfortable with formal training in a classroom setting - due in part to a bad experience with the public education system.

In contrast to many European nations, *the United States does not have a strong system for preparing non-college bound youth in making the transition from school to work.* While there are exceptions, the high school vocational education system in the U.S. has not done a very good job of developing work skills and habits of the non-college bound before they enter the workforce. The United States has nothing remotely comparable to the dual system of apprenticeship and vocational education in Germany. The U.S. does have a formal apprenticeship system; however, it generally serves workers in their early twenties or older, rather than recent high school graduates. Less than one third of one percent of the workforce is in an apprenticeship program.

American workers are also less prepared than their counterparts in Japan and Europe *for formal training* in terms of educational background. U.S. companies often discover that 20 or 30 percent of their workers need remedial education in reading and arithmetic before getting training on such subjects as statistical process control. Despite media accounts to the contrary, the problem is seldom illiteracy. Rather, workers need to upgrade their educational skills to cope with rising industry standards and requirements.

Finally, *training is often poor in quality or remotely connected to actual work needs.* Training often doesn't transfer back to the job. Equipment vendors often provide cursory training. Community colleges and other educational institutions have a mixed record in meeting industry needs. Some are excellent; others are not. Companies, especially small companies, have little time or resources to determine who really provides good training. Companies that have had bad experiences with training may not seek additional training.

Firms that are reluctant to train may also be reluctant to experiment with alternative forms of work organization. There are *some dramatic examples of company efforts to organize work in new ways.* By and large, however, the pace of change in the American workplace has been slow - too slow, in OTA's estimation. Indeed, one recent survey of larger firms found that only 13 percent of the responding firms had used any form of self-managed workteams. Another survey concluded that only about 5 percent of U.S. firms were dissatisfied with the skills of their workers. The reason: most firms used Tayloristic forms of work organization that do not require workers to use much skill in the performance of their jobs.

Now let me turn to *research needs.* Firms are not likely to change to more training intensive approaches unless they are convinced they will pay off. In such a situation, research, development and best practice diffusion are especially critical. OTA's analysis identified several priority areas:

First, *evaluation of training approaches.* Very few companies in the United States evaluate their training programs, and those that do seldom reveal their results. Efforts have been underway in the United States to develop better evaluation approaches. One promising effort, being pursued by the National Commission for Employment Policy and the National Governors Association, is to develop guide-lines that State Governments could use to evaluate their support programs for industry training. Forty four of the 50 states have small programs of support, involving total expenditures under S 500 million, in support of industry training activities. The guide-lines could be used to identify factors leading to success - and equally important - lack of success in specific training projects. A second area is *demonstrated of work-based learning approaches*. As I have mentioned, one of the key training challenges facing U.S. industry is to reach factory floor or nonsupervisory office workers who are uncomfortable with formal training in the classroom. For these workers, it is essential to find better ways to bring the benefits of structured training together with the informal process for training workers on the job.

Recently, *the U.S. Department of Labour established an Office of Work-Based Learning*. Among other things, this office is reevaluating traditional *concepts for apprenticeship* to see if they could be applied in non-traditional sectors. Apprenticeship in the United States is dominated by construction trades, with manufacturing a distant second. As I mentioned, less than one third of one percent of the American workforce is enroled in an apprenticeship program at any time. We have about as many apprentices today as 20 years ago, yet the workforce has grown appreciably. So far, the office has funded several pilot projects to demonstrate ways to adapt apprenticeship concepts in other industries. These include: training of child care workers for small business; development of structured on the job training for bank tellers; and development of upgrade training for aircraft manufacturing and health care workers. While the program is promising, funding levels are low.

Adult Learning is a third area where more research is needed. This is an especially critical need in the area of basic remedial education for adults. Recently, the U.S. Department of Education and the U.S. Department of Labour have been funding pilot projects and demonstration projects that deal with workplace literacy. The best workplace programs use materials closely related to work - an approach pioneered and used successfully by the U.S. military in improving the basic skills of recruits. A small research grant has recently been given for adult literacy research. A major expansion has been proposed in the U.S. Congress, but has yet to be undertaken.

Fourth: research, development and diffusion of *training technologies*. Information technologies have the potential to bring formalized training right to the worksite - thus helping to bridge the gap between formal and informal training. Information technologies can help merge training and performance support systems. Yet, at present, most existing examples of performance support systems are little more than glorified job aids. A great deal of development is needed to create the kinds of training that will live up to the potential of this technology.

Over the years, *the U.S. military has been a major supporter and developer of training technologies and learning research*. Finding effective ways to diffuse these technologies to the private sector would be useful. A program for training technology transfer was authorized in 1988 by the Congress in the Trade Act. So far, the needed funding has not been given.

In conclusion, there is *a need in the United States for stronger institutions to undertake research, development and best practice diffusion* in the fields of work organization, technology and training. To my knowledge, we do not have a sustained research and development effort comparable to the Work and Technology Program. While several research centres exist at universities, they tend to be fragmented. Often, funding for these programs will non be sustained for the protracted period of time needed to not only complete research but to assist in diffusion of results.

Ultimately, for efforts to succeed, *employers will need to be actively involved.* As I stated at the beginning my remarks have focused on areas where American firms and the American training system needs to improve. I did not discuss the many strengths of the American industrial economy. But in examining the system, there are, in this case as in every case, many areas where improvements can be made.

Industry organizations in the United States, by and large, *are not extensively involved in providing training methods and materials* to their members. Only about 6 percent of the industry trade and professional associations in the United States have training programs; and only 3 percent offer certified training. Employer commitment will be centrally important if a research and development agenda is to have much impact.

But *improving the American training system* will require much more than Research and Development: It will require partnerships among firms, workers, unions, and government. No level of government in the United States currently plays a major directed role in supporting efforts by firms to train their employees. Even the State Governments provide little support. For example, the typical industry training support program run by a State Government in 1988 only helped about 65 firms and a total of about 4,000 employees.

The Office of Technology Assessments report looked at a number of *government options for stimulating training*. Some options, such as small grants to firms for setting up training consortia or more technical assistance and support for apprenticeship, would build on existing efforts, and would not be controversial. But they also would not have much immediate impact on the training behaviour of firms. The most far reaching proposal - for a pay-roll training obligation similar to that employed in France - would be highly controversial, but would reach all firms. For much headway to be made, the United States will need to begin to see training as a kind of public good - like public education. This view is not yet widespread in the United States.

Plenary Discussion

Plenary Discussion of the Role of Labour Market Parties and Institutions Involved

K.W. West, Dortmund, Germany

Dr. Lübbert (BMFT):

I would like to make a brief comment on Mr. Bleicher's contribution: it is to support him on each of his eight thesis. This also applies to his criticism of research policy. I can do this without any conflict of loyalty concerning the present Minister for Research and Technology.

I am using two of the points Mr. Bleicher criticizes. In his sixth the-sis he states: "The first priority of technology policy is to maintain and improve the international competitiveness. Other aims such as the improvement of working and living conditions or the maintenance and the securing of our natural environment are, of course, propagated, but by no means sufficiently realized ."

The second critical remark was in thesis no. 8 which said: "If as part of government policy on technology, huge political markets are created for high-tech products, micro system technology etc, I ask: "Why does this not take place increasingly in the field of environmental and health policy?""

I am not sure, Mr. Bleicher, which period of time you are referring to in your consideration. But I have been working at the Ministry for Research and Technology for 20 years and as far as my knowledge of these 20 years is concerned, I would support your criticism whole-heartedly if you mean the period between 1970 and 1982. On the other hand, it is easy to prove that between 1982 and 1990 the amounts spent on the objective of industrial competitiveness have been constantly falling and that the amounts spent on the target of what we call "Vorsorgepolitik" (policy of precaution) - this notion means ecological and environmental research, health, "work and technology" - , have been growing more than proportionally. This means, that I support your political statements and I think, as we agree on political matters, we shall quite easily be able to agree these basic points.

Mr. S. Bleicher (IGM):

As there is so much agreement, I shall give a brief reply. I believe that we are all involved in a learning process. We, as a trade union, have learnt that in the shaping ("Gestaltung") of work and technology, the targets of humanization of labour and competitiveness are not contradictions. On the contrary: together with the employers, we have improved performance and products. At the end of this process a more competitive product was created that had also taken into consideration aspects of humanization. So there is no reason to change this policy.

As far as my criticism on environmental and health policies is concerned and their importance within research policy, I know from talks with Minister Riesenhuber, that a great deal has been done on environmental research. Considering the weak position of this policy in the research budget, considerable increases have been made. But we also have to talk about priorities in research policy. When I see what kind of sums are invested in space research, I think it is time to reconsider political priorities and a trade union politician has the right to discuss these matters.

Prof. Edquist:

I have a question to Mr. Lindholm. You were talking about over-mature industries and restructuring and you said, it never stops. This objectively means, that sectors and products and jobs disappear, but also hopefully that new ones will come instead.

Another of your main points was, that the new world of work, how working conditions could be improved, the importance of learning and so on.

My question or comment is: Could you connect these two? Maybe, the first one has more important impact on working conditions than the second one. New products, new industries certainly do not always lead to good jobs. We have for example electronic assemblies and so on. At the same time, old products and industries disappearing as well as very highly skilled and very rewarding jobs. I would like to hear your comment on this.

Mr. R. Lindholm (SAF):

You were not quite sure, whether it was a comment or a question. I think it was a good comment, because you have to speak about general trends in development. It becomes very much an attempt to describe typical or on average or common features and around these typical features there are of course, as you indicate, others.

First the process where overmatured businesses move away, is very painful and also geographically very unevenly distributed. Very often we can see mature industries disappear into - what we call - "one-company-towns". There are lots of people living there. They have their roots there, their houses and families. Suddenly they lose their jobs and there are no new jobs emerging. Of course, this affects in a very negative way working life and people. We can observe it everywhere in Europe.

In some cases very high levels of skills are becoming just inadequate, they are not used any more. In overmatured industries, there are many qualified jobs demanding educational training and practical experience for many years, before an employee can do his job.

But we must live with, and as I see, there is only one think we can do: to accept that, before regions with mature industries run the risk of losing their existence. In these areas strong efforts should be made to promote the development of new businesses, to promote the expansion of small businesses which may exist and thirdly, to promote that people in mature businesses can have a chance to start new businesses for example in entrepreneurial programmes etc.

But this is number one. We cannot say to ourselves that the re-structuring process is finished now and we have reached a new stable situation. Instead, we must be aware of the problem, that many of these processes are going to happen everywhere in Europe. We have to be aware of the fact, that we have to be active in that process and that we cannot come afterwards and try to cure. I think, that there is a very interesting process going on in the business community.

From the political side one can organize this by trying to offer good infra-structures. But the starting of new businesses must take place within the business community. I think we have a great responsibility in this concern, on the one hand those people from the research side who are present here today on the other hand those from the business communities and not to forget the trade unions. They all must try to develop policies and strategies speeding up constructive, positive transitions of the mature business community.

Contributions to Parallel Sessions

A

Economic Versus Human and Social Goals in Work and Technology Research

Developing Job Content Raises the Quality of Production

Monica Breidensjö, Stockholm, Sweden

There have always been *human reasons for banking on human resources at work*. But the level of work injuries and sickness absence shows that these arguments have not been sufficient. Labour law and legislation on the working environment have also been strengthened in many countries, but not even the legal arguments have been strong enough to bring about a constructive process of change at work places.

I would like to consider *three topics* which, in my view, will have to be right at the centre of the future development of working life. They are topics *on which research is needed*, but there is *an equal need for development work and practical change* on the spot, at workplaces, and in this process the employees themselves have a vital part to play.

1. Work Environment Improvements and Personnel Development: a Paying Proposition

The *occupational environment* and the economy have been topics of growing discussion in Sweden recently. I use the term occupational environment or working environment in the broad sense, so as to include all conditions at work with human consequences. *People* working in bad occupational environments *pay a high price* for doing so. But new research findings have shown that *companies and administrations too have a price to pay* in the form of sickness absence, personnel turnover and early retirement. But these expenses are not included in corporate calculations. No doubt there are several reasons for this but there are two main ones: *companies have not had models* for estimating the cost of "bad human resource management", and many have believed the cost to enterprise to be negligible, on the strength of public commitments through social policy.

Now companies work with visible things, and so the costs associated with the working environment have to be made visible and quantifiable. We need a better realisation of *the value of investing in personnel development*; minimum resource inputs are not good enough. Only too often this means, for example, that later on one has to call in consultants which is more expensive. Instead the occupational environment and personnel policy must be put *on the board room agenda* and connections between measures in these fields and operational results must be made clear.

The cost of personnel development has often been discussed without any similar consideration being given to the *benefit side*. Very often there are resources which can be committed to active personnel development measures, instead of passively accepting the cost entailed by high personnel turnover, a high rate of sickness absenteeism and so on.

To illustrate this point I would like to quote an *example from a Swedish municipality*. That municipality had 50 full-time and 10 part-time pre-school teachers. Their sickness absenteeism and personnel turnover were higher than for other employees - 25 per cent and 35 per cent respectively, as against 10 and 12 per cent.

As an alternative to just letting things take their course the municipality decided to try and reduce sickness absence and personnel turnover to the same level as for other employees. It was decided to invest the money which would then be saved.

The theory was that, if *personnel* were allowed to *develop and improve their competence*, this would make them more committed and confident in their role. Dissatisfaction would become a less potent cause of absenteeism and there would be a reduced failure rate. This in turn would mean less money spent on temporary staff replacements. Less personnel turnover means reduced hiring costs and less money spent on introduction and initial training. It was also anticipated that the staff would work more efficiently. People planning to leave are probably less efficient for a certain period before they actually do so.

The *development inputs* planned included, for example, field trips, education, travel, replacements for staff on education leave, more generous supplies of disposable materials, a re-organization of work itself, and certain alterations to the premises.

Where did the money come from for all this?

An estimate of the cost of advertising, interviews, introductory measures and initial training and all the rest of it showed that the *total cost of one employee leaving* was something like SEK 88,000.

By *reducing turnover* among pre-school teachers to the average for all its employees, the municipality could cut its annual expenditure by SEK 232,000.

If *sickness absence* also declined to the average level, expenditure would fall by SEK 270,000 a year. Then it would no longer be necessary to close pre-school facilities for a fort-night, which would mean SEK 300,000 additional revenue.

Then again, if *staff dropouts* were reduced by one employee per annum, this would save the municipality SEK 195,000.

Altogether these *savings* came to about MSEK 1, money which could be put into active development measures. The accuracy of these figures, of course, is open to question, but the whole point of cost-analysing the personnel situation is that it conveys some idea of how profitable investment in personnel development can be.

Of course, personnel cost-benefit analyses like these are not the whole secret of a good personnel policy. A more long-term strategy is also needed.

2. Personnel Development Becomes a Competitive Asset

The so-called *work strategy of Swedish labour market policy* is based on investigations in active measures instead of just doling out cash. The work strategy needs *to be developed* more and more *into a strategy of education and competence*. Temporary employment for the jobless is not enough. They have to be enabled to acquire the competence needed in order to hold their own in the labour market.

At the same time, future competition will be based increasingly on the quality of products and services. Retrospectively, the *division of labour* can have been a rational means of achieving efficiency and productivity. Developments along those lines have now reached a point where they are *detracting from efficiency*. And in any case, the quality and efficiency of service production - the real growth sector - cannot be controlled by sub-dividing job content.

A development of technology, competition and markets which is now taking place tells that *quality is assuming a much more pivotal role in production*. Quality as a factor of production means that activities are sustained more by the inputs of individual people than by complete technical systems. This being so, neither job definitions nor qualification requirements are appropriately defined today. Working life is going to be dominated more and more by intellectual production. The development of knowledge is never really predictable. It is a process of explorations, so that we cannot tell until afterwards what is right or wrong, good or bad.

The *element of intellectual production*, of course, is more conspicuous in some jobs than others. But it is present in the majority and, above all, it is getting more and more important all the time. Education, medical care, communications, financial and administrative services, industrial development and so on, all incorporate powerful elements of this knowledge-based process of exploration. Such work is characterized by the analysis of needs, by problem-solving, by individual adjustment and, in a word, by unceasing development.

The term *"intellectual enterprise"* has recently come to be used to denote companies *in which the employees' competence is the main factor* of production and the essence of net worth. Most companies and administrations of tomorrow will be intellectual enterprises. It is important, however, that research and development projects should cover the whole spectrum of working life, so as to avoid polarization between different sectors.

Working conditions in industry, and in services, too, have long been based on people's adjustment to the physical, technically determined flows of production processes. But today people are having to enter every stage of the process with *the opportunity and task of redirecting, adjusting, changing and improving things*. A decentralisation of decision-making and genuine participation in the working process have become essential prerequisites of the quality of production.

Which people, then must be allotted more development inputs, and what are the main lines of development? First and foremost, we must *broaden our concept of development inputs*, so that instead of concentrating mainly on key groups they will come *to include whole operations - all employees.*

It goes without saying that people who are not allowed to assume responsibility, to develop and to influence their work will be less committed to it than those who are able to make full use of their knowledge and experience at work. At the same time as *development at work* confers personal gratification and, moreover, should be looked on *as a democratic right*, it contributes towards the development of the enterprise. This has been more and more widely realized. It is now coming to be understood that the great mass of workers and junior salaried employees with one-sided jobs could, given *a different organization of work and a new personnel policy*, help to elevate the quality of activities.

But this challenge is accompanied by *many questions*. What are the new working models that need to be created? What changes are needed in order to give working life a developmental orientation?

At the same time as we have to create job content with the kind of balance which experience tells us that people will find engaging and positive, we also have to be careful not to create new residual jobs. *Job content must*, all in all, *draw on more of the employee's competencies.*

Measures to improve competence must not be taken at random or as an ordinary social policy measure for the personnel. Instead, *inventories of needs and strategic thinking* are called for, for the sake of both efficiency and a distributive fairness. In several surveys young persons have referred to opportunities for development as the main factor swaying their choice of occupation. This is a key issue in working life today.

In addition to vocational competence, though, employees must also be given an *opportunity of developing operational competence*. This can be done, for example, if they are allowed to perform various non-quantifiable tasks, cross boundaries into other operational fields and so on. This presupposes greater liberty at work.

3. Women and Technology - a Development Field

Development inputs need to put *more emphasis on women's roles and their experience and opportunities at work*. Up till now women have played a subordinate part in most activities, and their vocational identity itself has been fashioned by men. All that is now changing in some quarters. *Nurses* are breaking free of their role of doctors' attendants in order to create a caring profession for the promotion of health. This entails a new relationship to work, new developmental opportunities and, moreover, incentives for transforming activities everywhere in the caring sector.

Working life research has shown that *women organize work and solve problems differently from men*. There is a great deal to suggest that, if women's perspective and competence are made use of, the quality of working life can be improved from the viewpoint of employees and production alike. A very great deal could probably be gained if the distinguishing *characteristics of female leadership* - intuition, holistic vision and relations - could be *included in future development*. These qualities are particularly important for restoring the vitality and creativity of personnel - which in turn is crucial to corporate survival and, moreover, demands trust. The leadership exercised by tomorrow's managers will have to be delegated from below, rather than top-down.

Research into techniques and technology is an important field of working life research. Owing to the sexual segregation of working life, women and men have different functions and positions in relation to techniques and technology. But research in this field has mainly been geared to masculine fields. This only began to change with office automation. Since *offices techniques/technology operate in a female operational sphere*, it now became interesting to investigate whether women's relations to techniques/technology were different from men's.

In this field there are *many questions with a vital bearing on the development of working life* which have yet to be answered:

- What are the reasons for women not taking part in the development of technology and how are technology and work affected by this?
- What is the rate of technical change in the employment sectors of men and women?

One prerequisite for the optimum development of work in occupations and sectors where women predominate is a build-up of knowledge in the field of 'Women and Technology'.

Occupational Safety and Economic Efficiency - Procedures for Extended Economic Efficiency Calculations

Rainer Thiehoff, Dortmund, Germany

1. Introduction

"Extended" economic efficiency procedures are in no way only investment accounting methods with financial refinement, as might appear at first glance. Principally, they imply a planning concept which involves man at the work place in all his many-sided significance for the success of an investment.

The economic evaluation of in-company occupational safety and health measures remains an unresolved problem. Its effectiveness is, in the final analysis, to be judged against the background of market economy problem-solving in terms of efficient and low-cost organisation of production. This means that the anticipated cost and benefit effects of measures relating to occupational safety and health will, as with all other in-company activities, be evaluated on the basis of corporate objectives. Accordingly, such measures should be treated as company investment decisions.

The involvement of occupational safety and health in a company's economic functions is no longer concentrated solely on the cost side, as has been usual to date. It is *the attainable economic benefit of humane and healthy work places* which, to an ever increasing degree, interests planners and decision-makers in the companies. This applies all the more if man becomes a decisive factor for the economic success of an investment. For it depends increasingly on the human factor whether the new technologies can be mastered or not.

Extended economic efficiency procedures *assist planners in the systematic exposure of weak points* and in the consideration of non-quantifiable target contributions when assessing alternative investments. The greater transparency of the planning outcome afforded by revelation of the evaluation process and the subsequent ability to check the forecast utility values (work system values) also give the person who prepares the decisions greater competence in practical terms.

2. Deficiencies in Traditional Economic Efficiency Calculations

A characteristic of every business investment decision, and particularly of every innovation decision, is the need to weigh up profit possibilities and cost risks against one another, in spite of incomplete and uncertain information. The central problem for every economic efficiency calculation is the varying degree of certainty associated with the anticipated flow of incomings and outgoings. While the costs can normally be ascertained with adequate certainty, a *predictive assessment* of the benefit is very much more difficult. This applies from the uncertain market success of products through to the indirect effects on productivity due to improved working conditions.

While directly product-related planning data can be expressed in terms of monetary amounts or at least in countable units, the *effects of occupational safety and health measures* are primarily a *qualitative phenomenon* which usually manifests itself indirectly, e.g. through improvements in flexibility, quality or productivity. This "awkward" characteristic of indirect planning data explains why it is frequently the case that in investment plans no attempt is made to include in the decision-making calculation beneficial effects which cannot easily be quantified.

Clearly it is precisely measures of occupational safety and health which are at a particular disadvantage here. Procedures have therefore been developed which *go beyond the assessment possibilities of traditional economic efficiency calculations*. The usual dynamic investment calculations typically orient themselves towards the so-called capital value. The decision-making criterion for assessing the benefit obtainable from an investment is accordingly the sum of the future period surpluses discounted up to the present time. On the basis of reliable expectations, comparison is made only between those flows of incomings and outgoings which relate to production and are capable of being ascertained. No account is taken of areas of operation previous and subsequent to production nor of qualitative effects (assessment criteria which cannot easily be quantified) resulting from technical and organisational changes.

A further *deficiency* of traditional economic efficiency calculations is the over-emphasis placed on the *short-term profit motive*. The operating profit is indeed the most important aim of any company but it is never the only one, and it has a time-related dimension.

This can be seen, for example, in the problems of argumentation frequently observed in connection with the introduction of new technology. *Although flexible production technology* does not represent the most profitable investment alternative, it may be necessary for a company to install it. Short-term profit possibilities are dispensed with for strategic reasons, if survival on the market can be ensured in this way or presence on the market extended. In such a case, the desire to secure the company's long-term position predominates over the short-term profit motive.

3. Corporate Objectives as Decisive Assessment Criteria

Only against the background of corporate objectives and their individual weighting it is possible *to assess alternative investment projects* according to the most favourable combination of achievement levels for fulfilment of the various objectives. In practice, however, this takes place systematically only in very few cases. Consequently, account fails to be taken specifically of those risks and possibilities which arise from human factors.

A company which wishes to safeguard its profitability and viability in the long term needs a *"holistic" planning concept*. The latent, as yet unused productivity potential can only be tapped if one manages to link the individual factors of technology, organisation and work with one another in an optimum fashion. For this purpose, however, the one-dimensional investment planning which has prevailed to date has to be structured in a very much more comprehensive fashion. Assistance here is provided by the "extended economic efficiency procedures". These cannot be regarded as a first step in the direction of a holistic investment planning concept but they do indicate a practical way of taking account also of those assessment criteria which are difficult to quantify when making investment decisions.

Within the framework of extended economic efficiency procedures, consideration is given not only to the familiar profitability figures but also, basically, to all those *costs which are difficult to quantify* or to those costs which cannot be measured in monetary units, insofar as they are important for the assessment of an investment. On taking the example of occupational safety and health, however, it becomes especially clear which economic possibilities and which potential risks exist in that field. The expected consequences with respect to the following aspects, for example, frequently represent contributory factors which need to be considered when deciding upon the amortisation period:

- Flexibility of production,
- Non-productive and outrage times,
- Throughput times,
- Reject rates,
- Quality,
- Qualifications of employees,
- Work satisfaction and working atmosphere,
- Willingness to cooperate,
- Prestige on the labour, procurement and sales markets.

However, by considering assessment criteria which are difficult to quantify, it at least becomes possible to avoid expensive rectification measures for unsafe production facilities or for the elimination of ergonomic deficiencies.

Methods and instruments for extended economic efficiency procedures have now been refined to the point where even a general survey would go beyond the scope of this paper. The following remarks will be therefore be limited to *characteristic features* which have shown themselves to be essential during the course of practical application.

4. Characterisation of Extended Economic Efficiency Procedures

To make it possible to choose between investment alternatives, it is first necessary to have *objective-related criteria and data* on achievable levels of objective fulfilment. For this purpose, the entire work system has to be subjected to a thorough-going analysis. This is the first step in all investment planning procedures. Only in the second step are the results obtained to be assessed with regard to the various investment possibilities. The degrees of objective fulfilment are compared and finally condensed to form an overall judgment.

Extended economic efficiency procedures of an *essentially "analytical" nature* are employed for recording a company's economic efficiency figures, with the aid of various investigative methods, and for presenting them in a clear fashion. This ensures that consideration is also given to the non-monetary assessment criteria for an investment. A comparative study of the alternatives, however, is either not conducted or is left in the form of an unstructured draft to be dealt with by the people involved in the planning process.

It should be said that in practice the more serious problem is, specifically, *the assessment and selection of the best investment alternatives*. On the one hand, the company decision-makers demand an objective-related preparatory treatment of the relevant indicator results, incorporating the clearest possible recommendations. On the other hand, those responsible in the company for preparing the decisions are expected to refrain as far as possible from making a subjective assessment.

Instead of evading this conflict, the "assessment-oriented" economic efficiency procedures expressly include *the process of finding a consensus*. They make reference to the inevitability of subjective assessment influences, and this is certain to be the case whenever the knowledge of experts and affected persons is involved. It is even regarded as an information advantage, since the company decision-makers obtain valuable indications as to the future effects of an investment alternative.

The approaches adopted for assessing investment alternatives *differ fundamentally*. Procedures based on human capacity calculations attempt to set up functional relationships between measurable loads/strains and monetisable effects/consequences. Such causal relations are, however, on the whole only verifiable with extreme difficulty, and never without residual doubt.

For this reason, economic efficiency procedures which additionally make use of *methods of utility value analysis* dispense with this rather uncertain argumentation. Instead, they supplement the detailed conventional investment or cost-comparison calculation with a utility value analysis for those assessment criteria which are difficult to monetise or which cannot be monetised. By means of weighted point ratings (e.g. for the work-system value analysis), it is possible to render mutually comparable the levels of objective fulfilment for organisational solutions of varying suitability as regards human-related and health-related needs.

Practical experience with extended economic efficiency procedures in the industrial sectors of electrical engineering, mechanical engineering, iron and steel production, foundries, chemicals, and in the office and administration sectors, show that in practice the assessment-oriented procedures are applied much more frequently than those procedures which have a purely analytical orientation. A survey undertaken recently also establishes that an element of utility value analysis is always needed to ensure that these procedures can be applied successfully.

Viewed as a whole, extended economic efficiency procedures always arrive at convincing results when certain steps are pursued and certain methodological features observed. These "success factors" clearly play a major role as regards practical implementation. They will therefore now be outlined briefly, reduced to their essential points.

4.1 Characteristic Steps

- *Establishment and concretisation of objectives*
  The objectives pursued through an investment have to be determined and laid down as a basis for planning.
- *Situational analysis*
  By means of a survey of the actual situation, weak points are revealed and analysed with regard to the factors causing them.
  The initial data are established for the actual planning work. These also include the economic efficiency data which are customary in the company.
- *Formulation of alternatives*
  In order to develop an optimum solution, it is essential to formulate "conceivable" alternative investment projects which take account of technical, organisational and personnel-related variations.
- *Elaboration of assessment criteria*
  Those assessment criteria which are difficult to quantify have to be made "measurable" in order to identify the optimum investment alternatives with a view to concretisation of objectives.

- *Economic efficiency calculation and utility value analysis*
An investment or cost-comparison calculation and a utility value analysis are each drawn up on a completely separate basis for the planning alternatives.
- *Presentations of results*
The results of the economic efficiency calculation and of the utility value analysis are prepared in a clear form in order to make the results of the assessment easily comprehensible to the decision-makers.
- *Subsequent verification*
After they have been introduced to the planning, the economic efficiency characteristics and work system items need to be verifiable in order to allow comparisons to be made between the plan and its implementation, i.e. "learning from experience". This also ensures that realistic assessments are made as early as the planning phase.

The subsequent verifiability of the planning outcome is an extremely important criterion for verifying the competence of the planners. It is precisely the ability and willingness to reveal the assessment process which assists the decision-makers in arriving at their decisions.

4.2 Methodological Features

- *Team-work*
Planning is conducted in a team. For this purpose a planning team is formed from company experts and/or persons affected by the planning. This team meets regularly over a fixed period.
- *Planning which extends beyond individual subject areas or divisions*
Representatives from all in-company departments concerned - at all events from the planning, business management and production departments (stage, area, works etc.), but also from those company divisions which are previous and subsequent to the above, together with representatives from the occupational safety, ergonomics, personnel and central data processing departments, as well as from the works council etc. - all cooperate with one another in the extended project team. In this way it is intended that as many different viewpoints as possible shall be incorporated into the planning and assessment processes.
- *Methodological approach to the project*
The methodological instruments available to the project are used for making a structured approach within the intended procedural sequence (e.g. brainstorming, metaplan techniques, plenary and small-group work).

In many projects, the approach outlined here has proven its worth in practice. In many cases, however, an external consultant was engaged, who coordinated the overall process, or at least the team-work phases. A seminar guide has been compiled to assist the independent application of this extended investment planning concept or to facilitate appropriate training for personnel. This guide ("Better Planning") was prepared by M. Auch (IFAO Karlsruhe) and it forms a constituent part of a research report: B. Gottschalk, Wissenschaftliche Begleitung der Umsetzung erweiterter Wirtschaftlichkeitsrechnungen, Schriftenreihe der Bundesanstalt für Arbeitsschutz, Forschungsbericht Nr. 598, Dortmund 1980, Anhang I.

The new procedure for the integrative planning of technical innovations assists the "holistic" planning aimed at by the company. For the decision-makers, it is also worthwhile to incorporate as a preventive measure the knowledge possessed by experts and those affected by the planning, because provides valuable indications as to the future economic success of the various investment alternatives.

5. Advantages and Disadvantages of the Holistic Planning Approach

Extended economic efficiency procedures, however, also have *weaknesses*. The work and planning effort involved is clearly greater compared with traditional economic efficiency calculations alone. On top of this, extended economic efficiency procedures require co-operation between various company divisions and specialist disciplines, and particularly between cost accountants, organisers and planners, as well as ergonomists and occupational safety specialists.

Companies attach paramount importance to the *comprehensibility of the analysis results*. Company decision-makers justifiably demand that the manner in which a recommended decision is arrived at shall be transparent and that the result of the assessment shall be verifiable at all times. The acceptability of the analysis result depends directly on its comprehensibility.

The value of a multidimensional assessment procedure is also assessed according to whether the result can be made comprehensible to the company decision-makers. *This means that the quality of the argumentation* also has to be considered. If, for example, point rating procedures with comparatively large uncertainty are involved, the results of this part of the analysis will have to be clearly delimited from the monetisable assessment criteria. Finally it should not be overlooked that the form in which the assessment criteria are presented also contributes to their acceptability.

All in all, however, the *advantages* of extended economic efficiency calculations predominate. The greater planning effort involved in multidimensional assessment procedures is countered by *the reduction of uncertainty* with regard to the business-economic effects of a decision in favour of a certain investment. It is precisely the imponderables in the introduction of a new technology which increase the willingness of companies to give systematic consideration to the knowledge possessed by experts and those affected by the planning as early as the planning stage.

In addition, there is a desirable side-effect. The verifiability of the economic efficiency characteristics and utility value points *allows comparisons* to become possible between the plan and its implementation. This automatically ensures that realistic assessments are made right from the planning stage.

The initial benefit for occupational safety and health is the fact that such planning procedures ensure that the employee-related criteria are not simply forgotten, as they frequently are. The crucial advantage is that occupational safety and health is integrated into the investment planning as an objective and that investment resources - albeit sometimes only modest - are provided for the safe and healthy design of work places and the working environment. At all events, it serves to prevent the need for subsequent (and expensive) "rectificatory" humanisation.

6. Participation in the Planning of Investments

All those responsible for in-company occupational safety, and especially ergonomists and occupational safety specialists, are being called upon to involve themselves more than before in investment planning. The extended economic efficiency procedures support the arguments in favour of this, extending well beyond the avoidance of accident costs. The involvement of such personnel in investment planning opens up the possibility of considering important occupational safety and health criteria already at the planning stage of work systems.

The incorporation of their experience as experts and affected persons in weak-point analyses and in the determining of utility values enhances the rational character of investment decisions. *The investment-related residual* risk is reduced because the company decision-maker now has better information as to what to expect as regards those effects of different work systems which are difficult to quantify.

The idea of extended economic efficiency procedures is also gaining *increased international* recognition, and this not only in Europe. Examples of this are the second "International Production Management Conference" on management and new production systems and the international conference held in Finland on the subject of "Productive Working Environment - Cost-Benefit Analysis".

Workers' Participation - An Instrument to Achieve Productivity, Innovation with Human and Social Goals

Hubert Krieger, Dublin, Ireland

1. Participation in New Information Technology Can Support the Modernisation of European Enterprises

In order to remain competitive in world markets both now and in the future, the European Community needs to develop and apply new information technology. This technology is highly flexible and permits different forms of application in enterprises. To achieve optimum results and to make full use of its productivity potential, a skilled, co-operative and motivated workforce is needed and traditional hierarchies and forms of control in enterprises need to be revised. Much more involvement by employees and their representatives in the introduction and application of new technology can assist in making the modernisation of European enterprises a success.

The organisations of the European unions and employers and the European Commission have formulated their intentions on this topic. The union organisation, ETUC, favours extensive participation rights aiming at a model of industrial democracy. The European employers' organisation, UNICE, is open to all forms of workforce involvement, but according to existing national traditions and legal practices. The European Commission has adopted a role of facilitator rather than initiator of participation procedures. At Val Duchesse in March 1987, both employers and unions reached a common understanding that in the process of technological modernisation, employees and their representatives should be well informed and consulted in the process of technological change.

2. Foundation's Survey: Judging the Value of Participation and Estimating its Future Potential

Against this background, the European Foundation for the Improvement of Living and Working Conditions in Dublin carried out an attitudinized survey in all Member States of the European Community on the extent, impact and value of participation in actual situations, and on the future potential for increased participation of workforce representatives.

The main premise of the survey was that the involvement of the parties concerned in the introduction of new technology in European enterprises could contribute to a smooth functioning of these technologies, to an improvement of working conditions, to an improvement in industrial relations and to a more balanced distribution of costs and benefits of technical modernisation.

3. Largest Survey in Europe

The survey was designed to obtain factual information and to gather opinions from managers and employee representatives. The study concentrated on mechanical engineering and electronics to represent the industrial sector, and on banking, insurance, and retailing to represent the service sector. The interpretation of the results has to take account of these sectional limitations.

The survey was conducted between February 1987 and October 1988 in two waves. 3,663 Managers and an equal number of workforce representatives were interviewed, making a total of 7,326 respondents. This is the largest survey to be conducted on the topic of workforce involvement in the introduction of new technology in Europe. The data sampling and presentation was organised by Harris Research (London) and GFK (Nürnberg).

4. Intensity of Past Involvement in Technology Introduction: Higher Levels of Involvement are Comparatively Rare

Because of the distinct differences between each Community country, the study adopted a wide definition of involvement. Involvement in the introduction of technology was defined as "any participatory procedure or practice, ranging from disclosure of information to consultation, negotiation and joint decision-making, which formally or informally involve the parties concerned with the introduction of new technology in the discussion of decisions concerning the process of change".

The introduction of new technology can be viewed as going through the phases of planning, selecting, implementation and post-evaluation of technology; the first two phases are strategic, the latter two are operational. Data analysis concentrates on the strategic planning phase and on the operational implementation phase of technology introduction.

In the planning phase, workforce representative involvement is very low. According to two out of five managers there was either no involvement at all or employee involvement was restricted to the mere receipt of information; around 10 % stated that employee involvement was characterised either by consultation procedures or by negotiations/joint decision-making. Employee representatives produced a broadly similar response.

The pattern of involvement looks different in the operational phase of technology introduction: "No involvement" decreases sharply, information provision remains static at around 40 per cent, while consultation occurs in about 20 per cent of the companies. The percentage of respondents who indicate negotiation/co-determination practices rises to 16 - 18 per cent (see figure 1).

There is a notably high degree of consensus between managers and workforce representatives as to the different intensities of participation in both phases of technology introduction.

5. Content of Past Participation: Low Involvement in Management Concerns and Intensive Participation in Workforce Concerns

In substantive management functions such as defining market strategies and investment criteria, most management respondents (about 60 per cent) agree that there was no involvement at the time the survey took place. Between a quarter and a third indicate that participation was characterised by information provision. Consultation, negotiation and joint decision-making played only a very minor role.

The situation differs considerably in matters of traditional workforce concerns, from work organisation to health and safety measures. About half of all management respondents reported that consultation and co-determination procedures were in evidence (see figure 2).

6. Quality and Channels of Information: High Information Quality in Only Half of the Companies

All participation procedures depend on the relevance of information that employee representatives receive, and information provisions were the most often single involvement procedure cited. Half of all employee representatives received complete, timely and useful information. 13 per cent reported very poor information provisions on all three counts (see figure 3).

7. Positive Effects of Participation Clearly Prevail

The most general result of indirect participation on various issues is that employee representative involvement either had no impact at all or it had positive effects; negative repercussions were rarely reported.

The experience of participation has generally produced favourable effects as far as managers are concerned. It rarely slows down the decision-making process; industrial relations are improved; it assists in the quality of decision-making; it enhances the identification of employees with the goals of the enterprise; it pays adequate attention to the concerns of both sides; it facilitates the utilisation of skills and engenders a greater degree of acceptance of new technology in the minds of employees. The responses of employee representatives produced a broadly similar figure (see figure 4).

8. Future Involvement: Both Parties Want More, but Employee Representatives' Intentions are Much More Far-Reaching

At the European level, there is an overall positive climate towards increasing participation practices in the future. In the planning phase of technology introduction, more than half of all managers who did not involve representatives in the past intend to introduce some form of participation in the future and the percentage of managers favouring information as the best form of participation remains generally static. However, there is a significant increase in the numbers of managers favouring consultation and negotiation/joint decision-making procedures in the future (see figure 5).

As to the technology implementation phase, the non-involvement of employee representatives in the future plays only a minor role, but information, consultation, and negotiation/joint decision-making are favoured by almost equal numbers of managers with about 30 per cent favouring each option (see figure 6).

Employee representatives have much more far-reaching aspirations for the future; in both phases non-involvement is clearly a minority position and information provision is regarded as inadequate. Employee representatives clearly opt either for consultation, and particularly for negotiation/joint decision-making (see figure 7 and 8).

9. Survey Results in the Political Context

The political actors at the level of the European Community, the European Trade Union Confederation, ETUC and the employers' organisation, UNICE have differing views about the participation issue in the European Community. ETUC favours a model of a comprehensive co-determination right at all levels and in all issues of company concerns in legally binding or statutory form. UNICE is opposed to such policies and favours an open situation in which entrepreneurs and managers can practice any form of participation according to existing prescriptions and local and historical traditions. UNICE rejects EC-wide legal regulations on participation. Both sides have come to a compromise in the Val Duchesse Social Dialogue; information and consultation should become standard practice in all European firms.

The survey data reveals that the reality of workforce representative participation in many European companies falls well short of the standard practice envisaged in the Social Dialogue; this is particularly true in strategic matters of company policy such as the planning of technical change and traditional areas of managerial prerogative (investment, market strategies). Although consultation and even negotiation/joint decision-making sometimes occur on operational issues (technology implementation and workforce concerns), there are still considerable gaps between reality and the text of the Social Dialogue.

This situation is likely to change somewhat in the future. To some degree, management's intentions about the involvement of employee representatives have moved away from the practices of the past. This is particularly evident in operational company matters where we see quite a distinct trend to meet the provisions set at Val Duchesse. As to strategic issues, there is also a change of attitude, but its extent is rather limited. In both policy arenas there is positive change, but it is unlikely to result in attaining the Val Duchesse bench-marks of information and consultation in many European firms, particularly in strategic issues. These results indicate that the trade union model of industrial democracy finds little favour from management in Europe.

Even the majority of employee representatives - many of them union representatives at company level - are hesitant to opt for very far-reaching participation rights in the future. As far as areas of traditional managerial prerogative are concerned, only a distinct minority wants participation in a co-determination form. However, employee representatives generally opt for involvement practices which go well beyond what management is prepared to concede but which falls short of aspiring to radical forms of industrial democracy.

At the overall European level, relations between management and employee representatives in high-tech companies have so far been largely consentaneous and co-operative and our survey results lead us to suspect that despite the differences of opinion that exist between the two sides, this spirit of consensus is likely to continue in the future.

10. Country Differences: Five Specific Factors

The extensive throughout the last decade on the work implications of technological change shows that there are a number of country-specific factors which shape the opportunities for employee involvement in the introduction of new technology. They are inter-related to some degree, and strong forms of particular factors might well have a dominating impact on the outcome of the degree of participation in technological change that emerges. There are five principal factors which can help or hinder the opportunities that employee representatives have for participation. These are:

- management's dependence on the skills and co-operation of its workforce to achieve its objectives for introducing the new technology;
- management style and its attitude to participation;
- the bargaining power of organised labour to force management to negotiate or consult with its representatives in the absence of voluntary disposition to do so;
- regulations which lay down participation rights for employees or their representatives on a range of matters at enterprise level;
- the degree of centralisation of the industrial relations system that exists in the country concerned.

11. Denmark and Germany Lead the Country Ranking

The figures below give a breakdown by country of past participation in the planning and implementation phases of technological change as reported by managers in each country. The responses of employee representatives gave a broadly similar picture in both phases (see figure 9 and 10).

In both Denmark and Germany there are favourable factors in existence which assists employee participation; in both countries there are long traditions of co-operative management practices with trade unions underpined by a system of tripartite consultation and state support for jointly agreed goals between the two parties. Trade unions are accepted by management as partners in the planning of technological change and there is a stock of necessary goodwill between both sides of industry. Management in both countries use new technology to increase their competitiveness, to enhance product quality and improve customer service; in such cases we would expect a high degree of dependence by management on the co-operation of the workforce in problem solving etc. In addition, there are extensive provisions in both countries for training and retraining in the use of new technology coupled with effective means for re-deployment, time off for employees to perform their duties and with a premium placed on job satisfaction and adequate job design.

12. The Middle-Ranking Countries: Ireland, the Netherlands and Belgium

Just below Denmark and Germany there is a group of countries which also appears in the upper half of the ranking tables: Ireland, the Netherlands and Belgium. In the case of Ireland, the industrial relations system has become increasingly centralised over the past two decades and the Irish trade union movement has a fairly high degree of bargaining power. On the other hand, there is very little legislation promoting participation and Irish managers generally do not favour participation and has a low reliance on the skills and co-operation of the workforce.

In both the Netherlands and Belgium there is extensive consultation at national level before any employment legislation is enacted and both countries have well-established works council systems with extensive rights given to their members.

13. Countries with Lower Levels of Participation

As we move down the table and apply our explanatory factors to the lower-ranking countries, the factors become more and more unfavourable. In descending order these countries are the United Kingdom, France, Spain, Greece, Italy, Luxembourg and Portugal. If we consider each factor in turn we find that only in France and the United Kingdom there is a dependence by management of the skills and problem-solving abilities of the labour force; this explains the comparatively high level of consultation in Britain in the implementation stage of technological change.

Management style in all these low-ranking countries is generally unfavourable; in the United Kingdom there is evidence of a shift away from negotiation towards more consultation during the last decade and management are becoming increasingly paternalistic in their style. However, this serves only to increase the opportunities for employee representatives to be involved at the implementation stage of technological change. In France the Auroux reforms are working very slowly, and French management still jealously guards its managerial prerogative. In countries such as Spain, Greece and Portugal there is a legacy of a dictatorial past and management in these three countries are opposed to participation legislation which has been introduced by the new democratic governments and the trade unions are weak, poorly-resourced and inexperienced. Apart from the United Kingdom and Greece, all these low-ranking countries have trade unions which are divided on political or religious lines, and even where there have been recent efforts within the wider trade union movement to forge closer links and promote a greater degree of unity (e.g. Spain, Italy and to a lesser extent France), such efforts have had very little impact as yet on participation. In Greece there has been a recent split in the GSEE.

Whilst all these low-ranking countries (with the exception of the United Kingdom and Italy) have various forms of formal, institutional works council provisions which have emanated from legislation, in many of these countries the trade unions eschew participation and have a greater preference for subsuming technological change under collective bargaining.

The survey results for Portugal stand out compared to all other EC countries. This country has the *highest* level of "no involvement" in the planning stage of technological change and the *lowest* level of "no involvement" in the implementation stage. These unusual results are largely due to the very detailed and precise legislative provisions which define the rights for information and consultation for works councils and exclude major issues associated with technological change.

14. Participation in the Future: Increased Levels Envisaged by Both Sides in every Member State

Perhaps the most important result of our survey is that in every European Community member state both sides indicated in their responses that they wanted greater levels of participation in the future. Not only did this mean that the level of "no involvement" decreased substantially in the preferences of both sides for the future compared to what existed at the time of our interviews in 1987-88, but also both sides envisaged more negotiation and more co-determination in both phases of technological change (see figure 11 and 12).

There was a substantial shift in opinion among employee representatives - much more so than managers. In two particular countries, Denmark and Germany, there was a marked preference for major decisions on technological change to be taken either by negotiation or joint decision-making. Portugal again stands out as having the least shift in opinion (see figure 13 and 14).

It was not surprising that the preferences of employee representatives for the future envisaged much higher levels of participation in technological change than did managers; this was true in all countries. What is perhaps most notable is that this disagreement between the two sides is particularly evident in four countries - Denmark, Germany, Italy and Greece.

In Italy, over three out of every five managers envisage either no involvement at all or simply the mere provision of information on technological issues whereas employee representatives have much greater aspirations.

15. Practice of Participation in the Light of the Val Duchesse Joint Opinion

In comparing the results for each EC member state in our survey the benchmark of the Val Duchesse Joint Opinion on new technologies is that employee representatives must be, at the minimum, consulted about new technology in both the planning and implementation phases of its introduction into enterprises. Of course, levels of participation involving negotiation or joint decision-making are superior to the Val Duchesse provisions.

If we consider the views of managers on past involvement, we find that in the planning phase only in Denmark and Ireland did half of the managers concerned report that consultation, or stronger levels of involvement occurred. A middle-ranking group of countries which included Germany, the Netherlands, Belgium, the United Kingdom, France, Spain and Greece all had no more than 30 % of managers who claimed levels of participation which were within the Val Duchesse guide-lines. There was almost no compliance with Val Duchesse in the cases of Italy, Luxembourg and Portugal.

As we would expect, the compliance with Val Duchesse is better in the implementation stage. Yet here only in Denmark, Ireland and the U.K. do we find a majority of managers who state that the implementation stage of technological change is characterised by at least consultation or higher levels of involvement. There is a middle-ranking group of countries where the percentage of managers reporting Val Duchesse compliance clusters around 40 %. Below that the percentage range decreases down to just above 5 % (in the case of Portugal).

16. Improved Compliance with Val Duchesse in the Future

As for the future likelihood that the Val Duchesse guide-lines will be observed, three immediate conclusions can be derived from our data. Firstly, the most striking conclusion that can be made is that there is a significant shift of opinion among both sides in favour of a greater level of participation in all the countries surveyed. It is also evident that employee representatives would prefer a higher level of participation than their management counterparts; this is only to be expected. Thirdly, the pattern of preferences for the future, in terms of the ranking of countries, bears a close resemblance to that of the past.

There are five countries from our sample in the planning phase where over half of managers indicated that Val Duchesse guide-lines are most likely to be observed: Denmark, Ireland, the Netherlands the United Kingdom and Greece. A striking observation that can be made is that Germany is not included in this list of more highly-ranked countries; only two out of five managers in this country anticipate that Val Duchesse guide-lines will operate in the future.

In the implementation phase of technological change the position with respect to Val Duchesse guide-lines is much healthier. Here there is a similar pattern to that of the planning stage insofar as the future preference for strong forms of participation. A majority of managers in all countries except Italy, Luxembourg and Portugal envisaged levels of participation in the future which consisted of participation levels involving consultation or stronger forms. On the negative side, the levels of no involvement are still worrying. France, Greece, Italy, Luxembourg and Spain all have results which show that over 10 % of managers intend to introduce new technology in the future into their enterprises unilaterally without any employee involvement whatsoever.

17. Overall Positive Trend towards Increased Participation in all Countries but Great National Diversity

Perhaps the most important conclusion that emerges from our survey results in individual countries of the European Community is the sheer diversity in the forms of participation in new technology that takes place. Taking all our explanatory factors together, it is clear that the types of participation that exist in Europe, and which will emerge in the future, are a product of the way the contours of each Member State's industrial relations system have been shaped by wider political, economic, social and historical forces.

There is evidence that all countries of the Community are moving in different degrees towards a greater level of participation in the future. Our survey shows that the experience of participation by both sides engenders favourable attitudes towards more intense forms of participation in new technology.

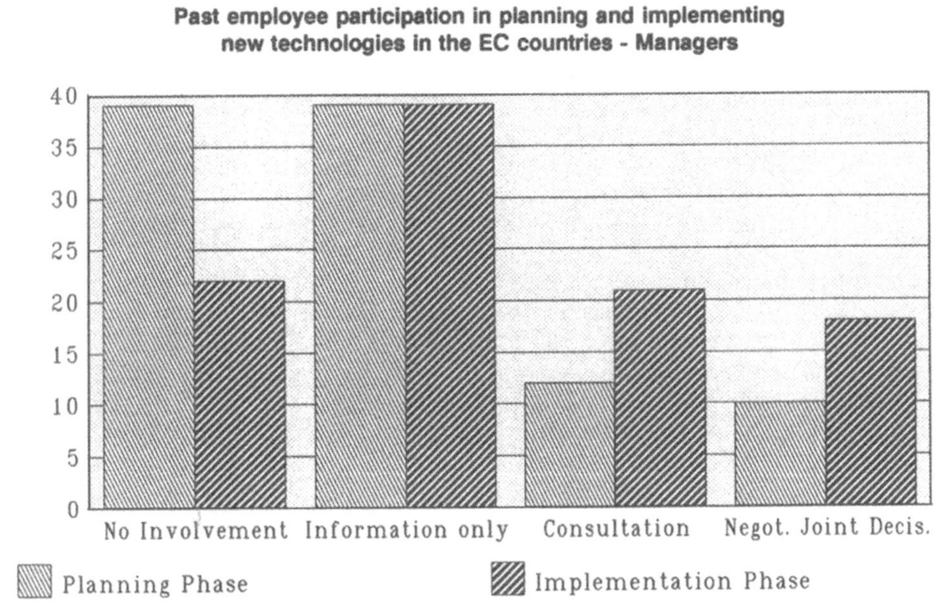

Source: Survey in all Member States, 1987-1988; 3 848 Managers

Fig. 1

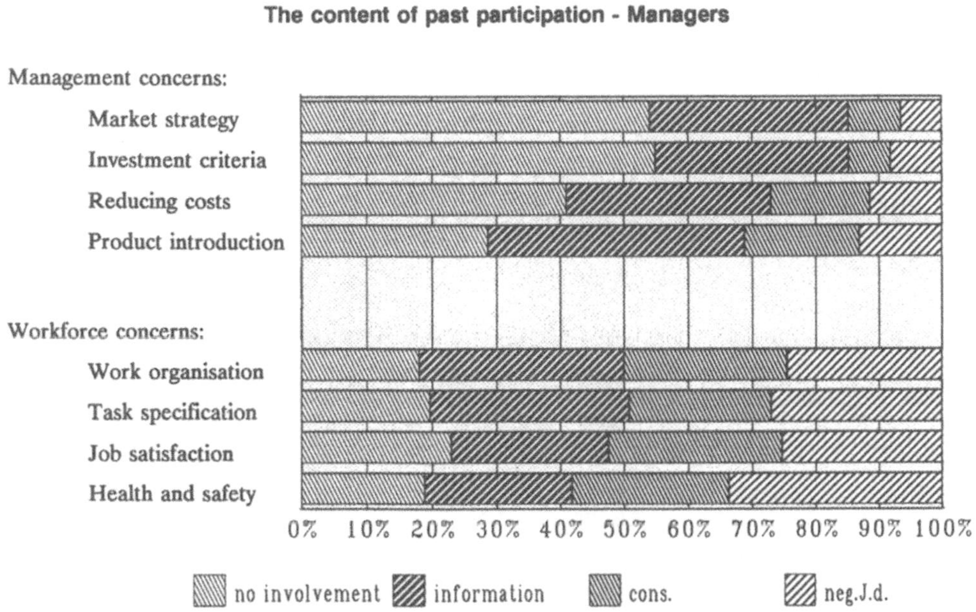

Source: Survey in all Member States, 1987-1988; 3 848 Managers

Fig. 2

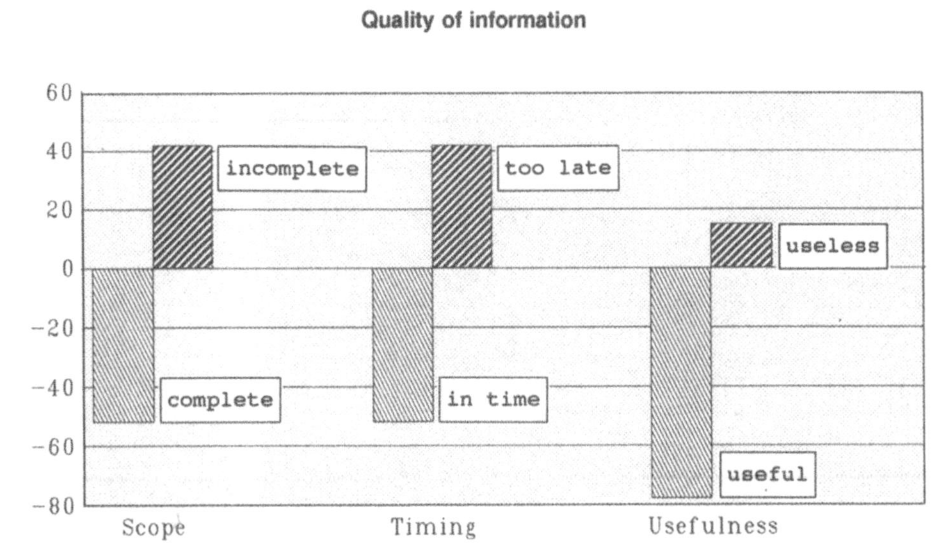

Source: Survey in all Member States, 1987-1988; 3 848 Empl. Reps.

Fig. 3

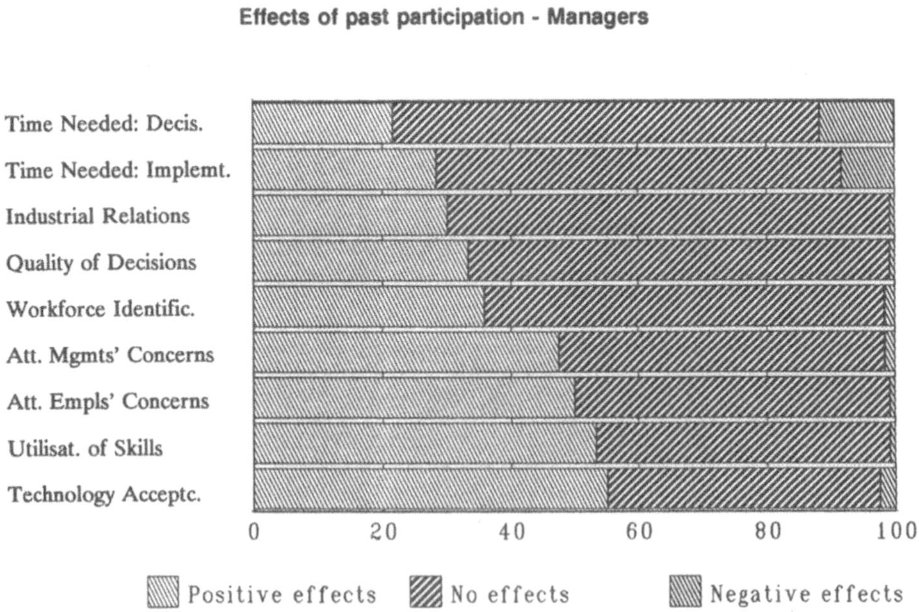

Source: Survey in all Member States, 1987-1988; 3 848 Managers

Fig. 4

Comparison of past and future involvement in the planning phase - Managers

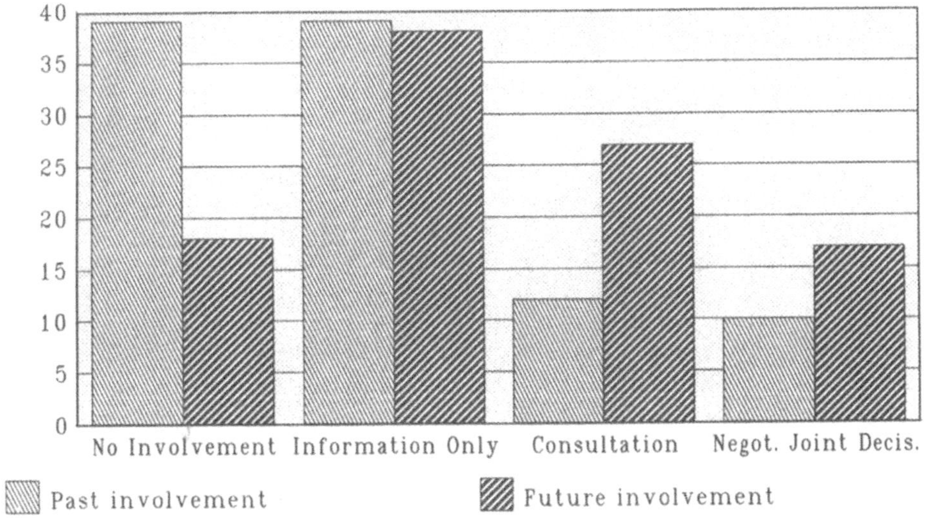

Source: Survey in all Member States, 1987-1988; 3 848 Managers

Fig. 5

Comparison of past and future involvement in the implementation phase - Managers

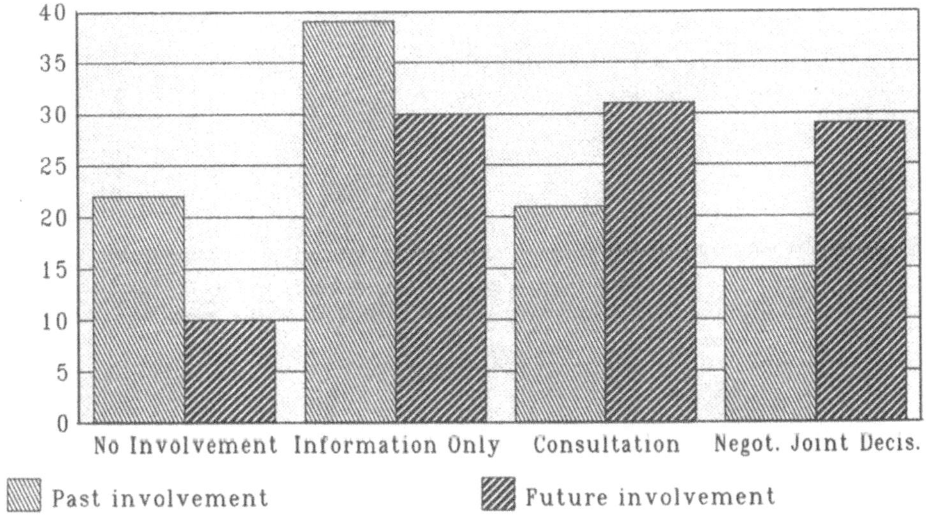

Source: Survey in all Member States, 1987-1988; 3 848 Managers

Fig. 6

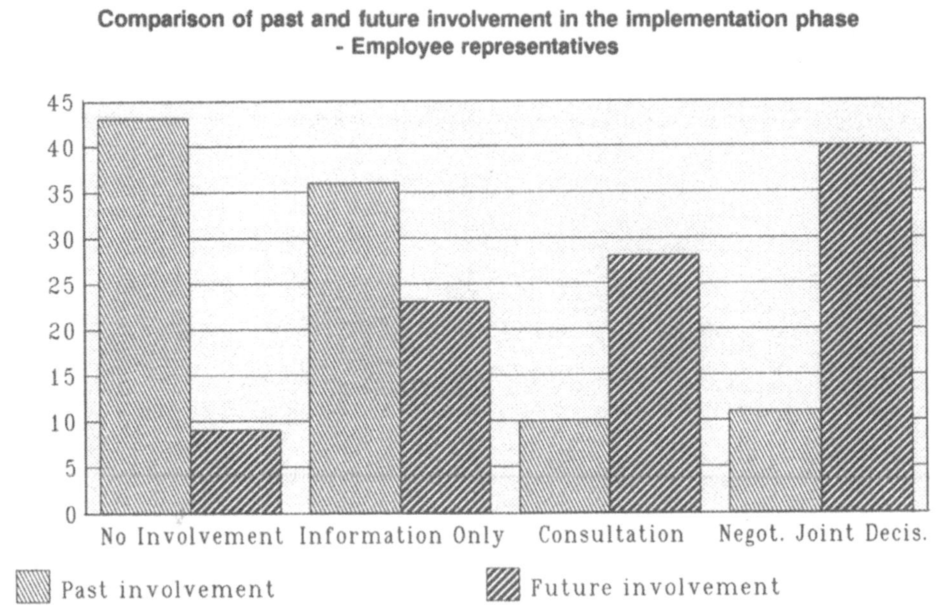

Fig. 7 Source: Survey in all Member States, 1987-1988; 3 848 Empl. Reps.

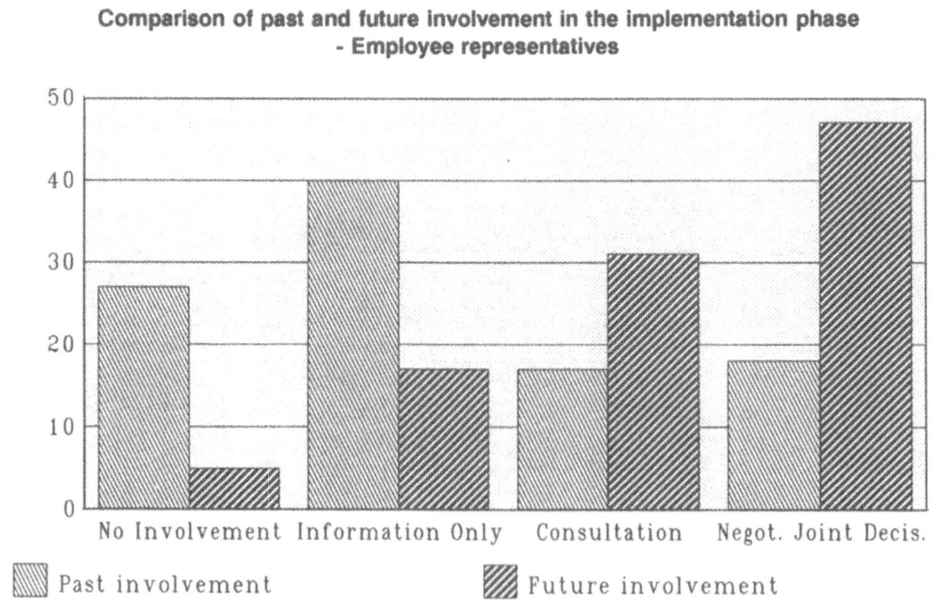

Source: Survey in all Member States, 1987-1988; 3 848 Empl. Reps.

Fig. 8

Past participation in planning for new technology in the EC countries - Managers

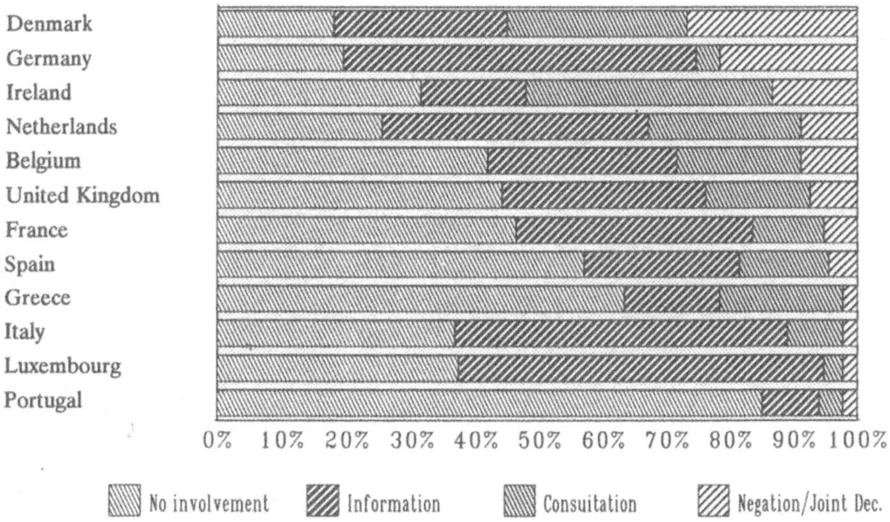

Source: Survey in all Member States, 1987-1988; 3 848 Managers

Fig. 9

Past participation in implementing new technology in the EC countries - Managers

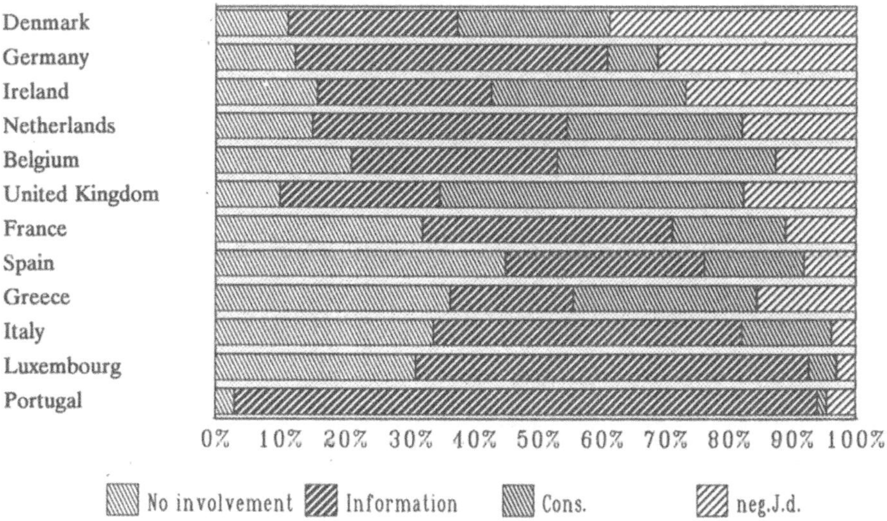

Source: Survey in all Member States, 1987-1988; 3 848 Managers

Fig. 10

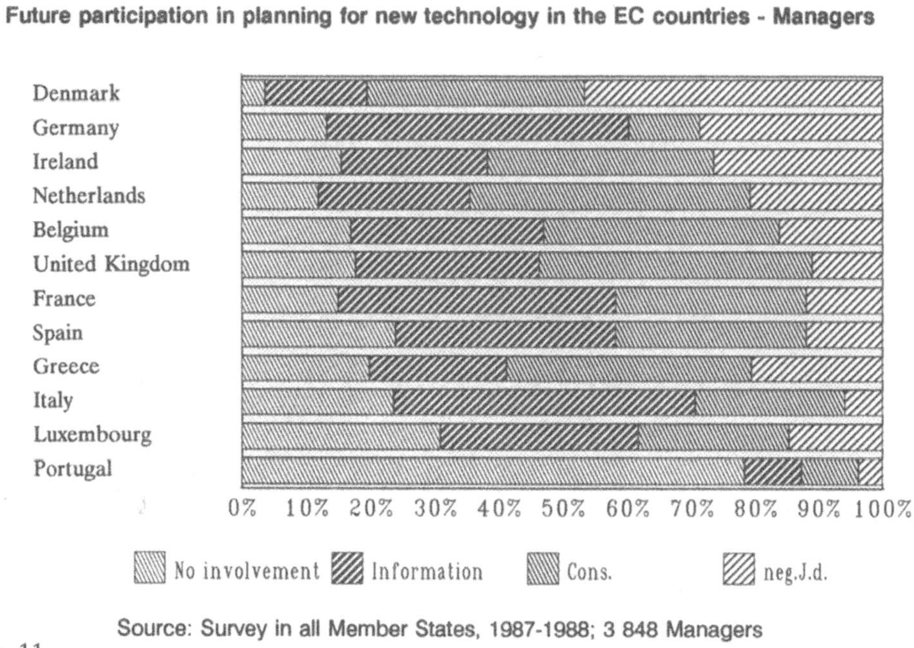

Future participation in planning for new technology in the EC countries - Managers
Denmark
Germany
Ireland
Netherlands
Belgium
United Kingdom
France
Spain
Greece
Italy
Luxembourg
Portugal
0% 10% 20% 30% 40% 50% 60% 70% 80% 90% 100%
No involvement
Information
Cons.
neg.J.d.
Source: Survey in all Member States, 1987-1988; 3 848 Managers

Fig. 11

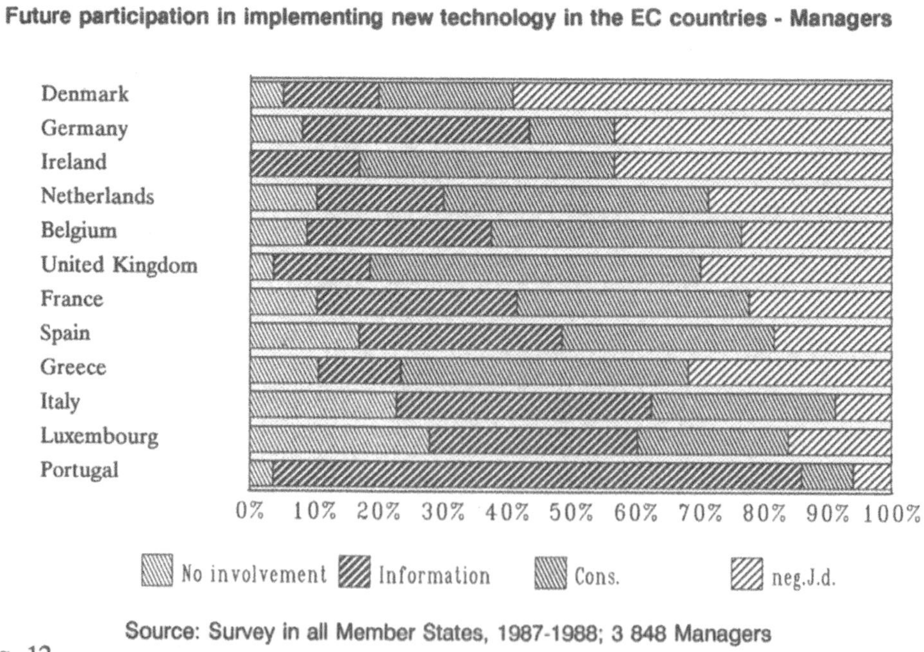

Future participation in implementing new technology in the EC countries - Managers
Denmark
Germany
Ireland
Netherlands
Belgium
United Kingdom
France
Spain
Greece
Italy
Luxembourg
Portugal
0% 10% 20% 30% 40% 50% 60% 70% 80% 90% 100%
No involvement
Information
Cons.
neg.J.d.
Source: Survey in all Member States, 1987-1988; 3 848 Managers

Fig. 12

Future participation in planning for new technology in the EC countries - Employee representatives

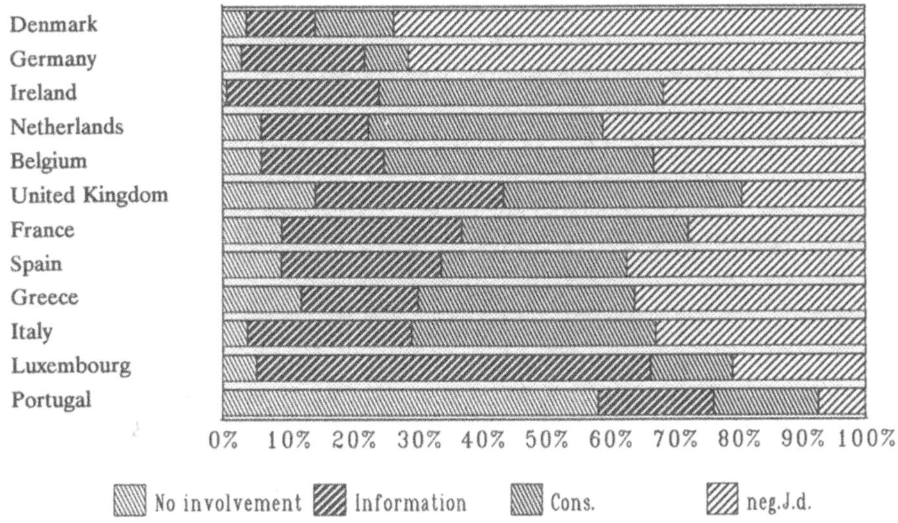

Source: Survey in all Member States, 1987-1988; 3 848 Empl. Reps.

Fig. 13

Future participation in planning for new technology in the EC countries - Employee representatives

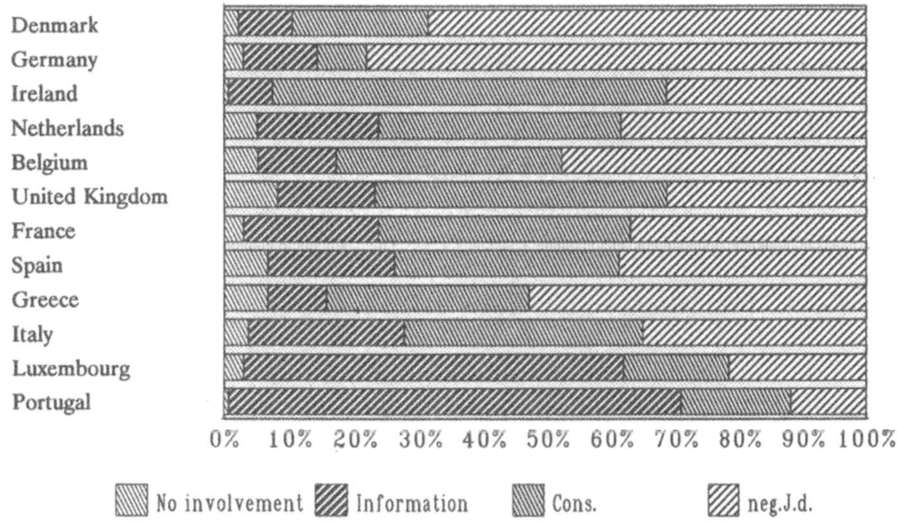

Source: Survey in all Member States, 1987-1988; 3 848 Empl. Reps.

Fig. 14

Productivity - Working Conditions - Environment - Search for an Integrated Concept

Klaus North, Geneva, Switzerland

Main issue of the *International Labour Conference 1990* was *"Environment and the world of work"*. In his report to the conference the Director General of the ILO explores the relationship between environmental issues and employment, development, working environment, industrial relations and training to mention only the most important questions.

This paper focuses specifically on how productivity and working conditions are related to environmental issues.

In *the Seventies* was *the "high time" of humanization of work* projects and the respective academic discussion. At that time enterprises provided playgrounds für researchers to try out new forms of work organisation. The enterprises, however, where not too convinced that these new organisational forms would contribute to increase productivity. In the eighties similar projects and discussions started in relation to environmental issues ("greening the enterprises").

Let us briefly examine what happened to the humanization efforts and what might also happen to the environmental questions. Having in mind the productivity of their costly equipment enterprises in recent years rediscovered *the not so "new forms of work organisation"* as a vehicle to make best use of their production resources. It seems that minds of management are changing and Taylorism is retreating. Will a similar evolution take place with the environmental issues? Developments of recent years suggest that *awareness in the enterprises* is increasing that *"pollution prevention pays"*. If management plunges into ecology will it forget humanization? Will resources be re-located and how are priorities of management changing? It would be a nice research project to answer these questions.

To tackle some of these issues it might be worthwhile to examine *which goals* of humanization, environmental protection and productivity *are congruent and which are conflicting*.

It is quite obvious that control of toxic substances, reduction of pollutants and hazards has a positive effect on working conditions and environment. Workers with increased responsibilities and decision latitude would - provided adequate training - have the chance to demonstrate their *environmental consciousness*. Humanization could also be counter-productive to the environment: substitution of processes and substances by less harmful alternatives to the worker might result in new environmental problems. Air conditioning, automation and the resulting increased energy demand might generate conflicts between humanization and *environmental considerations*.

While humanization is generally thought to contribute to productivity improvements we know quite well that the reverse has often been a challenge to humanization. I do not want to take up the humanization versus productivity discussion at this point but rather explore some *interrelations between productivity and environment*.

The productivity goal to make best use of resources may lead to reduced energy and material consumption and thus be in accordance with the careful use of natural resources. Pursuing productivity goals might also lead to pollution, noise, hazardous wastes etc.

These few examples have made clear, I hope, that there exists a *quite complex interrelation between productivity, working conditions and environment*. How can all three areas be adequately considered in the enterprises? There are of course many alternatives: joint committees, specialized departments, enterprise policies and strategies, training programs, design algorithms, research programs, data banks etc.

I would like here to think about the feasibility of *training industrial engineers also in environmental issues*. The industrial engineer is the main specialist to deal with productivity and work organisation matters in the enterprise. His suggestions on processes and technology also have an impact on the environment. Furthermore the design algorithms of industrial engineers (e g REFA 6-step method) should be amended with environmental considerations. In its first step "determination of objectives" the REFA 6-step method names objectives related to costs, humanization, organisation, and deadlines. Here a new category of environmental objectives should be added. For the evaluation of alternatives (step 5 of the REFA method) criteria related to the environment should be included.

As this conference is devoted to research and development strategies in the field of work and technology I will conclude my contribution by suggesting the following *research activities*:

- Analysis of the relative importance given to humanization and environmental issues by management,
- investigation of the interrelation between the goals of humanization, productivity, and environmental protection,
- exploration of the role industrial engineers can play to integrate all three fields.

Intangible Assets: The Capital Aspect of Cost-Benefit Analysis

David W. Budworth, London, United Kingdom

1. Introduction

In "Relevance Lost", Johnson and Kaplan(1) discuss the *inadequacy of current management accounting techniques* to deal with modern production methods. They point out that orthodox and widely-taught techniques were developed to suit the conditions which existed before the widespread introduction of computer-aided production. Starting with numerically-controlled (NC) machine tools, these now cover a wide range of techniques and a bewildering array of acronyms such as CAD (computer-aided design), CAM (computer-aided manufacture), CIM (computer-integrated manufacture), and just-in-time (JIT) manufacturing. All of these make profound changes to the way in which manufacturing industry operates, and they render invalid many of the implicit assumptions on which accounting, and hence costing and decision-making, are based.

One of the most spectacular ways in which the effect is manifested is in the *increase of overheads relative to direct labour* which modern methods involve. If overheads, instead of being 50% or even 200%, the sort of figure which would formerly have been regarded as normal, rise to 1000% or more, as can happen with modern production techniques, then the wisdom of basing a costing and, worse, a management and decision system, on direct labour becomes suspect, to say the least. For example, such a costing system inclines managements to put labour saving at the top of the list of desirable aims. The result of such shedding may, however, be merely to increase the overhead rate on such labour as remains, with no overall cost saving, to say nothing of the disadvantages, such as loss of experience and morale, which may accompany the labour shedding.

Although Johnson and Kaplan write largely from an *American point of view*, their observations are certainly valid in the UK context, where management accountants are busy devising new systems which allocate costs in accordance with the realities of current technology. I cannot speak with authority about the relevance of these issues to other countries, but I cannot imagine that there is none.

One of the more interesting of Johnson and Kaplan's observations is that *engineers played a very important part in the early development of management accounting*, but then allowed themselves to be edged out by professional accountants. My theme is twofold: current technical developments require there to be *more involvement of engineers and other technical people* in the development of the new accounting techniques which are needed; and the field of that *development* needs to be widened from management accounting and the shop floor *to company accounting and the strategic financial management* of the company as a whole.

Some recent research points in this direction, but more needs to be done, as I shall argue below.

2. The Management Accounting Problem

The techniques which are actively being developed *deal with the costs side of the cost-benefit balance*. The benefits side is less well developed. Conventional management accounting takes into consideration only the tangible benefits from the investment. Thus, the savings in labour, and perhaps in direct material costs, resulting from the purchase of a new computerised machine tool will be calculated and used to assess whether the purchase will result in an acceptable return on the capital invested.

The major benefits from the purchase may, however, be such things as reduced delivery times, faster response to orders, and reduced inventory and work in progress. A recently-completed project supported by the Joint ESRC-SERC[1] Committee (2) showed that enhanced perceptions of the company's capability in the eyes of customers was also rated as important by a surprisingly high proportion of investing companies.

Although engineers may regard these intangible benefits as sufficient in themselves to justify investment, accountants will be more cautious, and there *is a clear need for joint efforts to develop standard techniques of quantifying the factors involved*. The Joint Committee would certainly regard proposals to this end to fall within its sphere of responsibility.

3. The Corporate Accounting Problem

As the Bradford work has shown, a reputation for being in the forefront of technical advance may be a considerable help to a company in securing orders, and may thus be an important asset to it. Indeed, in an era in which capital equipment is increasingly standard across the developed world, the assets on which a company relies to make its living *are increasingly the intangible ones of the skills of its workforce at all levels, its collective knowledge of its customers* and how to satisfy them, and the *reputation* which it has in the market place for its skills in so doing.

But intangible assets are not, generally speaking, recognised on the balance sheet of a company. Just as management accounting fails to recognise intangible benefits, so does company accounting, as manifested in published accounts, nowadays governed by the *fourth EEC Directive on Company Law*, fail to recognise most intangible assets.

---

1) ESRC = Economic and Social Research Council, SERC = Science and Engineering Research Council.

There seems now to be a need to widen the perspective on this matter of intangible aspects of company operation. Just as the tangible capital assets can produce intangible benefits, so the intangible assets can produce tangible benefits such as improved profits.

We still have to find ways of achieving this end. There are *no established valuation and depreciation methods for intangible assets*, and some horror stories (3) from the past about the dangers of so doing. For this reason, analysts will usually strip out any intangible assets included in a balance sheet and make their analysis on the basis of what is left.

In consequence of the lack of recognition of intangible assets, the expenditure which builds them up has to be regarded as current expenditure. Expenditure on training; on innovation, of which research and development is an important, but often over-emphasised part; and on marketing, brand-building and so forth, is treated in the same accounting way as the purchase of flowers for the reception area. The different nature of such expenditures is sometimes recognised by calling them *"revenue investments"* in recognition of their intermediate status: although the prudence principle prevents their being capitalised, they are made with a view to producing future, rather than immediate, benefits.

Nonetheless, they are revenue items, and as such are deducted from the profit that would otherwise have been obtained in the year in which they are incurred. Management is thus presented with a problem, which is at its most acute when times are hard. Particularly if stock market pressures are high, there is a temptation to cut back on revenue investment in order to preserve current profit, even at the almost certain risk of prejudicing the future. There is *a need to find quantitative techniques to help decision-making in this area.*

4. Current Research

Even in more normal times and in the absence of undue external pressure, there is a suspicion that accounting techniques which ignore intangible assets result in *sub-optimum decisions*, judged in terms of the long-term development of the company. Although examples of over-investment are known, the *usual tendency is to under-invest in the future*.

One of the current Joint Committee projects (led by Dr. Tylecote at the University of Sheffield) is attempting to examine this proposition. The project is examining the effects of internally-applied financial performance criteria on decision-making. I am glad to say that it appears that the Joint Committee has been willing to be somewhat more adventurous in backing this work than some other funding bodies. Certainly the *Anglo-German Foundation*, a body for whose work I have considerable respect and admiration, decided to wait until results had appeared from the Joint Committee-supported work before considering a parallel application.

One of the reasons for the Foundation's caution might be that it is widely believed in the UK, and there seems to be good evidence for it, that here in Germany in particular the problem is less acute than it is at home (4). The unrelenting pressure from the stock market which characterises the Anglo-American system is less marked. The signs, however, seem to be that the German, and the somewhat similar Japanese, system, not to mention some of the others, will be pressed further towards the Anglo-American model. Dr Tylecote has therefore convened a *Steering Group for a Pan-European Project on Performance Pressures and Innovation*, for which European Community funding is being sought, and he is also seeking to extend comparative work to Japan.

5. Prospective Research

Whatever the development of financial markets in the respective countries, however, the *growing disparity* between the model of company operation which underlies current accounting techniques and the realities of modern life will affect industry in them all. The Board-level argument that we (the company) must spend money on training - or on R and D, or on brand-building - if we are to survive, which may well have been heard more carefully in some countries than in others, will increasingly be answered in all of them - either by other Board members or by the shareholders - by the question: how much?

That question cannot currently be answered at all coherently. The best answer is: as much as our competitors, but spending is only part of the story. The returns from it are more important. It is easy to waste money on advertising, on R and D, and even on training. Companies need to spend their money wisely. This is partly a matter of management, and I would be the last to deny that management instinct and flair can often be a better guide than badly-based arithmetical analysis. If *revenue investment* really is an investment, than it is not unreasonable to expect it to produce returns, and for those who are asked to sanction the investment *to be presented with evidence about what those returns* are likely to be.

There is a consequent need to develop the accounting techniques for intangible assets, and in particular to tackle the *problem of measuring both expenditure and returns*. There is considerable, but diffuse and rather unsophisticated, interest in this subject in the UK. The most systematic work has been done on brand valuations (5) and some bold companies have gone so far as to put these values on the balance sheet. This development is controversial, to say the least, and it is probably fair to say that the general view is that it is premature. Other specialists are interested in measuring returns on training, and on R and D.

Since these specialists were on the whole unaware of the interests of the others, I have made an attempt to bring them together to discuss their common concerns (6). There is, however, a need to develop things further against *a background of a common concept of what a company nowadays is.*

A commercial company aims to be *a self-perpetuating supplier of market offerings.* To do this, it has to offer goods or services which its customers want to buy, at a price which keeps it in business within whatever environmental or other constraints which the societies in which it exists choose to impose. In modern conditions, where the competition is increasingly global, technological change is rapid, and the environmental demands of society increase, there is *a constant need to update products and all other aspects of company operations.*

Just as machinery needs to be replaced as it wears out, so have outdated products, operating procedures, and so on to be replaced. We have *to recognise the depreciation of intangible assets*, and make it good, just as we recognise and allow for the depreciation of tangible assets. The various forms of revenue expenditure are interdependent contributions to this end. I have shown that *the concept of the product cash-flow history* can in principle be used to link expenditure to return, and that there is the potential for deriving quantitative guidance to management, investors, and those involved in national industrial policy (7).

There seems here to be a basis for developing accounting techniques which are relevant to modern conditions. To do so will demand co-operation between economists, accountants, marketing and training specialists, and engineers. The close co-operation of industry will be absolutely essential. Without access to real figures, collected in real companies in real conditions, no useful progress will be made.

I very much hope that the Joint Committee will move in this direction. Its terms of reference are well-suited to the task, and it has some researchers in its stable who are already making a start on some aspects of the problem. Their experience, most encouragingly, is that industry is surprisingly willing to let academics have access to its operations. *International comparative work would be helpful*, and there are signs that workers in other countries are interested in some of the same problems. Perhaps this meeting can help to forge links between us.

References

1) Johnson, H. Th. and Kaplan, R.S, (1987), *Relevance Lost: The Rise and Fall of Management Accounting.* Harvard Business School Press.

2) Price, D.H.R., Sharp, J.A. and Pike, R.H., (1990), *Investment in Advanced Manufacturing Technology: Benefits and Justification.* Project Final Report to the Joint ESRC-SERC Committee, University of Bradford.

3) In the UK, the collapse of Rolls Royce in 1971 is often cited as an example. Rolls Royce had capitalised development expenditure on the RB211 engine. Although not the main cause of its problems, this accounting treatment may have contributed to them.

4) Lorenz, Ch., *Investing in Success: How to Profit from Design and Innovation.* Anglo-German Foundation, 1979 was one of the early attempts to document the case.

5) Murphy, J., (1989), (ed), *Brand Valuation.* Hutchinson.

6) The Network on Intangible Assets of Companies has about 60 members.

7) Budworth, D.W., (1987), *Rewinding the Mainspring.* Technical Change Centre.

The Contribution of Practice Oriented Promotion of Research and Development to a Modern Evaluation of Work

Paul Oehlke, Bonn, Germany

Rolf Lindholm outlined a new world of work: not jobs but persons do matter renewing permanently their competence; only a small percentage of the workforce is directly productive whilst the service area is increasing.

Indeed, there are significant indicators for the enlargement of indirect functions becoming partially a precondition for further productivity growth which depends on the rapid integration of scientific findings into industry and trade. Though determining economic output and effectiveness, the rate and the distinction of the *diffusion and application of new technologies* are in turn *conditioned* by many economic and cultural factors, *especially a corresponding social and institutional environment.*

A skilled and healthy workforce, organizational flexibility and ecological awareness are becoming more and more prerequisites for the effectiveness and general acceptance of the generation and diffusion of new technologies. Therefore, private and public R&D activities have begun to incorporate the *development of human resources* in some European countries since the beginning of the seventies. The following remarks will inform on some basic issues of the German Research Programme "Work and Technology" superseding the programme "Research on the Humanization of Work" funded by the Federal Government since 1974. While in the past remedial solutions were in the foreground, the *present programme* puts *greater emphasis on a preventive and innovative design* of work and technology.

At first I would like to refer to some experiences which give evidence for a stronger evaluation of work; then I will portray the programme's arguments for a prevention research policy playing an increasingly important role and finally depict its practice-related orientation embedded in mechanisms of a cooperative programme management.

1. New Evaluation of Work

There is no doubt today that in many sectors of the economy a *breaking with the traditional principles of tayloristic organization of work* and technology based upon the strict division of labour would not only be technically possible but also be *reasonable from an economic point of view.* Especially in those industries, in which an increasingly diverse supply of products is demanded, such as in mechanical engineering, or in which product cycles are becoming shorter all the time, as in the production of information or communication equipment or in fashion-dominated branches, a need for *new forms of work organization with more tasks and holistic work concepts* and more *decentralized decision-making* structures is appropriate.

Experience has shown that *new technologies* (because of their inherent design possibilities) *can be combined with numerous management structures and organization models.* By chance this offers an opportunity to provide *different solution patterns for work organization and working conditions* enabling new technologies to be applied in accordance with human needs, and at the same time increasing efficiency and flexibility. Firms and branches which emphasize the functional flexibility of workers by ensuring internal mobility across tasks, flexible job boundaries and training facilities are more likely to successfully apply advanced manufacturing technology than those focusing on work control mechanisms and hiring and dismissal procedures.

According to such assumptions regaining strength during the eighties, *social research* found a *paradigmatic change of labour policy* in main sectors of industrial production - automobiles, machine tools, chemicals and even electrical industry. Due to the new technological potential for rationalization the engineering concept of the fully automated factory leaves room for a comprehensive utilisation of manpower. The companies' use of the skill, ingenuity and expertise of their employees is better paying off than restricting, pacifying and alienating their creative potential. Moreover, their special abilities of mastering new challenges by learning, cooperating and intervening have to be developed regarding high demands of quality, flexibility and innovation.

The new evaluation of work in modern production concepts derives partially from new constraints of international markets with sales possibilities for a large number of firms narrowing down. Renewal and variety of products, changing series, production in small batch sizes, assurance of product quality gain importance for the competitiveness of companies and create production problems which *traditional work structures are no longer able to cope sufficiently with.* On the other hand, the patterns of behaviour and the value concepts controlling them have changed radically. Burdens and deprivations accepted in the past as a matter of fact find acceptance any longer.

An outstanding example is conventional assembly line work or machine work with simple work contents being often much too inflexible to react quickly to the fluctuating demands in terms of time and materials. Demoralized by strenuous and unpleasant work conditions, the efficiency of the employees declines; numerous firms record an increase in the number of sick persons, of absenteeism and of defective products. These problems accentuate a growing sensitivity to the work environment which is attributed to a rising level of consumption and education resulting in correspondingly higher claims and expectations at the workplace. But, matching enriched work contents, participation in decision-making and, most important, in the shaping of technological innovation and organizational renewal calls for interrelated or *key qualifications* as well as for *work processes that promote learning* including active training methods.

Such forward-looking objectives contrast with traditional solutions which have been analyzed as patterns of structural conservatism deep rooted in managerial behaviour and organizational theories which are combined with concepts of a "value-free technological rationality" reducing the human being as a disturbing factor in production which has to be fragmented and precisely to be controlled. Hence, a great part of the workforce, often female, elderly and immigrant workers, has yet not overcome the prevailing division of labour including unqualified and unhealthy working conditions causally relating with various phenomena of stress and strain. They are still to be discerned in nearly all industries and services, above all in declining branches and in the broad range of small and medium-sized enterprises. The latter often lack the staff-related and organizational potential of how to work processes that might induce learning activities. A general rejection or, at least, a far reaching roll-back of tayloristic or fordistic methods has not taken place though the *demand for holistic innovation concepts and anthropocentric design strategies* is growing.

2. The Increasing Promotion of Research and Development

There are two main reasons for a *rising intervention of governments*: the *environmental damages* by the increasing use of resources and discharges of waste and the growing pressure of international competition with world-wide flows of capital, goods and manpower bearing chances and risks for the social and ecological state of our society. Therefore, the German programme stresses that the welfare of our nation being a *site of industry and services* depends on whether and to what extent we will manage to plan and organize work and technology in an innovative and preventive way according to human, social and ecological needs.

According to *its subsidiary function*, government is not inclined to assume industry's own task in this field, but will only provide supplementary services. The conditions on which such public support is needed refer especially to the *challenges of competitiveness* depending more and more on the streamlining of research, development and application of new technologies. Markets cannot adequately cope with this future-oriented task if the risk and cost of foresighted design of work and technology go beyond the individual company's cost-benefit calculations and planning capabilities. Due to a lack of well-tested models, principles and modes of procedure, many novel solutions appear to be too risky. *The Government is called to assist at two levels*:

- by means of setting *standards and technical rules* and by legislation it must contribute *to maintaining working and living conditions* as well as *the environment* to them being as free of impairments as possible;

- through *educational and social policies* as well as *the promotion of research and development* it must *improve the capabilities to cope with* the new demands in a situation of progressive fundamental *social and technological change*.

Therefore, the *German Research Programme on Work and Technology offers help to develop possibilities for preventive and skill-enhancing solutions* which can be used by in-firm and inter-firm target groups. As there obviously is not only one right solution but *many alternatives for the design of work and technology* in keeping with human needs, a sufficiently broad base of experience is required which may differ by industry, location factors, company size, product and service range and especially by employees' qualification profiles. The solutions have to be designed in a way that their results benefit *the greatest possible number of target groups* and can be accordingly translated into practice. By channelling the experiences and findings through associations and institutions in many ways, the *market functions* again are *strengthened*. In view of the harmonization of the European social area it is of strategic importance to transfer up-to-date and prospective design alternatives to the two sides of industry and to the organizations in the field of industrial safety, occupational science, initial and further training and education. This has to be done *in the interest of workers* as well as of *companies*, but also of *society as a whole*.

Most companies have not felt an urgent *need to build* up their own *efficient innovation potential in the area of personnel management and work layout*. The development observed so far has led to the conclusion that it might be sufficient to tackle problems of work and workers pragmatically within the scope of existing structures and regulations whenever they occur. But the possibilities to follow a method of pragmatic small-step innovations, widely practised and well tested up to now, are reduced by the spreading and networking of informations and communication technologies *emphasizing the systems character of in-plant structures and workflows*. The rapidly growing interdependence within and between companies makes it more necessary to design and plan processes of change and development on a long-term and comprehensive base. As the development and implementation of technology is pushed forward by industry and administrative bodies, dangers and less than optimum results of purely technical innovations can be observed. Hence, the new programme starts from the often proved experience that *economically successful innovations are characterized by also considering organizational, skill-related and health-oriented requirements* at the same time. The Federal Government has made this broad approach to innovation an integral part of its research and technology policy.

By the foresighted safeguarding and promoting of employees' health and qualifications the comprehensive innovation concept includes preventive effects. The shaping of technology and organizational renewal according to human needs, especially in the early phases of development, are crucial for the protection of health by reducing and warding off hazardous stress factors. Such future-oriented prevention of disease as well as job safety and vocational training opportunities are basic conditions for the individual to live an active life in our society and for maintaining workers' performance capability even to an advanced age. They contribute likewise to further developing the production potential of companies and the quality of services rendered by administrative bodies. Moreover, *preventive measures* bear not only economic advantages for the organizations involved but also for the various sectors of our system of social security, especially in the view of the rising expenditures for remedial requirements. Against this background the programme aims at *strengthening the preventive elements of our social system.*

The greater responsibilities covering the whole range from prevention to rehabilitation can only be accomplished jointly by enhancing the cooperation and competence in general. Under the conditions of a still accelerating social and technological change the education policy respectively vocational training is engaged in furthering interrelated or key qualifications. This has been, for instance, taken into consideration, when the vocational training of workers in the metal and electrical industries was reorganized by redefining their objectives of training and education. Thus, not only qualifications which are specific to a job or company are to be acquired, but also a general basic proficiency enabling a person to plan, prepare and approach a work task methodically. Moreover, in order to be able to identify and solve problems in the course of operations or to be able to help design processes or products to be developed not only a special technical attitude is required, but to an increasing extent, the skills and readiness to assume responsibility and to cooperate as well as a competent understanding of other departments of the company. For coping with the innovative demands we need, in particular, design solutions at the workplace that promote learning on the job and stimulate self-guided and informal learning. Thanks to the participation of the Federal Minister of Education and Science in the programme, new channels for the implementation of results via the education system are now being opened. The *incorporation of experiences and findings* derived from research projects into the curicula of schools and universities is, in the long run, absolutely *necessary for the practical implementation and effectiveness of the programme.*

3. The Translation of R&D into Practice as a Key Task

The practical orientation of the programme requires that *prevention research is playing an increasingly important role* in tackling the chances and risks of rapid change. Its main task is to contribute to an acceptable way of dealing with human resources and to creating living and working conditions that offer plenty of opportunities. The latter, in particular, are strongly influenced by the system of industrial relations, especially the autonomy in collective bargaining and the relations between management and works council laid down in the law on collective agreements and in the Works Council Act. Under this legislation it is the responsibility of the parties to a collective agreement as well as of management and the works council to plan and organize working conditions. By way of example, the works councils' right of codetermination and participation in social, personnel and economic questions essentially influencing the design of work and technology or personnel management and staff development and related framework conditions shall be mentioned here. Therefore, in the case of projects which entail the redesign of working conditions in the applicants' company, the formal application for support has to be accompanied by the written approval of the competent works' council. Accordingly *trade unions and employers' associations are proportionable represented* in the advisory bodies of the programme functioning also as a communication platform for scientists, associations and government.

Corresponding with the distinct system of industrial relations engraved into the operating of the programme, the basic idea behind the efforts of making research work more effective in practice has been that those using and implementing the results later on are involved not only in the definition of the objectives but also in the overall (steering) management of the programme and research process. Complex requirements have to be met by the project management team in improving the conditions which assure the mesh connection of research and practical conversion of results. Of crucial importance for creating awareness of science, industry and intermediary institutions is the arrangement of arenas fostering a dialogue between different actors: e.g. developers, manufacturers and future users, experts of occupational safety and health protection as well as of education and further training representatives of management and works council, social scientists and engineers. The commitment, participation and cooperation of those concerned at plant and workshop level are indispensable prerequisites for the design of work and technology according to human needs.

A practice-oriented concept of programme implementation which pays heed to all these principles cannot confine itself to general goals alone. For this reason, it has proved expedient to establish *priority areas* which provide a link between programme goals and the numerous individual projects. When defining and organizing such priority areas all competent scientific disciplines and social actors are involved that have an influence on the design of work and technology within the political, economic and social network of interrelations. In this way the *special needs of target groups for R&D results* are taken into account and deficits are identified which can be closed with the help of research and development. Intensive preliminary clarification on the part of the project management team and repeated consultation with experts and the advisory bodies are required before the draft for a new priority area can be recommended for publication. This process offers transparency and security with regard to promotion objectives, activities and conditions to all those who are interested and participating in the programme.

The present priority areas of the Work and Technology Research Programme include expiring activities, activities taken over from the former programme, updated and new activities. In the field of occupational safety and health the emphasis is shifted to a stronger orientation to prevention. As far as branch oriented and joint projects are concerned, R&D work is increasingly being based on the broad innovation concept as principle of design and orientation. Recent developments in technical and strategic areas, such as those characterised by artificial intelligence or CIM and JIT strategies, are playing a crucial role in the revision of the priority areas of design and application of new technologies in production, offices and administration. In the priority area concerning basic principles and cross-sectional issues, for instance, the effects of demographic trends will be observed.

To match the needs of the various fields and issues, a variety of promotion schemes and instruments is offered. Individual projects are related to a broad range of topics in different priority areas. But, there has been a trend to establish comprehensive projects, above all joint, branch-oriented and transfer projects. Joint projects are particularly suitable for small and medium-sized enterprises. They may include consultants, manufacturers and users. While these projects types are usually problem-oriented, the branch-oriented projects, such as in the forging, foundry, mining, printing, building, food industries or in road transport and commuter traffic, are dealing with a broad range of problems related to the specific working conditions. The special transfer projects are based on the experience that model solutions once developed and tested are not automatically transferable to other fields of application and that the interests of target groups have to be taken into account. The increasing complexity of issues taken up by the programme makes additional forms of cooperation, dialogue and common learning desirable.

Probably the most important effect of the diverse programme activities might be the acquisition of enlarged qualifications on a broad base of all people directly or indirectly involved. Scientists, entrepreneurs, experts in the companies and administrative bodies, works and staff councils, representatives of work-related institutions and many others are induced to deal in detail with the state-of-the-art in designing and planning working conditions and get a deeper knowledge of the increasingly complex systems and problems in production and services. As the organizations interested participated in the form of joint and branch projects and were able to carry out their own projects of implementation, the programme has contributed to finding rational answers to design questions and sometimes even to matching economic and humanization policy objectives and their practical application at the same time. Employers' associations offer meanwhile training programmes which include the acquisition of social competence and of designing capabilities. The Metal Workers' Union e.g. develops a broad range of activities to increase likewise the technical and organizational competence, thus improving the capabilities of workers to participate in the processes of planning and designing work and technology. These tendencies will have a growing impact on industrial relations: The centralised industry-wide bargaining system has been increasingly supplemented by decentralised negotiation processes giving a new reality to the legal rights of participation and codetermination.

In this context, the development of a scientific infrastructure and the *mutual opening up of science and industry* must not be underestimated. The systematic exchange of experiences and findings in various discussion circles, consulting panels and networks of social dialogue has resulted in a broader understanding of innovation and occupational safety and health. It is based on the accepted principle of a progress of productivity in accordance with human needs and has influenced the revision and further development of the programme. To make better use of the opportunities offered by new technologies and to avoid hazards, R&D measures have to be initiated in the early phases of the innovation process already. In doing so, the experience gained in other countries is to be included and, if appropriate, international cooperation in the field of research is to be proposed. By strengthening international communication the use of research funds can also be made more effective when agreements on setting priorities and on a mutual exchange of results and, at the end, cooperative projects can be reached. Such a preventive orientation, however, creates the risk of a widening gap between future-oriented objectives that are pressing ahead and a project reality and in-plant situation that are lagging behind. Therefore, new efforts have to be made in building up diverse and comprehensive networks of research, development, testing and transfer of results uniting the various actors in the field of work and technology.

4. Conclusion

The programme activities, though only in an experimental way so far, contribute to a new evaluation of work, which is urgently needed. The safeguarding of human and natural resources is gaining overwhelming importance. They have, certainly, ever been the base of all production and economy, but were partially exploited to an unreasonable degree. Nowadays emerge the outlines of the twin cores of future productivity growth which have to be progressively taken into account from the very beginning of the innovation process. Formerly a focal point of dynamic entrepreneurship, it gets democratic traits to an important degree corresponding strongly with the scientific character of the production process and of the work functions still remaining and gaining on strategic significance. Enlarging research departments and the spreading of quality circles as well as the whole range of R&D programmes and the private and public expenditures rising for elaborated tasks and requirements of R&D give evidence of this trend bearing a chance to incorporate human and ecological criteria in the development of the productive forces.

B

Integrated Concepts of Health Prevention, Industrial Safety and Ecological Requirements (including the European Dimension)

Integrated Concepts of Health Prevention, Industrial Safety and Ecological Requirements

Bo Tengberg, Stockholm, Sweden

Most people share the opinion that there is an urgent need to tackle the alarming ecological situation of the World by reaching international agreements. Agreements based on the understanding that pollution is cross-border, and, that the time-span is so urgent, that we cannot allow ourselves to get caught in endless discussions and, thereby postpone the necessary decisions being taken.

With few exceptions, *problems of environmental damage are directly related to production processes*. Either they are caused directly by pollution, or indirectly through hazardous chemical substances, transportation systems or industrial waste. That is why it is imperative to take appropriate measures at workplaces and, - during the Nineties -, *to develop preventive environmental policy measures in working life*.

Within the *Swedish Trade Union Confederation* we have, during the last ten years, adopted *an overall approach* as regards health and safety questions and the environmental issue. The reasons for us to involve environmental matters in the issues we traditionally deal with, is that we think it is necessary for the unions to take part in the adjustments which have to be made to solve the environmental problems.

Another reason is that we think it is the *workers who have the strongest motivation* to develop a production which can fit in with environmental demands to secure employment.

Thirdly, environmental adjustments of products, processes, maintenance and so on has *to be done with regard to workers knowledge and experience*. The know-how from the health and safety-area could be a resource also for environmental solutions.

In many cases i 'is easy to see the *connection between health and safety matters and environmental* ones. One of the most illustrative examples of the interrelationship between these two is to be found in the use of chemical substances as solvents, oils, fluids used in the engineering industry, acids and so on.

Regarding environmental policy and employment, measures taken in order to protect the general environment, *do not pose a threat to employment opportunities*. On the contrary, *not* taking preventive measures constitutes the real threat to employment opportunities.

Workers in some sectors of the industry regard a further growth of environmental pollution as a threat to their jobs. Unless, for example, dramatic cuts of Sulphur and NoX emissions are imposed, the employment rate in several industries will be affected negatively. As regards the forestry industry, for example, not only the quality of its products would deteriorate, but also employment opportunities for forestry, pulp and paper industry workers would decline. Other sectors of the industry that would be affected negatively are the tourist, agriculture and food industry.

For a long time it has also been commonly thought that there is a inevitable *conflict between environmental demands and employment*. As far as Sweden is concerned, we have not seen this conflict became a reality. Of course, environmental demands have probably accelerated some of the structural changes in Swedish industry during the late seventies and early eighties, but those changes would have occurred sooner or later.

For the future we have said that when a factory cannot solve its environmental problems in an acceptable way, we might have to accept that they must close down.

Industry and the society must, however, accept responsibility in order to prevent such situations as far as possible.

We accept of course the "polluter pays principle". Nevertheless in regions where people are very much depending on one or a few big industries with an important but polluting production it could be tempting at least for the local union to call for reduced environmental demands because of the risk of unemployment.

But *it shouldn't be necessary to compromise* with high environmental demands if some kind of regional support could be available, not to pay for the cleaning equipment, but to prevent unemployment. Training of the workers through labour market measures is one of the ways which can be used in such cases.

The most important contribution, however, concerns *the use of environmental policy as a job creator*. Society could play a more active role in supporting change and the spreading of new technology adapted to environmental demands. This could be important not at least among medium-sized and small industries where concern about environment is low. Demands concerning the environment will also create jobs in industries dealing with the "cleaning up" field. Data from the USA shows that every billion dollar invested in environment, as a result, beside the cleaner environment, also created 20 000 new jobs.

The situation in Eastern Europe and the immediate need for measures to clean up will give great possibilities for export and jobs in those companies which think up new technology in the environmental field.

But this will also give us an example of *the aspects of distribution*. Everybody can agree that health and safety problems mainly affect the workers. But health problems can be worsened by pollution, something we have seen around factories and in bigger cities.

The damages caused by pollution cost society billions and billions every year. To clean up and prevent pollution will be a big burden for society. If the burdens are not shared in a fair way, it could be impossible for the workers to accept environmental demands.

As I have mentioned in the beginning, it is necessary to involve the local trade union and the workers to get full support for environmental measures outside the factory. At the same time it is also necessary *to consider the problems concerning the working environment.*

The best way to do this is to take preventive measures at the beginning of the production process and solve inside and outside problems in the same environmental plan.

The opportunities to do this are often overlooked. There is a risk of a lack of confidence in industrial environmental policy from the trade unions if both sides are not regarded.

I also have to say that we not shall forget *the big resource 100 000 safety representatives* represents also for the outside environmental work. The know-how they have from health and safety work ought to be used in an integrated daily environmental activity.

Within the Swedish trade union we also think that there is a *need to extend the safety delegates' mandate*, so as to extend the rights to stop work which endangers health to work which endangers the general environment. There have not been any big problem with the use of these rights, which have been in force since 1974 in the health and safety field. The serious way the problem has been handled so far indicates that the law would be used with the same good judgement also in the environmental field.

However, the guiding principle for the Swedish trade union is, to *prevent environmental and health and safety problems from occurring* at all. In other words, techniques and devices for the purification of emissions, can only be accepted as temporary measures. Our aim must be to develop safe technology based on environmentally sound procedures.

Therefore, it is necessary to work out *an overall plan for the environmental issues.* It has to be indicated what preventive measures are needed, the schedule for this, control measures and so on. It is necessary to make *analyses of the entire production process*, not just analyse the stages at which pollution is produced.

The Trade Unions have to be actively involved, both in planning and control. That way the companies would be able to benefit from experience and knowledge of the workers.

Apart from this, we want *studies* to be carried out *on the impact of the different productions on the environment*, as well as analyses of the environmental problems that occur throughout the entire production process. That means, from the extraction of raw materials, to the transportation of these to industries, the environmental damage caused by the production processes, and finally - but not at least - how to dispose of resulting waste.

In the same way we want *to develop an environmental awareness in all professions* in a preventive strategy. Practical examples could be industrial process operating, transport planning, sorting waste in useful fractions and substitution of harmful chemicals.

In this way *local trade unions* will have the opportunities to influence the employers to develop products that are better adapted to environmental demands than the products of today. Clean products will be regarded as a must for industrial survival.

In the effort to create good working and environmental conditions, *workers can play an important role*, nevertheless it is the responsibility of the employer to secure healthy jobs and not to pollute the environment.

Provided, that workers are given the chance to develop an awareness of how the performance of their work effects indoor as well as outdoor environment, they *can share that responsibility*. Such a role, based on environmental awareness would for example imply:

- that a process operator, through accurate steering operations, could minimize the use of hazardous chemicals,
- that a forestry worker would know how to protect nature when foresting,
- that cleaning workers and garbage collectors would be able to give advice, at workplaces and in housing areas, on how to sort your garbage, that workers in shops would, for example, be able to advise about environmentally safe products.

For workers to be able to fulfil this role, *they must be trained* in health and safety *and* general environmental issues. It is also necessary that *the work organization at workplaces* has to be organized in a way that grants the worker an independent decision making function and, thereby safeguards the environment. These instruments do not exist in a traditional hierarchical structure of work organization, but is also important when you consider how important it is to use the knowledge and enthusiasm of workers when solving problems and to create a better environment for the common future. The work the Swedish Trade Union Confederation has done to form a program for this is just an example of what is going on both in local unions and in national as well as international federations.

Programme against Musculoskeletal Injuries

Peter Lidehäll, Stockholm, Sweden

1. Background

The *Swedish Work Environment Fund* in 1988 started its *Programme against Musculoskeletal* Injuries. Altogether MSEK 60 will be allocated to it during the time period ending 1993.

Of course we have had other programmes regarding work-related musculoskeletal injuries, and so far the Fund has, during the last decade, invested nearly MSEK 100 in *research and development related to work postures and workloads*. The major part of these grants has gone to medical institutions and to the National Institute of Occupational Health.

From this research we learned *how to diagnose musculoskeletal diseases and how to treat persons* suffering from such disorder. At the workplace, preventive measures were made in order to lighten work. Traditional human engineering provided check-lists and advice how to minimize work loads and find proper work postures. Skilful designers and engineers gave us technical equipment such as vertically adjustable worktables etc.

Now we know that *these measures were not enough*. Furthermore, our "correctly designed" work stations also meant more repetitive jobs and increased intensity. In our modern industry (and in public service as well) workers get tired and bored. Their motivation decreases as well as their performance.

The number of musculoskeletal injuries has risen steeply in recent years, especially among female workers. Today one-fifth of work-related injuries and *nearly 60 per cent of all cases of work-related illness in Sweden are musculoskeletal*. These problems often lead to very long periods of sick leave and cause a great deal of suffering.

Changing those conditions has become a very important issue in Sweden today, not only because of illness and suffering for the individual victims but because of the cost for the society and fear of future labour shortage.

2. The Programme

2.1 Policy

The current programme is action oriented and the basic concept is to support changes at workplaces, provided these changes are characterized by a holistic approach involving technological, ergonomic and organizational aspects.

According to the programme's directive, grants primarily are awarded to developing workplaces where ***physical strain caused by monotonous, repetitive work or inflexible work postures is regarded as the main problem.***

Those areas with extensive problems are of course given priority. ***The support*** is mainly given straight ***to concerned companies.*** But we expect those firms to contribute themselves to the projects and we also demand substantial plans regarding development activities.

Thus, support is given primarily to innovative projects. Everyone at the workplace is supposed to be involved in the process of change. This process is in fact regarded as more important than the change itself.

2.2 Initial Measures

In Sweden the ***labour market organizations*** in each business area ***have formed committees,*** which co-ordinate measures regarding improvement of work environment. These committees were assumed to provide active support to the programme as well as to the local projects.

Consequently the programme secretariate has approached the committees in order to ask them to privide suitable projects, or, in other words, identify companies where management as well as workers are seriously interested in developing work organization and job content. Furthermore they were assumed to make plans for their own support to local development activities and for their contribution to dissemination of results. Since several companies made their own applications for grants straight to the Fund, the committees concerned also were asked to consider those projects.

In order to establish programmes at industry levels some of the committees have been awarded grants for minor inventory studies.

2.3 Governing Principles for Awarding Grants

The programme can support

- *inventory of problems*, for example organizational or technical obstacles;
- *assistance from researcher or consultants*;
- *local activities* such as network-contacts, training etc;
- *dissemination activities.*

3. Projects

The programme is *focussed on relatively* few business areas, and is by now supporting *35 - 40 projects*, representing e g engineering industry, forestry, wood working industry, sawmill industry, fóod shop line, meat industry and garment industry. Support has also been given to development activities within the public and service sectors, eg dental care.

The industry programmes are normally based upon a series of local development projets regarding similar ergonomic problems. They are therefore connected to each other by co-ordinators and network organizations. Networks of projects and project participants also have been developed cross the business sector boundaries. Projects regarding documentation and evaluation have also been started within the programme.

Prevention in the Programme on Work and Technology - Contribution with Regard to Safety at Work in Practice

Hans-Jürgen Bieneck, Bonn, Germany

My lecture directly links up to the words of welcome which State Secretary Dr. Tegtmeier addressed to you. A conclusion to be drawn from his welcome address may well be that

- *German unification* of 3 October 1990 on the one hand realized a central objective of German politics,
- but on the other hand presented us with a variety of difficult tasks in the force field between Europe and German unification.

One would certainly be expecting too much of a government financed research programme were one to assign a decisive role to it in the framework of tasks of such a dimension. I am, however, convinced that *the programme on Work and Technology of the Federal Government might well play a significant role in this process.* We plan for example to use the programme explicitly in order to transfer our knowledge about occupational health and work organization at the company level to the former German Democratic Republic. The programme could also help to make it easier for the companies in the East to familiarize themselves with our system, e g by means of pilot projects aiming at a co-operation between persons and institutions responsible for occupational safety.

I do not want to dwell on this point any longer but shall now turn to our *general considerations concerning the programme on Work and Technology*. The Federal Ministry of Labour and Social Affairs considers this programme as an important support with regard to *the realization of a demanding objective in the framework of occupational safety policy.*

This objective is as follows: to further develop occupational safety to turn it into a comprehensive preventive occupational health system in dem companies.

I should like to explain to our foreign participants that the so-called "classic occupational safety" in the Federal Republic of Germany has its central tasks in the prevention of accidents at work and occupational diseases. These are the tasks primarily dealt with by the legislative bodies and by the supervisory organs. The vast area of "work related health hazards" is as yet neither subjected to satisfactory legal provisions nor to occupational safety experts deal with this area to an extent worth mentioning.

If one intends to change this situation it is of course possible to try to amend the legal basis accordingly. It is our experience, however, that laws based on good intentions have no effect at all unless one succeeds in concretizing the envisaged objectives to a sufficient extent, or if the application thresholds are fixed at a very high level in the legislative process; here I am for example referring to the "acknowledged findings of industrial science" which after almost 20 years of their legal establishment probably still play a rather modest role in company reality.

Another more realistic - and thus more promising - possibility may be to provide scientific findings and to promote their application - e g by means of pilot projects - thus making the intended objective generally acceptable in occupational safety practice. Once this has been achieved, a legal provision is no longer necessary, or else it is finally accepted as being reasonable, and politicians from all parties feel that they can agree to it.

The programme on Work and Technology - and the previous programme on the Humanization of Working Life - have, in my eyes, provided good examples for this alternative method.

I should like to mention that the interaction between research programme and the Federal Institute for Occupational Safety and Health where we are holding our meeting today, has always played an important role in occupational safety: The Federal Institute for Occupational Safety and Health is an ideal transfer agency for the programme as it has a firmly established and recognized position in the occupational safety "scene" of the Federal Republic of Germany.

When preparing the programme on Work and Technology the Federal Ministry of Labour and Social Affairs succeeded in having our intention of promoting and further developing occupational safety reflected in the programme's central objectives.

The relevant key-term in this context is "prevention".

The objective of prevention reflects our expectations with regard to the programme on Work and Technology, i.e:

- to contribute to *the improvement of practical occupational safety*, and
- *to identify* not only accidents at work and occupational diseases but also *other health hazards in the working environment, and to develop measures to prevent such hazards.*

The programme on Work and Technology is politically supported by *three federal ministries.* Tasks and objectives have thus been combined in a meaningful way; and we are participating because we consider the objectives of all involved parties as equal and unseparable in company practice. *The decisive key terms* in this context are

- *innovation* for the Federal Minister of Research and Technology,
- *prevention* for the Federal Minister of Labour and Social Affairs, and
- *qualification* for the Federal Minister for Education and Science.

In this context we see prevention as a target that can only be reached in a reasonable way in conjunction with the other objectives. We need a comprehensive approach as regards analyses and measures; after all problems do not occur as isolated in the companies as we would like them to for the purpose of drafting programmes and legislation. With regard to this programme I should like to say in conclusion that the participation of three ministries gives us access to the social institutions represented in these ministries: this is of utmost importance for the practical effects of a research programme.

I had explained that *a research programme can prepare and support legal developments*. With regard to topical developments in occupational safety three main areas are to be addressed:

- *EC legislation* on occupational safety;
- section 20 of *the Health Care Reform Act* (Gesundheitsreformgesetz), and
- our efforts *to further improve the implementation of the 1973 Safety at Work Act*.

EC legislation in the area of occupational safety has become increasingly important for the member states. We are therefore *obliged to harmonize German occupational safety* by translating EC Directives into national legislation. The new aspect is that this harmonization now also covers the principal elements of occupational safety.

Let me remind you that we have already translated many EC Directives into national law. Although this was done under the *heading of "removal of trade barriers"* it nonetheless meant that German legislation in the field of occupational safety was at least to some extent already influenced by EC legislation in past years. This includes eg dangerous substances, electrical equipment, agricultural machines and elevators. Irrespective of the view held until quite recently, eg that the EC moves very little, a quite considerable number of EC Directives have been enacted over the last years. It was, however, evident that the *concept of individual directives* together with the necessity to take unanimous decisions set the pace of which many said that it was too slow; this provided ammunition to the critics and sceptics of Europe; and it was easy, although not always justified, to be critical.

With *the entry into force of the Single European Act* on 1 July 1987 and a "new approach" to the harmonization *of technical trade barriers*, however, the situation has changed considerably.

- *In the field of occupational safety* the Single European Act introduced the principle of majority decision.
- *The "new approach"* meant a rejection of detailed regulations and a turning to

directives with general provisions the details of which are to be laid down in European standards.

As a result, *EC law-making as a whole has gained considerable momentum*; and this is true *for both areas* under consideration:

- *Technical quality requirements* are now harmonized - the catchword is "removal of trade barriers" - on the basis of Article 100 A of the EEC Treaty (*Internal Market Directives*). The most important example in this respect, and also of the "new approach", is the so-called "Machinery Directive" of 14 June 1989.
- *Occupational safety regulations* governing the ***handling*** *of technical equipment* at the workplace, for example are laid down in directives including minimum provisions in accordance with Article 118 A EEC Treaty (*Occupational Safety Directives*). The most important example in this respect is the so-called *"Framework Directive"* of 2 June 1989 with five individual directives which have been issued thereto up until now.

Against this background *the Federal Government* had suggested during its presidency in the first six months of 1988 that with regard to social policy, *the new instruments for European harmonization should first of all be applied to occupational safety*. Since the issues at stake in this field are relatively straightforward it seemed to be particularly suited to pioneer harmonization in the area of social policy as a whole. To this add that matters of safety and health are to a lesser extent part of an overall system than, for example of pension, health and unemployment insurance, of co-determination etc, where national traditions and funding models play a much bigger role making harmonization more difficult. The *Federal Government's view has been fully confirmed* in the meantime as is shown by the number of directives which have been issued so far.

I had already mentioned that we will have to *translate these directives into national occupational safety legislation over the next few years*. This will have to be effected by 1 January 1993 at the latest, ie by the entry into force of the Internal Market, and in case of some directives even earlier. This task must not be underestimated: Contrary to the individual directives of the past the more recent EC Directives are to regulate not only a limited range of technical problems but also fundamental issues of occupational safety and the institutional framework thereof.

Let me briefly elaborate by referring to two examples, namely *the Machinery Directive* and *the Framework Directive* both of which have been of considerable consequence.

The most important directive issued on the basis of Article 100 A of the EEC Treaty is *the "Council Directive of 14 June 1989 on the approximation of the laws of the Member States relating to machinery"*. The aim of this directive is to remove technical trade barriers to ensure the free movement of goods within the Community. Technical trade barriers may be caused, for example by different occupational safety regulations and standards in the individual member states. The directive lays down *essential safety requirements* and includes *provisions on a declaration to be given by the manufacturer* to the effect that his products are in conformity with these requirements; in case of certain machines *compulsory type examinations* are prescribed and *labelling requirements* are laid down. These general safety requirements are being put in concrete terms by *European standards*. On the whole the Machinery Directive has had a *marked impact on our German system* which has been established on the basis of the Equipment Safety Act to provide for the safety of machines.

Directives may serve *as instrument to remove technical trade barriers*, and in the meantime a large number of individual directives has been adopted in order to make use of this possibility. These individual directives concentrate on hazardous substances on machines to be used in agriculture and forestry and on various other categories of technical working equipment and installations, such as pressure vessels, gas-consuming installations, lifting gears and conveyors.

As we all know *this type of directive has to be translated into national law* as it stands, i.e. *upward or downward deviations from the required level are not permitted.*

In contrast *the occupational safety directives on the basis of Article 118 A EEC Treaty* lay down *minimum provisions only* which means that the *member states may adopt more far-reaching regulations*.

Of central importance among the new directives is *the "Council Directive of 12 June 1989 on the introduction of measures to encourage improvements in the safety and health of workers at work (Framework Directive)"*. It provides the framework for further specific individual directives and regulates general principles of occupational safety, above all

- the *responsibilities for the employer*,
- *general obligations* of employers and employees which are to be established,
- *obligations with regard to consultation*, co-participation and instruction and
- the obligation of the *public authorities to exert a supervisory function.*

One might say that the Framework Directive is *the "basic law" of occupational safety in Europe*. As you probably know this Framework Directive has meanwhile been *supplemented by five individual directives* which deal with

- workplace requirements,
- the use of equipment at the workplace,
- the use of personal protective gear,
- the manual handling of heavy loads,
- work at VDUs.

The Commission has declared that it intends to work out quite a number of *further directives in the field of occupational safety*. In so doing it will increasingly deal with *matters relating to working hours* and *protective legislation for workers as a whole*. It can therefore be assumed that in future EC legislation will influence our national legislation on occupational safety more and more extensively.

For the time being *it is hardly possible to evaluate the more recent law-making activities of the EC* with concluding effect. As you probably know German experts in occupational safety pointed out quite early that the need to come to a compromise at EC level and to adjust the levels in the member states could lead to a lowering of the high level of occupational safety prevailing in Germany in order to bring about the necessary harmonization with the standards in the other member states. This means invariably that in some countries the standards will have to be levelled up for this purpose; the extent to which they will have to be levelled up will depend to a considerable degree on the quality of the new European standards. This is why it has come to be a *strategic importance for the German Standardization Institute DIN to participate in European standardization activities.*

On the whole it can be said that *the EC directives will give a positive impetus to the further development of occupational safety legislation in the EC*. For example, public service employees are covered by the scope of these directives which is not yet a matter of course in the field of occupational safety in the Federal Republic of Germany. Particularly interesting for the future is, however, that the *EC Directives reflect a comprehensive approach to occupational safety. The concept of health* as used by the EC includes, eg health impairments caused by mental strain, stress, bad organisation of work, monotony, etc. These are *stress factors which have not yet been taken into account by German legislation on occupational safety*. With the translation of these directives into national law the EC member states will introduce occupational safety regulations which will cover a broader range of work-related stress factors and hazards than the legislation in force. This is a perspective which can only be welcomed in the interest of a comprehensive workers' health protection.

Let me mention in passing that *the EC is planning to set up an occupational safety agency* which would inter -allied take part in activities to supervise the translation of EC Directives into national law. If realized, these plans would lead to the creation of another supervisory instrument at the European level besides the European Court of Justice which would certainly give the Commission greater possibilities for enforcement.

I already pointed out that *a comprehensive health concept is used in the EC directives.* This is not so much the result of a concrete definition of this term but rather follows from the problems mentioned in the directives and to be solved in the fields of work organisation, stress and mental strain, for example. I further elaborated that *the Federal Republic of Germany is obliged to translate these EC provisions into national law* - as binding legislation and/or minimum requirements. This means that the scope of problems to be regulated by *German occupational safety legislation will have to be extended*, a process which is comparable to the more comprehensive tasks given to company doctors by virtue of the Safety at Work Act. Here I should like to add that these issues will also be subject to the participation rights of the works council under the *Works Constitution Act* once they are regulated by law.

When looking at the development of occupational safety in the Federal Republic and also abroad we realize that there is indeed movement: *the traditional concept of occupational safety* which focused on the prevention of industrial accidents and occupational diseases has *been further developed in the meantime*, and not only recently. Incentives were given by companies on the basis of their practical experience but also by researchers. The occupational safety experts of the companies act on the legal basis of the Safety at Work Act, and they have often achieved interesting results. In *the field of occupational health care*, for example, it seems that the reservoir of feasible activities is far from being exhausted. In a series of articles published in the "Bundesarbeitsblatt" we currently present various concepts in this respect.

The legislator has taken measures partly to encourage and partly to safeguard this development:

- The 1972 Works Constitution Act introduced the concept of *"acknowledged findings of industrial science"*.
- Under the 1973 Safety at Work Act company doctors were inter -allied given the task to examine *"work related diseases"*.

Section 20 of the 1988 Health Care Reform Act authorized health insurance funds to take part in the prevention of *"work related health hazards"*.

These examples illustrate that we were *moving towards a more comprehensive health concept* also *in our legislation* on occupational safety when the new EC Directives made a point of introducing such a concept, too.

The Federal Labour Ministry has begun to prepare the translation of the EC Directives into national law. The legislative procedure will be initiated upon completion of this preparatory work. It will be at this time at the latest that the public will start to discuss these interesting aspects of EC law-making in the field of occupational safety. I think we should constructively follow up the incentives given by the EC- not to mention the fact that we are actually obliged to put them into practice - to improve preventive health protection at company level.

Finally I should like to sketch out a *few of our activities in the field of occupational safety at company level*. Their aim is to further improve the implementation of the Safety at Work Act. For this purpose we have worked out proposals in several working parties including occupational safety experts with practical experience, and these proposals have recently been presented to the interested public. The aim is to discuss and realize these proposals, for example in accident prevention regulations and further training concepts *The proposals deal with the following matters*:

- *Improving the training of safety* engineers, safety technicians and safety experts with the qualification of a master;

- *criteria for the employment of such experts* at company level;
- *possibilities to extend* the provisions of occupational health care and safety technicians' *services to small and medium-sized companies* and
- *qualification standards* for the provision of occupational health care.

I am glad to say that those who deal with occupational safety in practice have given considerable support to these activities so that I assume that we will make progress in the field of occupational safety at company level. Here, too, *the programme "Work and Technology" helps* us to make headway *by means of pilot projects* and the development of working aids for occupational safety experts at company level.

In this international gathering I will refrain from elaborating *the possible strategic importance of s. 20 of the Health Care Reform Act* which entered into force only recently, namely in 1989. Turning to our German participants I should like to refer to what has been said in this respect elsewhere and limit myself to the following:

S. 20 authorizes the health insurance funds to play an active part in the prevention of work-related health hazards; for this purpose they are to co-operate with the bodies responsible for occupational safety matters. This is an attempt to create a *legal basis for a co-operation between the occupational safety scheme and the general health care system* in the Federal Republic of Germany, a new development which has also been strongly supported by the Federal Institute for Occupational Safety and the programme "Work and Technology".

Let me conclude by saying that *the research programme "Work and Technology" could fulfil functions* as a result of being embedded in legal developments in such a way as has been described in relation to occupational safety by way of example which go far beyond those of a "normal" research programme. We hope that it will be possible to make use of the opportunities which have emerged.

Dangerous Substances Database of the West German Federal States

G. Albracht in co-operation with A. Walter, K. Bott, K. Schneeweiss, Wiesbaden, Germany

1. Introduction

Some 100.000 chemicals contained in over a million preparations and products are sold throughout the world. Because of *the deluge of dangerous substances* (Fig. 1), information about them plays an important role in enforcement and the further development of dangerous substances monitoring. That is particularly true because exposure to dangerous substances *is one of the most important occupational health hazards*, as highlighted by the fact that two-thirds of the rising number of 45.700 industrial disease notifications, or 31.000, were due to the effects of dangerous substances (2).

The plethora of chemical accidents in the past few years has shown that *officials* (Trade Supervisory Authority [1], fire-brigade, police and the agencies responsible for occupational safety, public health and protection from exposure to pollution) *need rapid and reliable data about dangerous substances* in order to minimize risk and avoid acute hazards, so that they can take preventive measures and react quickly and competently.

In 1982 the Government of the Federal State of Hessen proposed establishing a centralized federal register of dangerous substances. When that suggestion met with no response Hessen worked together with three other West German states to develop their own model for a dangerous substances database and submitted it to the federal authorities and the other states. After unanimous approval of the resolution by the Sixty-second Conference of Employment and Social Affairs Ministers, the remaining states joined Rheinland-Pfalz, Baden-Württemberg and Niedersachsen which hat participated from the outset.

---

1) The standard translation of the German concept Gewerbeaufsicht is "Trade Supervisory Authority". In order to draw attention to the fact that the emphasis in this context is on the supervision of the maufacturing rather than the retail process, the less common term "Factory Inspectorate" has also been used.

2. Dangerous Substances Database Of The West German Federal States

The resolution asked the *Federal States Committee for Occupational Health and Safety (LASI)* to develop a *Dangerous Substances Database* for the states. It also established the *Specialized Group on the Dangerous Substances Database* in September 1986. The Specialized Group includes all West German states, the Central Organization of the Industrial Employers' Liability Insurance Associations (HVBG) and the Federal Office of Occupational Health and Safety (BAU). Other participants in the meetings of the Specialized Group include representatives of the Federal State Working Party on Environmental Information Systems (BLAK), the Information and Communication System on Dangerous and Environmentally Harmful Substances of Nordrhein-Westfalen (IGS-NRW), the Federal Environmental Protection Agency (UBA) and the Federal Health Agency (BGA) (Fig. 2).

In the LASI resolution of 23 and 24 January 1989 the Specialized Group on the Dangerous Substances Database was assigned to Subcommittee II on Dangerous Substances of LASI. That committee supervises the Dangerous Substances Database of the Federal States (GDL, Gefahrstoffdatenbank der Länder) within LASI. A *clearing-house collects the data* that have been prepared, compiles them and makes them available to the members of the Specialized Committee in either disk or tape form. Sources include legislation, rules and regulations, reference works, original literature, other databases and information from manufacturers. The databases are currently kept at central offices of the Specialized Group members. It has been agreed that in future the Federal Office of Occupational Health and Safety (BAU) will act as clearing-house.

3. System of Information on Dangerous Substances

Unlike many other databases, the *GDL will be linked with enforcement data* (operator files, plant and workplace files, Substance exposure and occupational disease files) which makes it an essential tool for enforcement and monitoring by the Trade Supervisory Authorities. The GDL (fact files) thus forms the cornerstone of the states system for the exchange of information on dangerous substances among the federal states and HVBG (3) (4). In connection with the substance exposure file it should enable the authorities to obtain rapid, reliable information about dangerous substances in order to avoid risk or to take action in specific cases when accidents occur. The aim is to optimize information on substances for enforcement and monitoring authorities as well as to link substance information (substances, preparations and products) and enforcement data. Used in that way it becomes *an important foundation for planning concerted actions and regulations* to control certain substances. Moreover it helps to answer inquiries and provide advice to industry. Fig. 3 summarizes the use of the Dangerous Substances Database by the Trade Supervisory Authority. There are plans to make the Database available to interested third parties in the medium term.

Fig. 4 shows the connection between the Industrial Cadaster and the Dangerous Substances Database.

4. Substance Data Standard

The substance data standard forms the basis of the work of the Specialized Group by establishing the content of the Dangerous Substances Database. The *content* is based on *the information required for notification of new substances according to the German Chemicals Law (ChemG).* The substance data standard (5) is a rigidly formatted data description that forms the basis for the exchange of data in the Specialized Group.

5. Identification of Substances

Identification of substances is the most important aspect of the Dangerous Substances Database, because it involves allocating a ***distinct substance number*** to each different substance (mixture). That substance number is used independently of any other substance characteristics as a linkage criterion between the individual characteristic areas. Fig. 5 lists the primary identification characteristics used for substances. Fig. 6 provides a summary of all characteristic areas.

6. Query for Names

To recognize whether a given substance is stored in the Database under another name, all *substance names with synonyms* are kept in a separate characteristic area. The possibilities for convenient query depend on the hardware and software used. For purposes of query, the spelling of the substance names is standardized in Hessen by changing all capital letters into lower case and by eliminating umlauts and any special symbols. The output on substance names then appears in the original form, ie with capital and lower case letters, umlauts and special symbols.

7. Integration of other Databases and Areas of Specialization

The GDL goes beyond the area of occupational health and safety and is also important for environmental questions, planning and monitoring. Matters from various related subject areas can be integrated without difficulty into the Database by adding other specialized subject areas to the existing date and information, for example the dangerous products, security data sheet and police and fire brigade information files (Fig. 7).

8. Hardware and Software

The federal states operate the Dangerous Substances Database using different hardware and software. In Hessen the Database, like the IGS-NRW System, is kept on a UNIX computer (Nixdorf/System TARGON 31) as relational database (DDB/4) for the Social Affairs Ministry of Hessen at the Hessian State Environmental Protection Office (Central Office for Occupational Health and Safety). Rheinland-Pfalz uses networked personal computers (MS-DOS) and the DBASE IV database.

The user surface (screen mask) of the Hessian Dangerous Substances Data Bank was adapted to the interests of the various user groups (Trade Supervisory Authority, fire brigade, etc). Thanks to the menu-driven programme operation a substance can be queried at the screen without difficulty, even by non-experts (Fig. 8). The various information areas can be called up at any time using function keys (soft keys).

Because fo the relational structure it is easy to increase the database by adding additional characteristics or tables.

The memory capacity for the current level of approximate 100 000 substance data records is about 500 MByte including the database and the operating software. The details of the database structure, the user functions, the mask configuration and query possibilities are found in the IGS Status Report of November 1989 (6).

9. Data-, Information- and Communication Linkage

The GDL will include data from Federal databases as well as those from other federal states, the Central Organization of the Industrial Employers' Liability Insurance Associations (HVBG) and industry. The HVBG, IGS-NRW and the Federal Office of Occupational Health and Safety (central clearing-house) have agreed to a mutual exchange of communications, information and data (Fig. 9).

Figure 10 shows the most important points of that agreement, such as exchange of substance data free of charge, the establishment of the substance data standard of the federal states as a binding basis for the exchange of date and prohibiting commercial use of the data. The substance data standard which is binding on all partners thus forms the basis for a *large-scale exchange of data*. It is drawn up with the consent of the Specialized Group on the Dangerous Substances Database and evaluated in co-operation with the participating organizations. They establish priorities and assign tasks in *an annual working programme* covering collection, data handling and other technical details.

The Information and Communication System on Dangerous and Environmentally Harmful Substances of Nordrhein-Westfalen (IGS-NRW) uses and combines data from federal databases and those of various federal state authorities, the chemicals industry and particularly the GDL. The feasibility study (7) that preceded the project analysed 37 different data sources and concluded that the data of the Dangerous Substances Database could make a significant contribution to establishing the IGS. On the suggestion of the chairman of the Specialized Group on the GDL and the Federal States' Working Party on Environmental Information Systems (BLAK) a seminar on the status of the "Development of Harmonized Substance Data of the Federal Government and the Federal States" (8) was held in Berlin on 28-29 November 1988. It concluded that the Dangerous Substances Database of the Federal States is one of the most practically suited for the work of the federal state and monitoring authorities and because of its structure it allows additions at any time. An analysis of 16 different substance database projects, both in West Germany and abroad demonstrated that various databases are modelled on the data structure standard of the federal states. Various database projects of the chemical industry and the automotive industry also use the dangerous substances concept of the federal states as a guide.

Co-operation between IGS and the Federal Minister for the Environment, Nature Conservation and Reactor Safety has permitted an exchange of INFUCHS date and other dangerous substances files of the German Environmental Protection Agency and other international dangerous substances databases (Fig. 11).

The IGS database concentrates on topics that are nearly identical with those of the GDL (Fig. 12). The *area of measurement methods* is an interesting addition that *should be integrated* into the GDL database in the future. To test the IGS, pilot projects are being carried out in close co-operation with the various users in the fields of occupational health and safety, protection from exposure to pollutants, water quality, fire prevention, accident and catastrophe prevention and monitoring of the transport of dangerous products by road.

By combining different databases and data sources (such as LIS accident files, regulations, Merck PC lists, Dräger substance database, AUER Technikum), the IGS now has about 7.000 substance data records and 13.000 product data records with a total of about 100 characteristics. In addition, in conjunction with the German Environmental Protection Agency large numbers of safety data sheets are being input into the system. Fig. 13 summarizes the data stock contributed by IGS. The addition of data relating to environmental medicine is planned. IGS is based on a relational database that uses the UNIX operating system.

10. Data Stock of the Dangerous Substances Database of the Federal States

Both the content, i. e., the amount of data per dangerous substance and the number of dangerous substances and preparations included and the range of functions of the GDL are being increased. Fig. 14 shows both existing and planned date stocks. A fair division of labour and effective completion of the Dangerous Substances Database are ensured by the productive team-work of the GDL, the Central Organization of Industrial

Employers'Liability Insurance Associations, the Federal Office of Occupational Health and Safety and the Information and Communication System on Dangerous and Environmentally Harmful Substances of Nordrhein-Westfalen as well as co-operation with the Federal Environmental Protection Agency and the Federal Health Office. The aim of all participants is to provide information about the most important dangerous substances in the shortest possible time. In its final version the database system should be able to furnish information about more than 50.000 substances.

List of Abbrevations Used

	German Name	English Translation
ARGUS	AltstoffDB registr. gesundheits- und umweltrel. Stoffe (BGA)	Existing substances data base of substances affecting health and the environment
BAU	Bundesanstalt für Arbeitsschutz und Unfallforschung	Federal Office for Occupational Health and Safety and Accident Investigation
BGA	Bundesgesundheitsamt	Federal Health Agency
BK-Datei	Berufskrankheitendatei	Occupational diseases file
BLAK	Bund/Länderarbeitskreis Umwelt-Informationssysteme	Federal/Provincial State Working Party on Environment Information Systems
CAS		Chemical Abstracts Service
ChemG	Chemikaliengesetz	Chemicals Law
D A B A - WAS	Datenbank für wassergefährdende Stoffe	Databank of Water Pollutans
ECDIN		Environment Chemicals Data and Information Network
EC		European Community
EINECS		European Inventory of Now Existing Chemical Substances
GDL	Gefahrstoffdatenbank der Länder	Dangerous Substances Database of the Provincial States

GGVS	Gefahrgut-Verordnung Straße	Regulation on Transport of Dangerous Products by Road
HVBG	Hauptverband der Gewerblichen Berufsgenossenschaften	Central Organization of the Industrial Employers' Liability Insurance Associations
IGS-NRW	Informations- und Kommunikationssytem gefährliche und umweltrelevante Stoffe des Ministers für Umwelt, Raumordnung und Landwirtschaft Nordrhein-Westfalen	Information- and Communication System on Dangerous and Environmentally Harmful Substances of the Minister of Environment, Regional Planning and Agriculture of Nordrhein-Westfalen
INFUCHS	Informationssystem für Umweltchemikalien, Chemieanlagen, Störfälle (UBA)	System of Information on Environmental Chemicals, Chemical Plants, Accidents (UBA)
IRPTC		International Register of Potentially Toxic Chemicals
IUPAC		International Union of Pure and Analytical Chemistry
LASI	Länderausschuß für Arbeitsschutz und Sicherheitstechnik	Provincial States Committee for Occupational Health and Safety
LIS	Zentralstelle Störfall-VO und gefährliche Stoffe in Nordrhein-Westfalen, Ministerium für Umwelt, Raumordnung und Landwirtschaft	Central Office for Accidents Regulation Substances in Nordrhein-Westfalen, Ministry of the Environment, Regional Planning and Agriculture
LWA	Landesamt für Wasser und Abfall des Landes Nordrhein-Westfalen	Office of Water and Waste of Nordrhein-Westfalen
RTECS		Registry of Toxic Effects of Chemical Substances
TUIS	Transport- und Informationssystem der Chemischen Industrie	Transport and Information System of the Chemical Industry
UBA	Umweltbundesamt Berlin	Federal Environment Protection Agency, Berlin
UN		United Nations
VCI	Verband der Chemischen Industrie	Chemicals Industry Association

List of Illustrations

References

1) Albracht, Gerd, (1989), *Dangerous Substances Database of the Provincial States*. In: Die Berufsgenossenschaft, Volume 10/89.
2) Occupational Safety 1988, Report on Accident Prevention of the West German Federal Government.
3) Hessian Dangerous Substances Database, Report of the Social Affairs Ministry of Hessen, 23 March 1987.
4) Dangerous Substances Information System of the Industrial Employers' Liability Insurance Associations (GESTIS), Published by HVGB (Central Organization of Industrial Employers' Liability Insurance Associations e.V.), Sankt-Augustin.
5) Standard of the Dangerous Substances Database of the Provincial States (Data Structure Standard), August 1989 version.
6) IGS Status Report, November 1989, for Ministry of the Environment, Regional Planning and Agriculture of the Provincial State of Nordrhein-Westfalen.
7) Feasibility Study by Nixdorf AG (company) on establishing a system of information and communication on dangerous and environmentally harmful substances 1989. lient: Minister of the Environment, Regional Planning and Agriculture of the Provincial State of Nordrhein-Westfalen.
8) Progress Seminar on the Development of Harmonized Substance Files of the Federal Government and Provincial States on 28-29. November 1988 in Berlin. Working file prepared by Freie Universität Berlin, Research Centre on Environmental Policies, for Federal Minister of the Environment, Nature Conservation and Reactor Safety.

Dangerous Substances

		Number
• Europ. Invent. of Exist. Chemicals	approx.	100 000
• even if only 10 % of them relevant for occupational health and safety	approx.	10 000
• VCI survey: Substances with more than		
• 10 tonnes/year	approx.	4 600
• 1 000 tonnes/year	approx.	1 000
• Preparations (Mixtures for specific applications)	›	1 000 000

Fig. 1

Provincial States Committee for Occupational Health and Safety (LASI)

Headquarter in Hessen — **Subcommittee II Dangerous Subst.** — **Decision-making body approval of annual programmes, etc.**

Clearinghouse — **Specialized Group on Dangerous Subst. Database**

- West German Provincial States
- Central Organization of Industrial Employers' Liability Insurance Ass.
- Federal Office of Occupational Health and Safety (BAU)
- Federal Environmental Protection Agency
- Federal Health Agency (BGA)
- Representative of Federal/Provincial State Working Party on Environmental Information Systems (BLAK)
- Information and Communication System on Dangerous and Environmentally Harmful Substances of Nordrhein-Westfalen (IGS-NRW)

Fig. 2

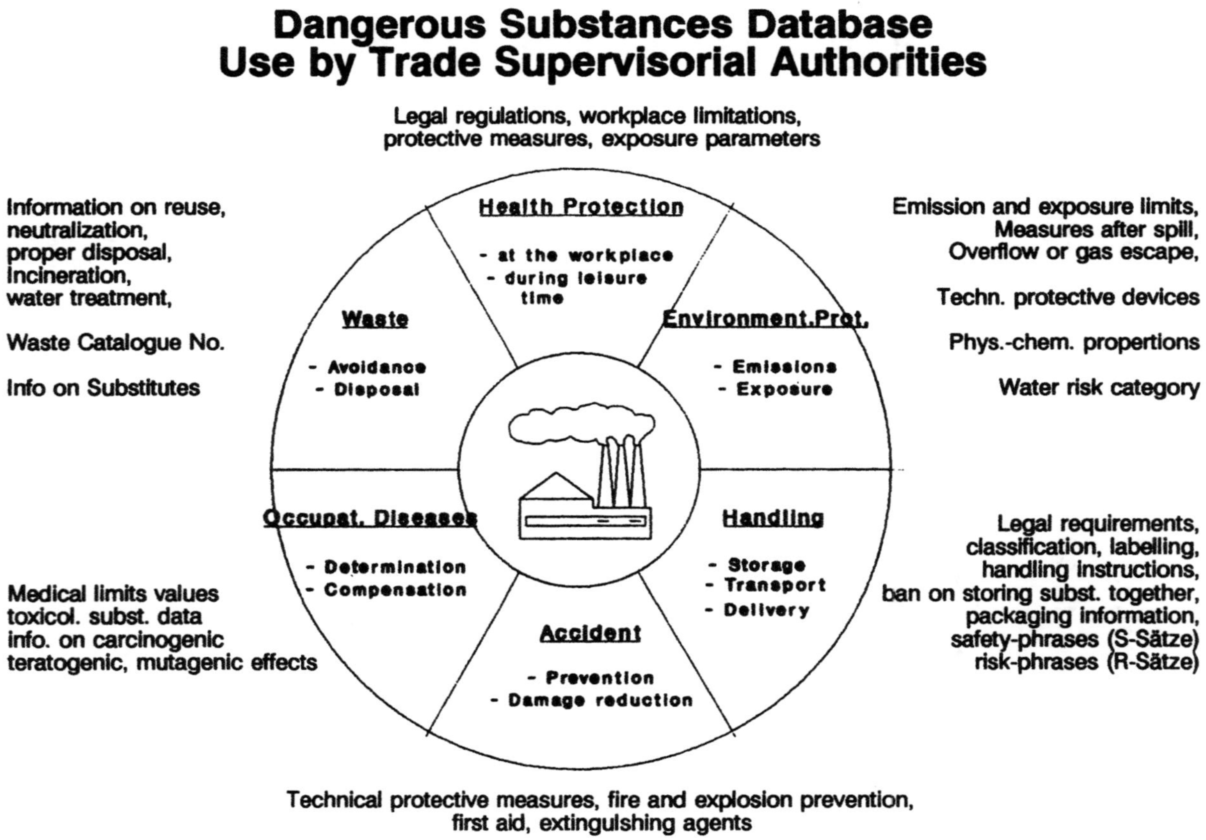
Dangerous Substances Database
Use by Trade Supervisorial Authorities
Legal regulations, workplace limitations, protective measures, exposure parameters
Information on reuse, neutralization, proper disposal, incineration, water treatment,
Waste Catalogue No.
Info on Substitutes
Emission and exposure limits, Measures after spill, Overflow or gas escape,
Techn. protective devices
Phys.-chem. propertions
Water risk category
Health Protection
- at the workplace
- during leisure time
Waste
- Avoidance
- Disposal
Environment.Prot.
- Emissions
- Exposure
Occupat. Diseases
- Determination
- Compensation
Handling
- Storage
- Transport
- Delivery
Accident
- Prevention
- Damage reduction
Medical limits values
toxicol. subst. data
info. on carcinogenic
teratogenic, mutagenic effects
Legal requirements, classification, labelling, handling instructions, ban on storing subst. together, packaging information, safety-phrases (S-Sätze) risk-phrases (R-Sätze)
Technical protective measures, fire and explosion prevention, first aid, extinguishing agents

Fig. 3

THE DANGEROUS SUBSTANCE DATABASE WITHIN THE SUBSTANCE INFO. SYSTEM OF HESSEN

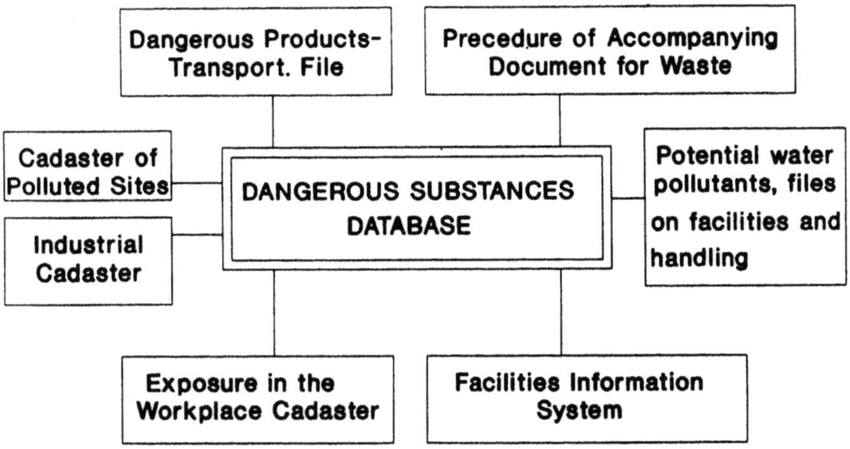

Fig. 4

IDENTIFICATION CHARACTERISTICS

- CAS - No.
- EG - No.
- UN - Nr.
- RTECS - No.
- EINECS-No.
- Molecular Formular
- Name
- Substance group key *
- Chem. Characterization

Fig. 5

** Being prepared or examined*

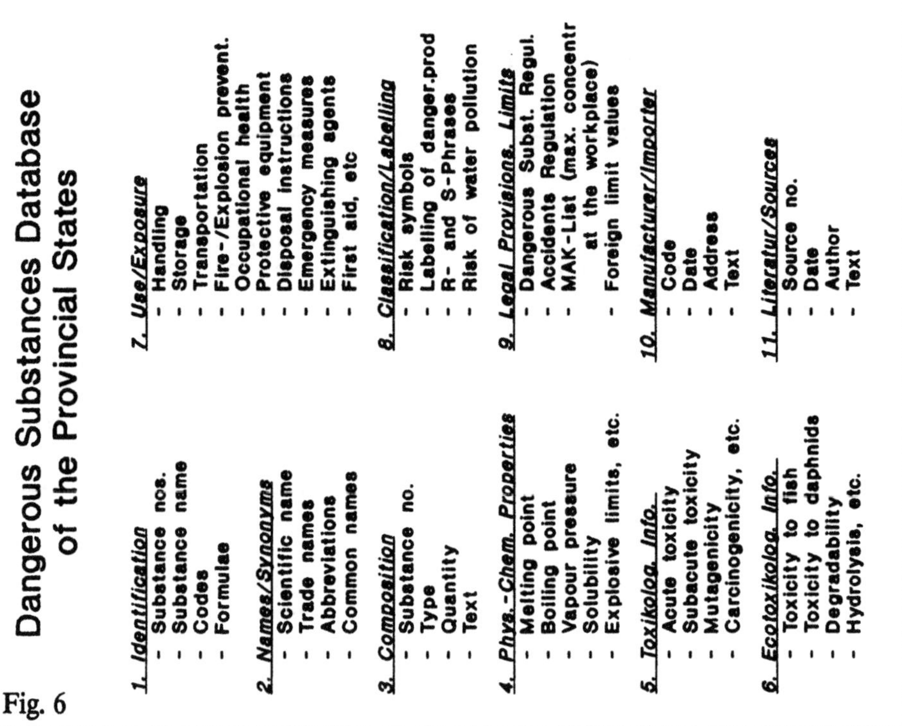
Dangerous Substances Database
of the Provincial States
1. Identification
- Substance nos.
- Substance name
- Codes
- Formulae
2. Names/Synonyms
- Scientific name
- Trade names
- Abbreviations
- Common names
3. Composition
- Substance no.
- Type
- Quantity
- Text
4. Phys.-Chem. Properties
- Melting point
- Boiling point
- Vapour pressure
- Solubility
- Explosive limits, etc.
5. Toxikolog. Info.
- Acute toxicity
- Subacute toxicity
- Mutagenicity
- Carcinogenicity, etc.
6. Ecotoxikolog. Info.
- Toxicity to fish
- Toxicity to daphnids
- Degradability
- Hydrolysis, etc.
7. Use/Exposure
- Handling
- Storage
- Transportation
- Fire-/Explosion prevent.
- Occupational health
- Protective equipment
- Disposal instructions
- Emergency measures
- Extinguishing agents
- First aid, etc
8. Classification/Labelling
- Risk symbols
- Labelling of danger.prod
- R- and S-Phrases
- Risk of water pollution
9. Legal Provisions, Limits
- Dangerous Subst. Regul.
- Accidents Regulation
- MAK-List (max. concentr at the workplace)
- Foreign limit values
10. Manufacturer/Importer
- Code
- Date
- Address
- Text
11. Literatur/Sources
- Source no.
- Date
- Author
- Text

Fig. 6

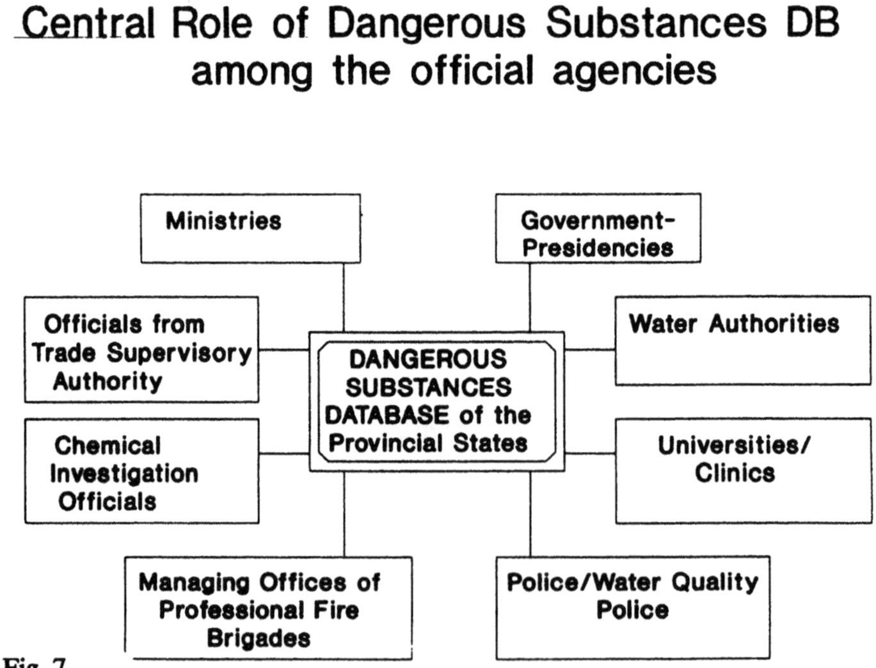
Central Role of Dangerous Substances DB
among the official agencies
Ministries
Government-Presidencies
Officials from Trade Supervisory Authority
Water Authorities
DANGEROUS SUBSTANCES DATABASE of the Provincial States
Chemical Investigation Officials
Universities/ Clinics
Managing Offices of Professional Fire Brigades
Police/Water Quality Police

Fig. 7

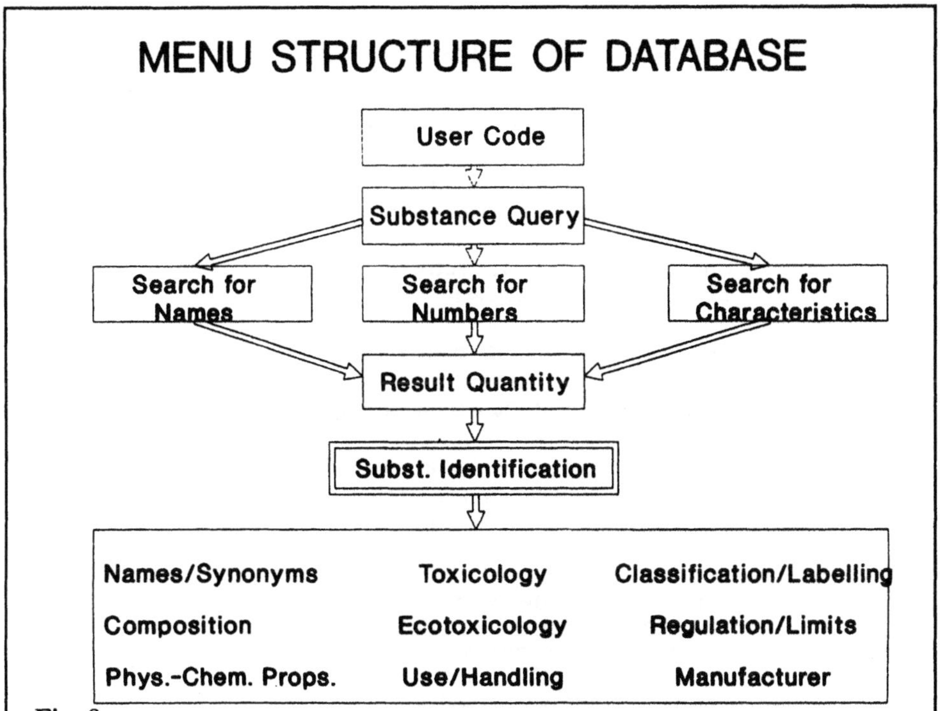
MENU STRUCTURE OF DATABASE
User Code
Substance Query
Search for Names
Search for Numbers
Search for Characteristics
Result Quantity
Subst. Identification
Names/Synonyms
Toxicology
Classification/Labelling
Composition
Ecotoxicology
Regulation/Limits
Phys.-Chem. Props.
Use/Handling
Manufacturer

Fig. 8

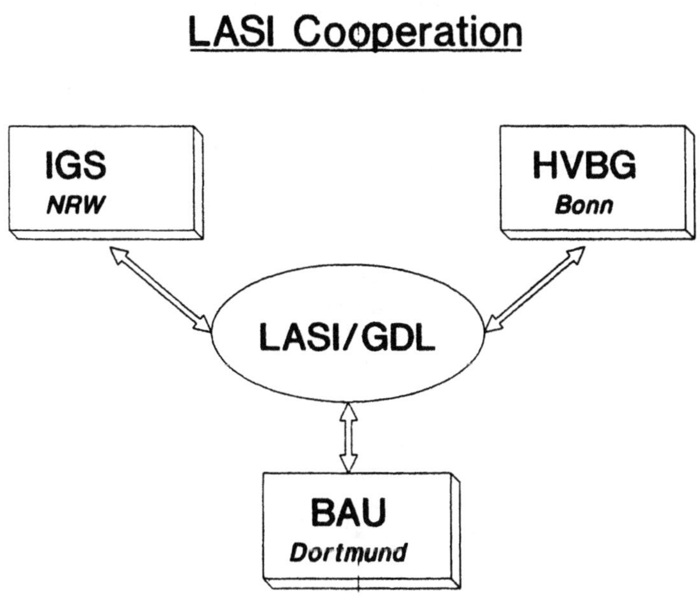
LASI Cooperation
IGS
NRW
HVBG
Bonn
LASI/GDL
BAU
Dortmund

Fig. 9

IMPORTANT POINTS OF AGREEMENT

- Exchange of substance file data, free of charge
- Substance Data Standard of the Provincial States: binding basis for the exchange of data
- Cooperation in the user software, exchange of screen masks
- Data to be examined for consistency by IGS
- Commercial use of the data is prohibited
- Division of labour in collecting, managing and updating data; partners may work with third parties, but may not disclose fact or operational data without the consent of both partners
- Data subject to laws protecting data privacy or that contain industrial and commercial secrets are labelled as confidential and may be provided only for monitoring within the legal scope and for (preventive) medical purposes.

Fig. 10

COOPERATION WITH IGS/NRW

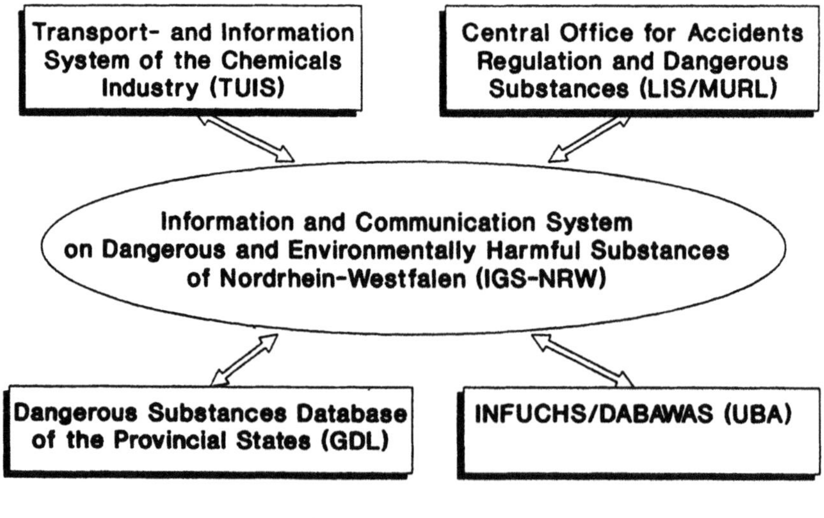

User: Trade Supervisorial Authorities, LWA, Medical Officers of Health, etc.

Fig. 11

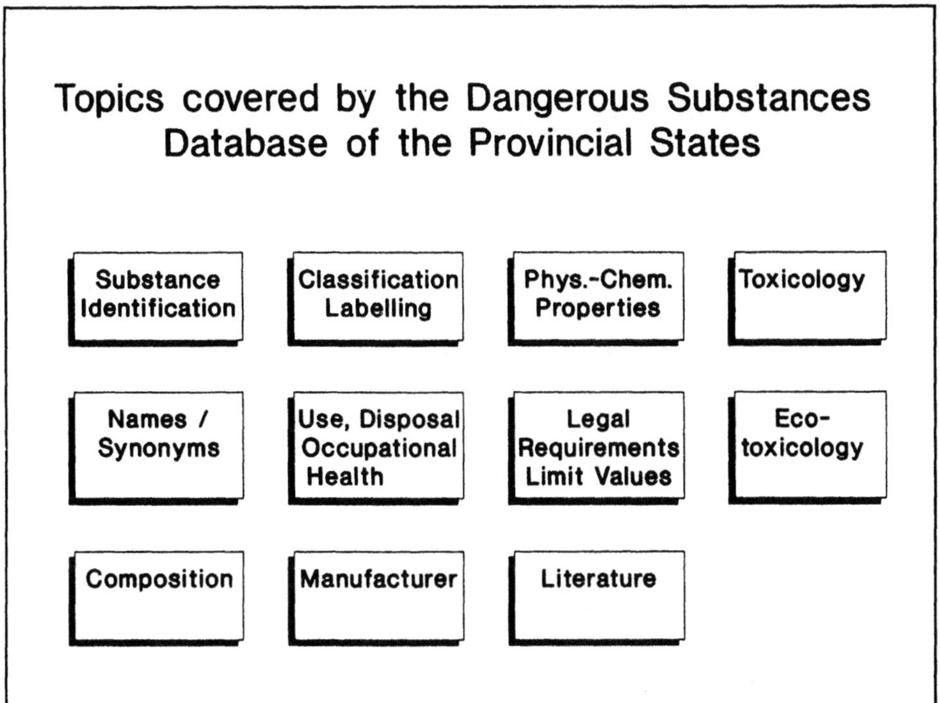
Topics covered by the Dangerous Substances Database of the Provincial States
Substance Identification
Classification Labelling
Phys.-Chem. Properties
Toxicology
Names / Synonyms
Use, Disposal Occupational Health
Legal Requirements Limit Values
Eco-toxicology
Composition
Manufacturer
Literature

Fig. 12

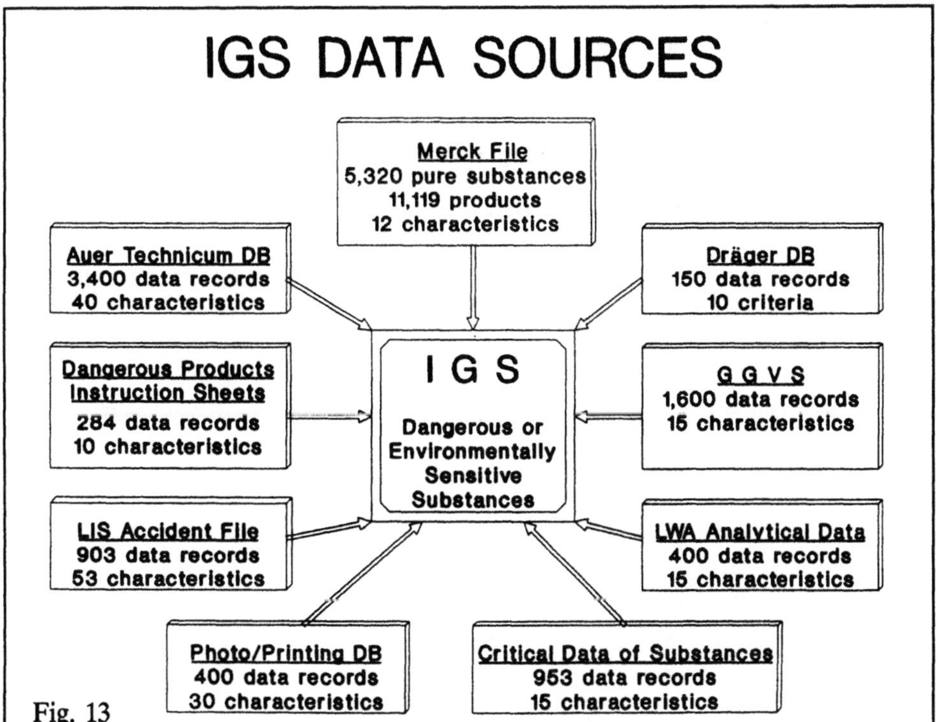
IGS DATA SOURCES
Merck File
5,320 pure substances
11,119 products
12 characteristics
Auer Technicum DB
3,400 data records
40 characteristics
Dräger DB
150 data records
10 criteria
Dangerous Products Instruction Sheets
284 data records
10 characteristics
I G S
Dangerous or Environmentally Sensitive Substances
G G V S
1,600 data records
15 characteristics
LIS Accident File
903 data records
53 characteristics
LWA Analytical Data
400 data records
15 characteristics
Photo/Printing DB
400 data records
30 characteristics
Critical Data of Substances
953 data records
15 characteristics

Fig. 13

DATA STOCK of the Dangerous Substances Database (GDL)

- **EINECS** - Existing Substances	**100.000**
- **RTECS** - Toxic Effects	**90.000**
- **Collection by Specialized Group**, with info. on: phys.-chem. properties, legal provisions, limit values, **use/handling,** ecotoxicolog. information	**1.800**
- after sample taking (300): continuing collection and exchange of preparations	
- **VCI - Safety data sheets**	**4.600**
- **Gas Atlas**	**300**
- also planned:	
- **New applications Chemicals Law / BAU**	**800**
- **DABAWAS data / UBA**	**20.000**
- **ARGUS data on existing substances / BGA**	**3.000**
- **ECDIN / IRPTC data of EG**	**60.000**

Fig. 14

Human Work in Advanced Technical Environments and Anthropocentric Production Systems in the Context of the CEC Monitor/FAST Programme

Werner Wobbe, Brussels, Belgium

1. Introduction

Within the framework of its MONITOR programme, the Commission of the European Communities *operates its FAST activities* (Forecasting and Assessment in Science and Technology), a research programme set up for consultation on *the priorities and orientations to be considered in formulating its science and technology policy.*

When the second FAST programme ended, it put forward *a proposal for a European Community R&D programme on "Human Work in Advanced Technical Environments"* end 1989 with a view to developing anthropocentric technologies. This proposal intended to broaden the scope of the Community actions from its main concentration on IT products and modernisation of old industries. The main message of FAST was that the tryadic competition between USA, Japan and Europe, by concentrating on technological development itself, was not sufficient for the welfare of Europe. Cultural differences in the European Member countries support different implementation modes of specific technologies. Adoption of this policy would therefore increase imbalances in the Community instead of making it more coherent. Furthermore, it was questionable whether the innovation and application modes of traditional technologies were as efficient as always claimed. Therefore, *better adapted technologies*, which *take into account the skills of users and new organisation modes*, thus moving away from the traditional Tayloristic factory, present a way to cope with future developments. This is a *typical FAST approach*; namely, *first exploring the socio-economic developments* and *then following up with the science and technology policy* to be adopted in the light of these developments.

The following outlines the FAST programme in general, together with certain research results.

2. FAST Objectives

The FAST programme, launched by the European Communities in 1979, is *a tool for developing long-term orientation for R&D* at Community level and *contributes to the definition of new strategic priorities for Community policies* associated with scientific and technological developments. FAST has two main tasks:

- *a research task:* the analysis of scientific and technical changes in its many dimensions - economic, social and political.
- *an organisational task:* the *strengthening of the basis of current* European *forecasting and assessment thinking* by encouraging the creation of ad hoc cooperative networks between researchers, civil servants, industrialists and concerned social groups on a European scale.

The main orientation of *the FAST approach is towards the social and individual changes* which are being brought about by the application of new knowledge and technology. A technology driven analytical approach (technological development first, then afterwards consideration of its impact on the economy and society) is often not the most pertinent and useful approach to serve the interests of policy makers, who expect to derive from technology assessment and forecasting a better understanding of ongoing and potential complex societal changes and clear indications for options and strategic choices.

Since 1989, the FAST forecasting activities are a *reoriented follow up to the previous FAST programmes*. It has been allocated 11 mio ECU (1 ECU = 2,07 DM) for the research activities, to cover personnel, scientific fellows, conferences, research networks, etc. The aim is to provide the Commission with global analyses and long term projections in relation to the Community's major objectives for the 1990s, namely the creation of a single European market and the strengthening of economic and social cohesion within the Community, in the light of worldwide economic and social developments.

3. Conclusions Drawn from Fast Research

More than two hundred studies have been carried out in the second cycle of the FAST programme, which have contributed to orientations synthesized by FAST. Results regarding the industrial future are summarized in the following (FAST, Volume II, Prospects for Human Work, Industrial and Organisational Strategies). The FAST studies point out that employment problems and changing employment conditions are caused less by technological developments than by *structural changes in the reorganisation of our industrial society, termed Meta-industrialisation*. FAST does not therefore tend purely to favour the concept of reindustrialisation, which policy makers have supported over the past few years, but recommends that the strategic alliances between producer services and industries and their suppliers be strengthened.

The fully automated manless factory is not expected to materialise, but new forms of divisions of work and cooperation by *computer integrated manufacturing* will result. It is important that the potential of information technologies is used in this context to adapt to the abilities of its users.

To stabilise *Europe's technological position* in the world, it is recommended that besides supporting the information and communication technologies in the ongoing Community programmes, more attention should be paid to new materials and light technologies.

In addition to the scientific and technological objectives, by establishing R&D programmes, the safeguarding of European *manufacturing knowledge and its human resources* has been given a priority position.

R&D for industrial application, should, in future, pay much more attention to the human aspects. Technologies should be such as to assist man to achieve his practical ends, rather than regarding him as a servant of technology. Such an approach demands the development of *computer and human integrated systems (CHIS),* and promises to be a much more fruitful approach in generating productive, robust and controllable technologies than a purely technical one.

Industrial restructuration has to be faced and adaptation measures for innovation have to be developed as follows: the traditional industrial structures are changing with regard to the principles of traditional mass production, including a high degree of division of labour, dedicated automation, product life cycles, large production facilities, hierarchical enterprise organisation. The economies of scale are in transition towards economies of scope or variety. This has serious implications for enterprise organisation, technology, human resource questions, industrial cooperation schemes. It leads to *new forms of industrial structures.*

One element of this new industrial structure is *meta-industrial cooperation*. It means that core enterprises are operating with a periphery of subcontractors for goods and industrial services. This presents changes which could create serious problems, particularly for small firms. Business services are of increasing importance. Telematic facilities are a strategic means to aiding this process, given that the cooperation between small enterprises will depend on access to information of large firms, on long term relationships and publicly provided infrastructures.

New forms of manufacturing - the *"flexible specialisation pattern"* - will continue to increase. These developments increasingly favour small production units. In terms of world competition, they seem to be becoming a dominant manufacturing pattern for Europe, able to adapt rapidly changing markets and delivering complex and quality advanced products. These production units are based on programmable technologies, skilled workers and intelligent organisation and knowledge based management. Key elements are skills, factory and work organisation. To aid this process in Europe, anthropocentric technologies, computer and human integrated manufacturing (CHIM), flexible forms of organisation (orgware and orgknow), as well as training measures have to be developed.

4. Human Work in Advanced Technical Environments

As a consequence of these orientation pointers, *in November 1989 FAST had proposed a R&D programme for anthropocentric technologies for the Community*. A working group lead by M. Cooley (Cooley et al.) was charged to give a first outline which was further developed by FAST (FAST Proposal). The particular reasoning of the programme initiative is as follows:

Productivity, economic success and social welfare in the beginning of the 21st Century will be dependent on a number of strategic factors in which technology has to be carefully embedded. Already by now, flexibility, changing markets, the role of SMEs, environmental concern, the use of new materials, education, ageing of society and the role of young people, misinvestment, vulnerability of "big" technical systems, have led enterprises and public bodies to reconsider the question of scientific and technological development, application and diffusion.

It has become evident that productivity and competitiveness will increasingly depend on the link between the skills and ingenuity of humans and advanced and complex forms of technology. New R&D programmes have therefore to be based on the "valorisation" of the human factor (Hingel 1990, Commission 1987) in production environments by designing, developing and applying technologies which can be called *"anthropocentric technologies and systems"*. The key focus of an R&D programme would be:

- how to optimally achieve higher productivity and competitiveness in manufacturing, and
- how to avoid misled investment.

This would deal with *an issue* which is to be found *on the agenda of advanced enterprises today*. Experiences with complex information technology based systems have not led to the results expected. 20-50% of investments have not succeeded in reaching their aims and are often taken out of deployment.

Such an R&D action will fill an important research gap, i.e. fundamental and applied research devoted to the identification and definition of the scientific, technical and organisational prerequisites with the view to *designing and developing highly advanced technical production systems that conform to human, social and environmental requirements.*

Its subject is a *new generation of anthropocentric technologies and systems* that will enhance human skills and competence. Although it has been conceived and designed in Europe to respond to the variety of social structures and cultures within European societies, this proposal is also open to cooperation with non-Community research groups, particularly from the less developed countries.

4.1 Specific Reasons Demanding a New Approach

Several socio-economic and technical reasons exist which support the development of anthropocentric technologies and systems.

The socio-economic reasons reflect the particular nature of the European production structure, and are related to problems posed by SMEs, shifting markets, changing demands, broader product variety, environmental factors, education and demographic development. Technological reasons are concerned with flexibility, vulnerability and misled investments.

4.1.1 Socio-economic Reasons

- SMEs: Contrary to the situation in the United States and Japan, much of European industry is characterised by SMEs. Indeed, in some Member States the entire industry is composed of SMEs. In general, SMEs provide around 75% of total employment and 65% of GNP in the Community. All too often, publicly funded production technologies have started from organisational structures and production modes of large firms and have been reduced in size afterwards to be adopted in SMEs. The holistic organisation and the skill base of small production units have to be considered as a starting point for technological development and design.
- Markets: Demand for industrial consumer goods is shifting from steady expansion to stagnation. The growing wealth in industrialised countries has led to higher prices, and more individualised and customised products. The European industrial core might, therefore, be better off to specialise in those areas which require order bound manufacturing, and hence experience in flexible specialisation based on the flexible production systems and skilled work force at their disposal. Long production runs and mass production have given way to forms of production with a broad scope of variety and with high quality features.
- Environmental and material problems: The 90s will be marked by a growing demand for processes which involve the lowest possible release of eco-impacting materials to the environment and the highest possible degree of recyclability, as well as the use of materials together with the highest achievable energy efficiencies. This implies not merely new forms of materials, but new forms of production and new product ranges which facilitate repair, recycling and refurbishment. There are grounds for believing that anthropocentric technologies will actually facilitate production processes and product ranges which display more energy and ecologically desirable characteristics.
- Demographic Development: According to UN's projections, the population of Europe over the age of 60 will increase by 40% between 1989 and 2000. The traditional development of technology rejects existing skills and creates a demand for

skills that do not exist. This process of skill transformation in industries has serious implications for demographic change.

- Education: The tendency for education to concentrate on narrow, specialist areas is counterproductive and must give way to more holistic forms. The concern should be with education rather than training. Above all, education should be the transmission of a culture which values "proactive", sensitive, creative human beings. The education of engineers and managers has to be such as to equip them to design systems and produce organisational forms which are skill enhancing.

4.1.2 Technological Reasons

The concept of "Total Automation" runs into several difficulties, rendering its long term success doubtful:

- Flexibility: Firms allowing the strategy of "total automation" would suffer from relative inflexibility with respect to alterations of batches and process innovation. The more complex automation systems become, the more inflexible (including maintenance and break down) they seem to be.
- Vulnerability: Indiscriminate attempts to replace humans by machines have led to serious difficulties with machine dependent systems which are vulnerable to disturbances and which frequently lack robustness and flexibility. Disruptions, large scale accidents, systems breakdowns, low capacity usage are different forms of unexpected vulnerability. They are mainly caused by inherent problems in the design of complex systems, deterioration of norms, ignorance of human communication systems.
- Misled investment: Worldwide investment in flexible automation and CIM is considerable (in 1977, 35 billion were invested in automation technologies and this sum is estimated to increase to 77 billion in 1991). However, economic problem solving applications do not seem to lie in bigger and more complex systems, but in new organisational concepts, and smarter technologies calling for a new link between the skills and ingenuity of humans and the advanced forms of technology.

5. Are All Production Cultures Receptive to Anthropocentric Technologies?

Although the advantages of anthropocentric technologies seem to be plausible, it has to be considered that technology design and its implementation have a strong cultural bias (Warner, Wobbe, Brödner, 1990). For analytical reasons, we distinguish between at least two production cultures: *a technocentric production culture and an anthropocentric one.* They could be characterised schematically as follows (see Figure 1).

The Technocentric Pathway	**The Anthropocentric Way in Society**
- White collar dominance in manufacturing - The belief that only engineers could solve manufacturing problems - Productivity raises by technology - Low level of blue collar training - high level of white collar education - Heavy influence of military automation - Bureaucratic and centralised control - Fordistic paradigm	- Broadly educated human capital at all levels - Acceptance of blue collar expertise - Productivity achievements by organisation - Dispersed markets with different cultural flavours - The flexible specialisation paradigm - Cooperative and decentralised patterns of control - Service paradigm

Figure 1

It is the peculiarity of Europe that it is composed of very different cultures, including production cultures with strong relationships to national histories and traditions. FAST has therefore set up an assessment exercise to explore in more detail the factors which are fostering or hindering the implementation of anthropocentric technologies. This research exercise started in 1990 and the results are now available

6. The Fast Assessment Exercise on Anthropocentric Production Systems (APS)

The Objectives

The study aims at assessing development perspectives, during the next 15-20 years, of anthropocentric production systems. This term is used to describe a particular principle of technology design. Anthropocentric systems are human centred systems based on human competences and holistic organisations (low division of tasks, flat hierarchies). Anthropocentric production systems are considered to be an instrument to improve the production performance and European competitiveness at the beginning of the 21st C Century.

The Structure of the Research Work

The core of the research work is concentrated on two areas:

- the assessment of prospects of APS in all Member countries
- the requirements of their technical developments and diffusion.

The core activities are complemented by two further activities: a world comparison of production modes and a Summer school. (See Figure 2).

1. The Core Research Activity

The **Assessment of prospects of APS** is a socio-economic exercise in eleven European Community Member countries established to analyse the different (national) production systems and cultures with the view to comparing to what extent they offer different conditions and responses to the design and development of APS. Besides such an overall assessment, a case study is being carried out on three "small" countries (Ireland, Portugal, Greece) to explore to which extent and under which conditions anthropocentric technologies could play a role in their development process.

The **technical focus** of the core activity is split up into two parts. The first part will study selected advanced experiences with APS in different Member countries. The second part will be an attempt to define the technical development needs that are required for the further development of APS in Europe.

2. The Complementary Activities

Complementary to this core research, two further activities are being carried out:

- A world comparison of production modes and APS is scheduled for the USA, Japan and Europe. In this case, FAST tries to valorise an existing network, CAPIRN, by expanding their activities to cover the question of anthropocentric production systems.
- A first test of the outcome of the core research has been carried out by a Summer school on APS which was held in Vaals (NL) bordering on three countries (Belgium, Germany, Netherlands). It was jointly prepared with the European Foundation for the Improvement of Living and Working Conditions.

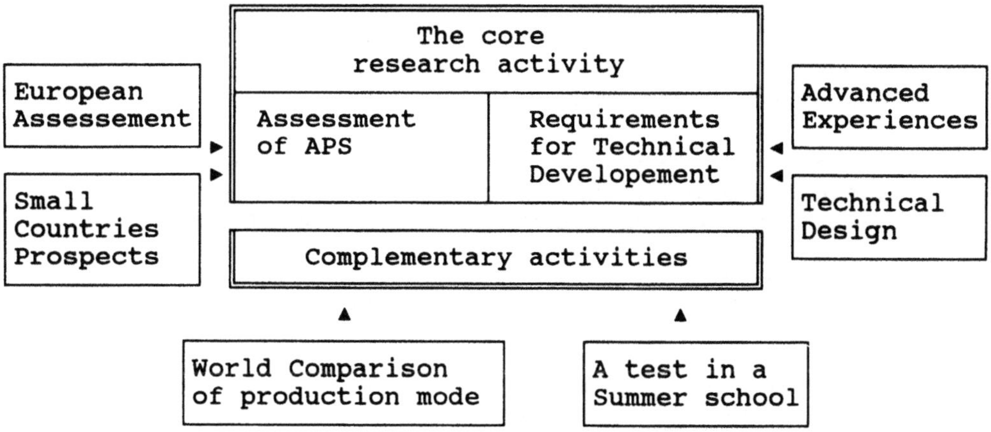

Figure 2

Organisation and Management

To organise the *core research activity* a network (see Figure 3) has been set up. We then chose a *main contractor* for coordination and the final report (the Wissenschaftszentrum Nordrhein-Westfahlen - Institut Arbeit und Technik (IAT). The expertise of the Institute lies in both areas, as progress in the research network points out: in the socio-economic and technical field. The expected results should highlight national production cultures as well as give recommendations for European research and development activities.

In conjunction with this the European network is offering a broader reflection on a world scale. The activity on "World Comparison of Production Modes" is carried out by the CAPIRN network (see Figure 4). Results from the IAT and CAPIRN network have been broadcasted at a conference in Gelsenkirchen, end September 1990.

To help the Commission implement the research activities, an *Advisory Experts' Board* has been set up. It is composed of experts chosen on a personal basis from industry, trade unions and public administrations.

In addition, FAST has established an *in-house* research capacity, thanks to *three fellows* who have sound experience in this research field: J. Schwarzkopf, University of Bremen (D), O. Danielsen, University of Roskilde (DK), R. Bennati, FIOM (I).

1. **Overall coordination and final report:**
Institut Arbeit und Technik des Wissenschaftszentrums Nordrhein-Westfahlen (IAT) - Franz Lehner, S. von Bandemer, J. Hilbert, J. Hennig
2. **The core research activity**
"Assessment of Prospects of Anthropocentric Production Systems (APS) in Europe and Elaboration of Requirements for their Technical Development and Diffusion" a. **Assessment of APS** **Member of the research group on the European Assessment of APS** **Coordination : IAT, D** **(B) P. Berckmans, E. Eysackers, Stichting Technologies Vlaanderen, Brussels** **(NL) B. Dankbaar, MERIT, Maastricht** **(F) D. Linhart, CNRS, Paris** **(DK) L. Rassmussen, Technical University of Denmark, Lyngby, B. Clematide, P. Banke, Danish Technological Institute** **(I) O. Marchisio, Studio Giano, Milano, F. Garibaldo, Centre for Mind and Technology** **(UK) T. Charles, A. Roulstone, Staffordshire Polytechnic, Stoke-on-Trent** **(E) O. Homs, CIREM, Barcelona, F. Mana, Institut Catala de Technologia** **Prospects for APS for small countries** **Coordination : Nexus, IRL** **(IRL) S. O'Siochru, B. Dillon, Nexus, Dublin** **(P) I. Kovacs, A. Moniz, A. Mateus, CESO, Lisboa** **(H) Z. Papadimintriou, University of Thessaloniki** b. **Requirements of Technical Development and Diffusion** **Analysis of advanced experiences of APS** **Coordination : K. Henning, D. Brandt, HDZ, Technical University of Aachen, D** **Technical design Proposals for APS** **Coordination : P. Kidd, Cheshire Henbury Management Consultants, UK** **(IRL) J. Browne, University College Galway** **(UK) T. Husband, Imperial College of Science and Technology, London** **(D) G. Seliger, Fraunhofer IPK, Berlin**

Figure 3: Detailed Composition of the Research Teams and Networks

<u>World Comparison of Production Modes and APS</u>

An already existing network, the CAPIRN network (Culture and Production International Research Network) carried out a research project on the impact of different national production cultures on the design and development of programmable automation in the USA, Japan, Germany and Italy.

Network coordinator : F. Rauner, University of Bremen (D)

Principal regional coordinators are:

F. Rauner, K. Ruth, University of Bremen (Europe)
R. Gordon, University of California, Santa Cruz (USA)
Y. Masuda, Tokyo Keizai University (Japan)

Three workshops/conferences have been held by CAPIRN :

- Santa Cruz December 1989
- Tokyo May 1990
- Gelsenkirchen September 1990

They will specifically address issues related to APS.

<u>Summer school on APS</u>, together with the European Foundation for the Improvement of Living and Working Conditions, Dublin

Coordination : D. Brandt, HDZ, D

Figure 4: The complementary activities

7. Final Remarks

The assessment research on anthropocentric production systems is now coming to an end. FAST expects the bulk of the reports to be delivered by end 1990 as well as a summary report. All the documents will be available and mailed upon request by FAST as soon as they are printed.

The FAST proposal on "Human Work in Advanced Technical Environments" has provoked a lot of discussion within the Member States, as have the content of the report (Cooley et al.) and the research activity itself. However, the outcome is still uncertain as these views, which try to integrate people, organisation and technologies, are different from the mainline thinking of technology developers and science and technology administrators. This is not surprising, as all long term reasoning needs to undergo a discussion process to promote awareness raising. This has led MONITOR/FAST to get in close contact with the concerned services in the C.E.C. with regard to new research and applied assessment areas on anthropocentric production systems. This could ensure that in the forthcoming Community Framework Programmes on Research and Development the whole or parts of the proposal might be adopted. The Community

Research Framework Programme which has just been established has, to a certain extent, already opened up an area in the BRITE/EURAM Specific Programme on work environment. Other Specific Programmes are now more receptive to social science related questions, such as in the "Environment" Programme or the Programme on "Human Resources and Mobility". Prospects therefore exist in the medium term for European Community research, which gives more scope to integrated concepts for technology, work and organisation.

References

Commission of the European Communities, (1987), *The Regions of the Enlarged Community, Third Periodic Report on the social and Economic Situation and Development of the Regions of the Community*. Brussels.

Commission of the European Communities, (1988), *Research and Technological Development Policy, European documentation*. Brussels/Luxembourg.

Cooley M. et al., (1989), *European Competitiveness in the 21st Century - Integration of Work, Culture and Technology.*

FAST, *The FAST II Programme (1984-1987), - Results and Recommendations*. Brussels:

Volume 1: *A Synthesis Report "Science, Technology and Society - European Priorities"*. December 1988, 151 p.

Volume 2: *Prospects for Human Work, Industrial and Organisational Strategies*. July 1988, 214 p.

Volume 3: *L'Europe des Services - Un virage à réussir.* November 1987, 89 p.

Volume 4: *Communication in the Europe of Tomorrow : Beyond Technology*. July 1988, 160 p.

FAST, (1989), *Proposal for a Community R&D Programme on "Human Work in Advanced Technical Environments"*. Brussels.

FAST, (1988), *Small Countries, Science and Technology and EC Cohesion - Issues and Opportunities*. Brussels.

Hingel A., (1990), *Diversity, Equality and Community Cohesion*. Brussels.

Warner, M., Wobbe, W. and Brödner, P. (eds.), (1990), *New Technology and Manufacturing Management : Strategic Choices for Flexible Production Systems*. John Wiley, Chichester.

New Preventive Concepts in Occupational Safety and Health

Karl Kuhn, Dortmund, Germany

1. Development in Social Policy

Over the last decades *government social policy* has undergone considerable qualitative and quantitative development. The group of persons covered, the estimated risks and the level of benefit have been expanded with the results of a comprehensive and differentiated system of social security. Expenditure for the individual areas of social security (e g financing of pensions and the health system) is growing, and the inadequacy of retrospective corrective interventions has become evident. In particular with this in mind, it is now acknowledged that the *preventive elements in our social system have to be strengthened*. This not only has economic benefits in the long run, but the prevention of illness, the safeguarding of the workplace etc are prerequisites for the individual to enable him to lead a dignified and active life in our society.

The historically grown and very *different approaches in the various areas of our social security system* (health, invalidity, old age, workplace) should, therefore, *be more closely interlinked* with regard to their preventive functions, thus furthering the interest of a comprehensive security. It will be possible to organize the social security system more effectively in the future if, in particular, the preventive functions are more closely interlinked.

The demand and need for preventive solutions will continue to rise sharply in future, thanks to changed boundary conditions. Thus the demographic composition of the population will change considerably. The proportion of older people in the overall population will increase sharply. There will, therefore, be fewer young people on the labour market in the next few years. It thus becomes considerably more important to preserve the full work capacity of all working people up to a relatively high age. This sets new requirements for employers, the education system and the labour market and for those responsible for funding the social security system. These requirements concern *the creation of suitable forms of work and technology* in industrial plants, so as on the one hand to maintain, to adapt and develop further the necessary qualifications, and on the other hand to preserve the health of working people for many years.

This demands new concepts for health preservation, the health care of those whose work capacity has changed, *and further qualification*. Only one third of all employees reach pensionable age while still working. Circulatory diseases, rheumatic complainants, psychiatric disorders and tumours dominate the picture of early retirement through reduced work capacity. Plant-based and non-plant-based health protection must take this into account by means of appropriate preventive measures. The growing proportion of women in the working population requires that greater account be taken of the situation of women when designing workplaces. These developments present challenges with regard to occupational safety and health. A retrospective reaction to accidents and occupational diseases is no longer called for, but instead *the preventive design of working conditions* in terms of precautionary measures and provisions.

2. Instruments for Preventive Action

The plant-based and non-plant-based *occupational safety and health protection* is strongly *characterized by experience.* The empirical knowledge is generally based on a quite specific logic which characterizes the accumulation of knowledge. *This logic* can be described in terms of the following stages of action (see picture 1):

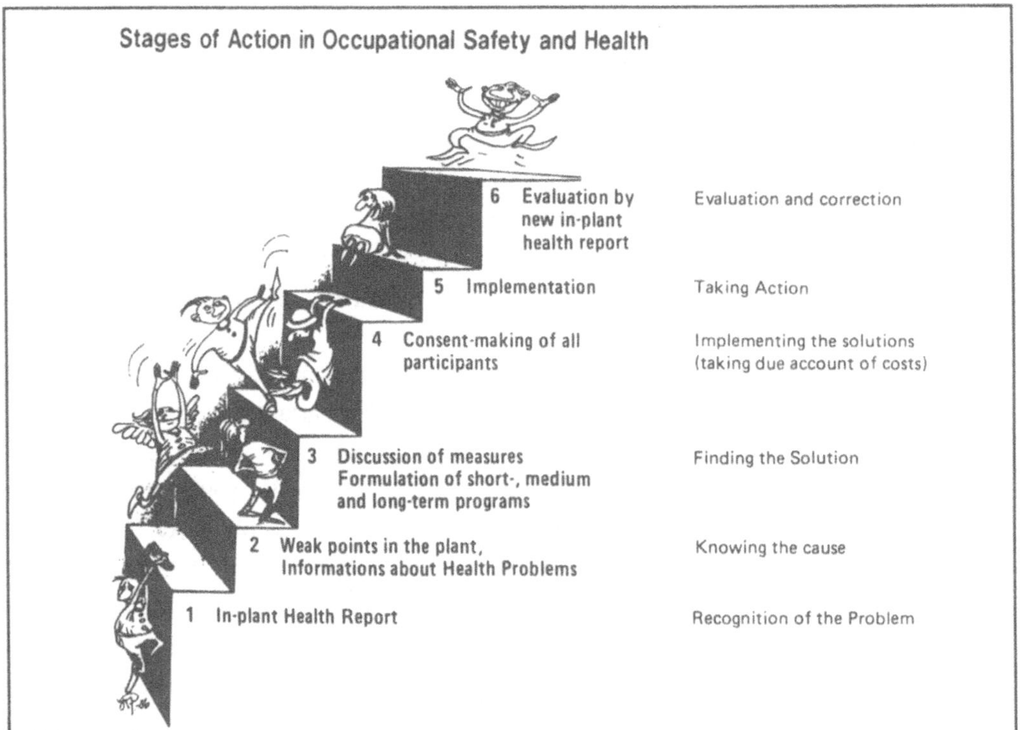

Picture 1

1. *Recognition of the problem*: this initial step already involves major problems, since a nspecific basis of information has to be available to recognize the problem.
2. *Knowing the cause*: if the first step has been taken, the way to step 2 is generally relatively simple. As soon as the problem has been recognized, the causes are also nearly always evident.
3. *Finding the solution*: as a rule a solution has to be found which is acceptable. For this purpose it is also necessary to have knowledge, as to who can take the subsequent measures. The lack of good solutions can be a major obstacle.
4. *Implementing the solutions* (taking due account of costs): this step is normally very painful, because the solution not only has to be implemented in the plant, but it must also be economically feasible.
5. *Taking action*: once this step has been reached, one is generally very successful.
6. *Evaluation and correction.*

This sequence of steps can also be interpreted *as a basic conceptual approach for overall preventive occupational safety and health protection.* For preventive measures always assume knowledge of the causes, conditions and relations which are significant for the development and course of health damage. For the plant as the place of prevention, this means that working conditions have to be established which have an effect on the development and course of health impairments and diseases. One can, with Auguste Comte, describe this new direction for occupational safety as follows:

- savoir pour prevoir; *recognition of risks and hazards*,
- prevoir pour prevenir: *prevention by taking action.*

To date it has generally been a very great problem in the Federal Republic of Germany to give a *plant-related empirical basis for the entire area of work-related diseases*. This has now changed. We have the new situation with, 20 of the Social Code V (Sozialgesetzbuch V), a substantially extended prevention commission. With the introduction of the Social Code V (SGB), the Health Reform Act (Gesundheitsreformgesetz) has included illness prevention and the promotion of health in the catalogue of benefits of the statutory health funds. For example, according to 20 SGB V, the health funds (Gesetzliche Krankenkassen) have to investigate the causes of health hazards and health damage and take action to eliminate them (see picture 2). Furthermore, they would help in the prevention of work-related and environmentally-related health hazards. With this provision, there arise new and varied potential forms of co-operation between plants, those actively involved in occupational safety and health protection and, in particular, the health funds and employers' third party liability associations (Berufsgenossenschaften).

In numerous research projects, *the Federal Institute for Occupational Safety and Health has developed* over the last few years *methods for evaluating the information on work incapacity patterns* available at the health funds. The aim was to develop a pragmatic approach to collecting information on the plant-based illness pattern. The most recent result of these endeavours is the development of the instrument of *the "in-plant health report"*. With this is will be possible in future to present the work incapacity pattern on a plant basis with the help of a few parameters. This will be done in such a way that reliable empirical information will be obtained for the management personnel, occupational safety personnel and works and staff councils. All this information is available in anonymous form. With the help of the in-plant health report, not only is a broad empirical basis concerning the illness pattern in the plant provided, but it also gives an opportunity to conduct an intensive in-plant discussion on sickness and health in the plant.

It has thus been possible successful to instrumentalize the plant-based information base in such a way, that intensive communication and co-operation is ensured in *initiating measures* and, in particular, measures *relating to plant-based health promotion.* The plant-based information base may become the core of an in-plant discussion and decision-making process, leading finally to

- a possible new orientation for the *industrial medical service*,
- the possibility of *assessing health promotion measures* developed internally or "imported from" outside, specifically with regard to how much they help solve the problems involved,
- the possibility of *releasing the initiative* of personnel managements, works councils and company doctors and relating it to jointly recognized problems,
- the possibility of assigning to *the occupational safety committee of plants* a new function of *in-plant "health reporting"*, by also treating the development of the sickness situation as a measure of the success of actions taken by those responsible for health matters in the plant.

Small and very small plants can be merged in terms of sectors and regions and also analysed in a health report.

It has now become evident that these beginnings have introduced *a major communicative dimension* into the plants and also developed a communicative dimension outside in the social security institutions. This gives the impulse to initiate and put into motion both - behaviour-related and condition-related preventive measures. Those, involved inside and outside the plant, find that in the plant, as the place of action and practical implementation, they occupy *completely new roles*; it can be assumed that thus routine occupational safety will be permanently and positively influenced, by virtue of the fact, that it is possible to discuss health-related questions in a wider context.

It may, therefore, also be possible to draw in-plant occupational safety and health from its semi-neglected corner. *The occupational safety and health* of the future must be anchored in the line organization of the plants as *a major part of in-plant management functions*, and it must be present especially *in all areas of in-plant controlling* as a natural part of the planning and decision-making processes. Only in this way, in future it will be possible for occupational safety and health to meet the challenges mentioned at the beginning.

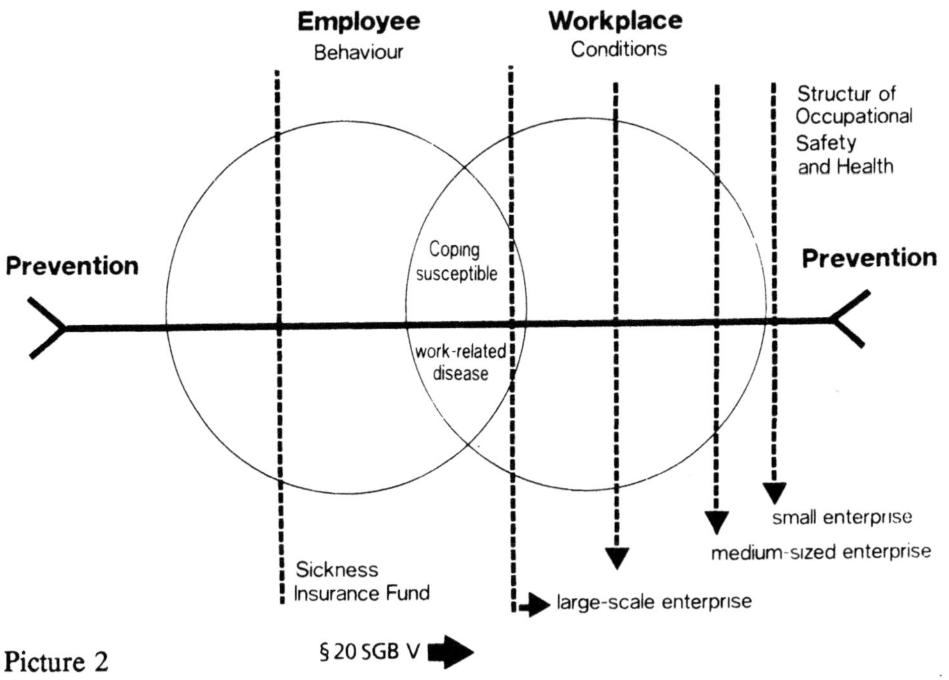

Picture 2

Strategies for Improving Occupational Health and Occupational Safety in Medium-Sized Enterprises - Approach of the RKW - the Productivity Center of the Federal Republic of Germany

Ingra Freigang-Bauer, Eschborn, Germany

The *RKW* - as Productivity Center in the Federal Republic of Germany - has the *overall task to compile and process information regarding strategies and objectives to improve productivity*, as seen from the individual and overall economic viewpoint, and submit them to company managements, works councils, union officials, employers' organizations and to politicians.

One of the essential goals of the RKW is to put the elaborated data at the disposal of *small and medium-sized enterprises* since these, as a rule, are neither personally nor financially in the position to work out *programs continuously* and systematically and to transform them into factors of increased productivity. A very important function of the RKW is to offer possibilities for discussion of socio-economic development perspectives, of occupational and social problems and their solutions.

The RKW is composed of the Main Office in Eschborn and 16 Regional Groups (refer to illustration 1). The Regional Groups execute the important field work including consulting, in-plant and field personnel training, symposia and congresses.

The *section Occupational Health and Occupational Safety* is part of the Department Arbeits- and Sozialwirtschaft.

The main objective of this department is to work out the human aspect in the economy as an economic factor and to appraise it regarding its importance for the individual and the overall economic productivity goals. The program for the nineties, e.g. comprises the topics "New technologies and their effects on the enterprises and the employees" or "Development of work conditions in the area of product-related services".

Following these outlines of the RKW tasks, I wish to point out *the specific approach the RKW takes* within the area Occupational Health and Occupational Safety which, I wish to stress, has been newly defined and formulated at the beginning of this year.

The *deficits, especially small establishments do have*, regarding Occupational Health and Occupational Safety as well as the importance this subject has for the RKW need not to be emphasized. Our objectives can be broken down in two sections:

1. Since *the working capability of employees*, and thereby the overall economic productivity, is greatly jeopardized by so-called civilization damages, such as myopathia,

skeleton, heart and circulatory diseases, which are quantitatively far higher than those damages caused by occupational illnesses or occupational injuries, the RKW concentrated its efforts on the *general "Occupational Health Promotion Programs"*. This, of course, *includes Occupational Safety*.

2. As a *future-oriented, preventive corporate goal Occupational Health Promotion* must be incorporated in various corporate fields, e.g.:

   - the humanizing of working conditions;
   - the organisational and personnel development.

These broad perspectives characterize the strategic approach of the RKW. Prevention of behavioural diseases must hereby correspond to the improvement of the working environment.

These indicated objectives can, however, only be reached if the *specific situation of small and medium-sized enterprises* is realized and if the measures taken, comply with structural conditions. For, although two thirds of all employees are active in small and medium-sized industrial enterprises, the knowledge about working conditions and health status of this group of employees is relatively low. A great number of business branches of various sizes is characterized by small and medium-sized establishments so that general statements are hardly possible. *Health reporting*, broken down in size, branch and region, *does not exist* at all the moment.

Although one should avoid generalizations when talking about small and medium-sized enterprises, there are *a few general and proven results* in this area:

- *Socio-industrial studies in recent years* clearly showed that the innovation and qualification aspect in medium-sized enterprises frequently differs from that in large-size corporations. If occupational health promotion is discussed as innovative corporate objective, the following studies are relevant and must be considered when developing a marketing concept "Health".
- In many small and medium-sized enterprises, a *disproportion of financial expenditures for the technical infrastructure and expenditures for the human capital* is evident. An average of only 5 per cent of all employees of medium-sized enterprises in the metal and electrical industries participate in organized development programs annually (1). In an RKW study "CAD/CAM Project Management", supported by the Federal Ministry of Research and Technology (BMFT), it clearly showed that medium-sized enterprises very rarely engage external consultants who offer more than technical advice. Often "prejudices" towards external advice can be found. There is an *extensive abstinence from external advice* (2).
  Considering this background it can hardly be expected that medium-sized company managements will qualify employees in the sector occupational health promotion.

- Furthermore, studies proved that *long-term objectives and plans are rarely in existence* in medium-sized enterprises. This is true as well for the areas of product innovation, marketing and personnel development (3).
  It is therefore doubtful whether through the publication of results, as e. g. "expanded efficiency" or "re-organization of work structures", the consciousness of the importance of the human capital can be deepened on a wide base.

These results, mentioned as examples, clearly demonstrate that the task to carry out "preventive occupational health and occupational safety" in small and medium-sized enterprises can only be solved by *enhancing motivation and by means of problem-solving strategies*. Such strategies must take into account the basic deficits of such enterprises and thus consider

- the lacking objective planning and future-orientation of company managements and
- the in-house qualifications'deficit, especially in the others than technical areas.

This task is even more difficult in as far, as there are wide differences in the individual business branches of this size regarding work conditions. Here the span of work load is extraordinarily large (4).

Although the envisaged field of intervention, that is "small- and medium-sized enterprises", is not of an uniform character and furthermore difficult to tackle, the Productivity Center is concentrating on *some specific tasks in the Nineties*.

1. Main Office Level

The section "Occupational Health and Occupational Safety" in the Main Office of the RKW will presumably have *an empirical problem study*, carried out on the actual occupational safety and the occupational health actions in small and medium-sized enterprises in the old and new federal states of Germany within the next year. In one or two selected business branches an analysis will be made in which corporate areas having information gaps or where innovative progress in the above area is prevented. Such research tasks are only realized by the RKW where a specific information gap is evident which is not covered by other institutions.

Moreover, the subjects of occupational health promotion and programs for their realization will be treated in *platform discussions* where employer organizations, personnel managements and employees' representatives are represented.

2. Regional Group Level

As mentioned in the beginning, the Regional Groups through their Consultant and Development Programs have direct contact with medium-sized enterprises. In 1991, a Regional Group will carry out *a model conception "health promotion in small and medium-sized enterprises"*. Following the in-house qualification of the RKW colleagues, and after a marketing strategy has been formulated for this area, the contents shall be successively presented to the corporate management level in the form of seminar components.

Focal points of the continuing corporate consulting and development will be problems such as absenteeism and drug problems of the staff. It is anticipated to achieve a sensibilizing process and to carry the motivation for the innovative realization of "health programs" into the management levels.

It is intended to develop *an overall consulting strategy* whereby technical and economic viewpoints, as well as organizational and personnel aspects, will be taken into account.

Co-operation with other health organizations - but also with other institutions, such as environmental protection - can be useful.

3. Transfer Projects

On a third level - in co-operation with the RKW Regional Groups, the Bundesanstalt für Arbeitsschutz and governmental institutions - we intend to *transfer the results of human working conditions* into enterprises eager for innovations. When working out recommendations, the special structure of the medium-sized enterprises must be reflected. Such basic "transfer interventions" can and should flow back to the consulting and personnel development process. I do hope that I will be able to submit a report on the first successful transfer very soon.

References

1) Friedrich, W., (1987), Innovation and Qualifikation in Klein- und Mittelbetrieben, Teil I: *Innovations- und Qualifikationsverhalten*. unveröffentliche Untersuchung im Auftrag des Rationalisierungskuratoriums der Deutschen Wirtschaft. Köln.

2) RKW-Kurzbericht, (1990), *CAD/CAM Projektmanagement gegen die Macht der Gewohnheit*, Eschborn: RKW-Verlag.

3) Besselmann, K., Friedrich, G.W., (1987), Innovation und Qualifikation in Klein- und Mittelbetrieben, Teil II: *Strategien zur Verbesserung der Innovationsfähigkeit*. unveröffentlichte Untersuchung im Auftrag des Rationalisierungs-Kuratoriums der Deutschen Wirtschaft. Köln.

4) Weimer, S., (1983), *Arbeitsbedingungen in Klein- und Mittelbetrieben*, Eschborn, RKW-Manuscriptabdruck.

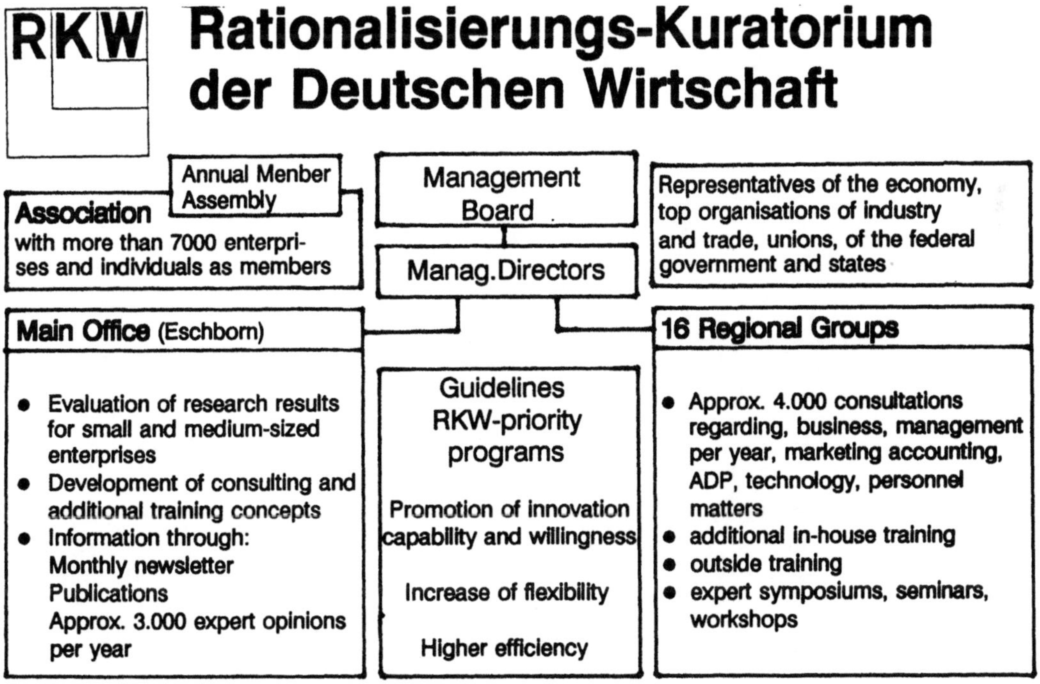

Illustration 1

C

Holistic and Innovative Concepts of Work and Technology

Policy Considerations for Anthropocentric Research and Development

J. Friso den Hertog, Limburg, Netherlands

1. Summary

There is a growing awareness of the importance of integration technological and organizational innovation. This integration is needed to foster both the effectiveness of the firm and the quality of working life. It has been argued that such integration calls for an investment in anthropocentric R&D. In the last decade a number of projects were carried out which can be considered as crucial experiments in anthropocentric R&D. These projects from sociotechnical engineering in a laboratory context to comprehensive organizational change in a real life business context. This paper disputes that at this stage, policy strategies are needed in order to leave the margins for both technological and social research as well as the business organization. A number of considerations are presented for the design of such strategies. This paper is part of a study commissioned by the FAST directorate of the EC.

1.2 As soon as the Process Design Begins ..

In 1968 an internal consultant of the Dutch Philips organization made a survey of thirty 'work structuring' or 'job redesign' that had been projects started within the firm at the beginning of the sixties. One of his basic conclusions was that *"work structuring becomes relevant as soon as process design begins"* (Van der Does de Willebois, 1968). Some years later the company's Central Works Council (COR/Philips, 1973) of the firm came to the same conclusion which was based on a survey compiled of 54 projects. The council made a plea in its report for a *preventive approach to work structuring*. That means a concern for the quality of work at the moment that new products, processes, control systems and factory layouts are shaped, instead of the removal of inferiour jobs once these core structures have been introduced. It represents the shift from 'selecting out' what is undesirable towards 'designing in' what is desirable. The work researchers within the company (Den Hertog, 1978; Van Assen, 1980) stressed that job design could be regarded as a derivative of the design of the product, the process, the control systems and of the organizational structures (figure 1).

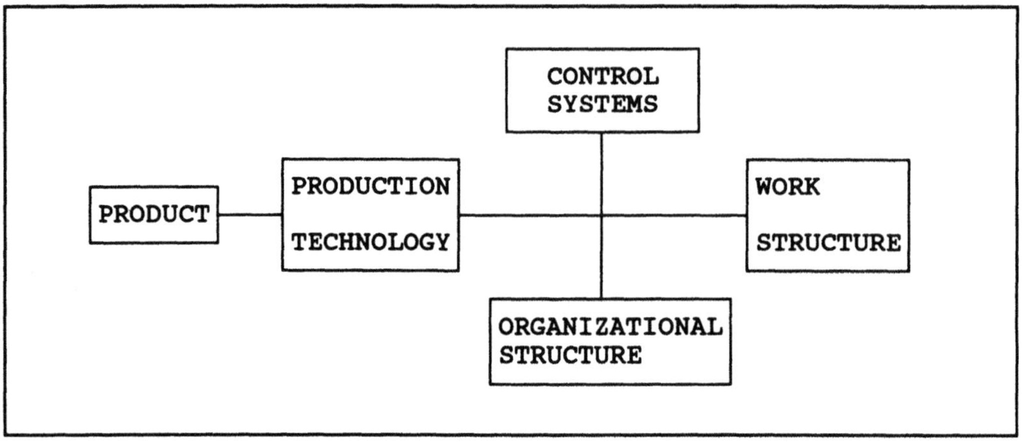

Figure 1: Work Structure as a Derivative

During the seventies, *the link between technology, work and organization* was increasingly stressed. If we want to push job design "out of Marginalia", argues Davis (1976, p. 76) we have to tackle the technological and organizational core structures. In this respect he stressed that technology and organization are, to a large extent, no enforcing exogenous factors, but a product of human choice. This argument has been brought forward by many others since the seventies and has become a point of departure in "the Quality of Working Life (QWL) movement" (cf. Jenkins, 1983). Since the late seventies a form of reasoning (in the opposite direction) has also been advocated: New (information) technology only becomes effective in organizations when job-related criteria are being introduced from the onset of the design. This is particularly true in situations where the socio-technical consultants and researchers have underlined the necessity of an integrated approach of production system design. Enid Mumford's *ETHICS approach* (Mumford & Weir, 1979) is well-known for its effort to bring this integration about. Mumford, Rosenbrock (1983) and Cooley (1989) can be regarded as pioneers in this field. Working groups of national and international professional organizations, like IFIP and IFAC, can also be regarded as important facilitators in this development.

It has become increasingly recognized that such a development has *to be based on the accumulation and integration of new social, economic and technical knowledge*. The reason is simply that we have to invest at least a fraction of the enormous amount of money spent in purely technological R&D into humanizing advanced manufacturing technology (Clegg & Corbett, 1987), or into what others call (Brödner, 1990): *anthropocentric design*. As we enter the nineties we notice an increase of the verbal support for *human-centered or anthropocentric design of technology*. This support is no longer restricted to that of individual design experts or individual firms, but is also shared by policy agencies on the regional, national and international level (see Den Hertog & Schröder, 1989).
In this respect, this paper argues that *policy-making has to enter a second stage*. It is no

In this respect, this paper argues that *policy-making has to enter a second stage*. It is no longer sufficient to underline the importance of this new field of R&D, *effective strategies to guide future efforts must also be developed*. The paper delves into the policy context of anthropocentric design, briefly sketches a number of different forms of R&D projects and offers considerations for future policy-making. This study regards anthropocentric R&D as a temporary concept covering a broad range of initiatives focused upon the objectives sketched above. It is considered in the sense that the final aim of anthropocentric R&D is to design effective production systems in which technology, work and organization are integrated as self-evident and equivalent elements. This paper can be regarded as an interim report of an explorative study, commissioned by the FAST directorate of the European Commission, and builds upon an earlier survey on social research for technological change issued by the Dutch government (Den Hertog & Schröder, 1989).

2. The Context

The key notions regarding the integral design of production systems date back from 20 to 25 years. During the initial 15 years, it was especially *the social scientists that raised the issue*, at a time that business was expanding and the labour market was tight. Research (e.g. Den Hertog, 1978; Van Beek, 1964) indicated that the high level of absenteeism and employee turnover resulted from monotonous work and inadequate industrial relations. Not only was it the power relations, that played a role here, but the cultural turn, which became manifest in the late sixties, also had a considerable impact on the meaning of work in Western societies. The *new legislation, in which quality of work and participation were introduced* as formal criteria for business policy, can be seen as a result of this development.

Particularly in the beginning of this development, *the sociotechnical approach* contributed considerably to finding alternatives to the deep-rooted production bureaucracy. For example, the Norwegian and American experiments in the processing industry (Emery & Thorsrud, 1969; Walton, 1972) and the work of Enid Mumford (Mumford & Weir, 1979) at Rolls Royce and some British banks. A number of large companies, such as Volvo, Philips, AT&T and Procter & Gamble, have also played a central role in this development in which the quality of working life was a key concept. In the late seventies and early eighties this movement fell into an abyss. *The causes* of this can be found *in three particular factors*.

In the first place, the unilateral solutions: autonomous groups and the dominance of the social scientists - compare the Scandinavian critique of employers and Trade Unions to the socio-technical projects (Gustavsen & Engelstad, 1986; Engelstad, 1990; Agurén, 1989) and the German critique to the job redesign projects carried out in the context of the Humanization of Work programme (see Dankbaar, 1987).

In the second place, the socio-economic climate in which all attention was directed to apparently more important issues, such as employment and economic survival (Buchanan & McCalman, 1988). During the eighties, the relationship between work, organization and technology received renewed attention. A reviving economy and new scarcity in parts of the labour market have certainly contributed to this renewed attention, but there was another factor: the growing attention for technological innovation as a motor for economic growth and regional development (Smits & Leyten, 1991). Due to the stream of publications about the difficult implementation of new technologies (e.g. Jaikumar, 1986; Nadler & Robinson, 1987) there was growing awareness that the effectiveness of new technical systems does not only depend on the level of investments and training of personnel. *The industrial context* has changed considerably, which is also due to competition from the Far East: Companies are expected to develop and deliver a larger diversity of goods and services with a higher quality, less stock and lower cost prices. Gradually, it has become clear that the quality of the product is strongly dependent upon the quality of work, and that flexibility is not only a question of machines and logistic systems, but also a question of flexible and transparent organization. As a result, the approaches and concepts developed in the seventies have once again been restored. Stefan Agurén (1989), Advisor to the Swedish Employer's Federation, SAF, points out that in the early seventies these very same ideas, which had been rejected, as they were considered by employers as being socio-scientific hobbies, now offer the key to effective technological development. A precondition is that they are geared to the business development. In other words, design of production systems should take place from an integral perspective, or in the terms used by the *Swedish development programme* (Utbolt, 1989):

- *a holistic perspective* balancing the goals of the business, of the personnel, technology and work organization; by .
- an utilization of technology directed at the *maintenance and development of the knowledge and skills* of the individual workers, and
- an active and meaningful *participation of the workers and their representatives.*

2.1 Towards Anthropocentric R & D Policies

During the eighties, this development was reinforced by policy initiatives in North and Western Europe. *Regional and national governments* in France, Germany, the Netherlands, Norway and Sweden launched R & D programmes in which the humanization of the advanced technology is regarded as a central objective. This objective can also be recognized in parts of *the FAST and ESPRIT programmes of the European Community.* These programmes bring together a range of different actors:

- researchers and consultants from different disciplines (technologists, industrial engineers, economists and social scientists);
- employers' federations and managers;
- workers and their representatives.

Figure 2 shows the remarkable convergence of ideas. These ideas are presented under different readings with almost the same meaning: humanizing advanced technology, human-centered technology, anthropocentric design, humanizing technology, human factors in systems design and constructive technology assessment. The common denominators in these concepts are:

- the aim of *matching human needs* (in terms of job security, safety, health, personal development, qualifications, etc.) with organizational and societal needs (employment, economic growth, organizational effectiveness, etc.);
- safeguarding this match by intervening in an *early design stage*;
- *participation* in the design process of the people who have to work with the technical systems; and
- the *development of knowledge*, models instruments and methods.

What is interesting about this development is that the pleas for a new approach to the design and implementation *originate from different fields of policy-making*:

- *Regional policy:* strengthening the social and technical infrastructure of problem areas (e.g. the Sotech programme (presently: Technology and Work in North Rhine-Westphalia and the SAR initiative in the Port of Rotterdam (Buitelaar, 1989);
- *Technology policy:* heightening the effectiveness of technological development programmes by incorporating work and organizational design criteria (e.g. in the framework of the German production engineering programme (the "Fertigungsprogramm");
- *Science policy:* reinforcing the scientific infrastructure (e.g. the Dutch TAO programme);
- *Labour policy:* safeguarding qualifications and improvement of the quality of work life by early participation in decision-making by workers involved. These objectives can be found both in the laws on working conditions and workers' participation as in government-sponsored programmes in countries like Norway, Sweden, Germany and the Netherlands;
- *Business policy:* contributing to the business development by integrating technological and organizational innovation.

Arbeit und Technik (Work and Technology, Germany)

In this programme, research and development are based on an **innovation** philosophy which is oriented to new **extensive design solutions** considering technical, organizational, qualifying and health-related aspects in their complexity and interaction. At the same time **questions of environmental protection** are taken into account insofar as they are related to the planning of design of work and technology (Research, Pornschlegel, H. 1990, p. 2).

MDA (People, Computers and Work)

Sweden has invested considerable resources in research and development work within the field of computer technology. By international standards extensive research has been carried out in the area of work environments. However, human aspects have received little attention with regard to developments in the area of computer technology and work environment research has had difficulties in keeping pace with these developments. Thus, there is an expressed need for systematic interdisciplinary research where researchers in computer technology co-operate with behavioural work environment researchers (MDA: Pornschlegel, H. et al. 1986).

TAO (Technology, Work & Organization, The Netherlands)

A balanced development of new product and process technology calls for attention for social criteria in the design process, for social impacts following implementation, and for policy-making with respect to design, implementation and use. Consequently, new knowledge and experience is needed to establish the effects on work and organisation, design alternatives and to keep the implementation on the 'right track', (Technologie, Arbeid & Organisatie, 1986, p. 3).

FAST II (European Community)

Priorities for immediate action: The promotion of an R&D community programme on Human Work in advanced technological environments. This programme should aim at designing anthropocentrically oriented systems which will:

- improve manufacturing technologies, their productivity and reliability;
- reduce the vulnerability of complex technological systems;
- apply new methods for socio-technical organization;
- develop communication abilities (FAST MONITOR Programme, 1989, P. 30);

Technology, Employ, Travail (TET)

It is no longer sufficient to safeguard the technological strength or to be effective in sales. What really matters is the most effective way of linking of technological, social and economic systems along the whole chain of production and commercialization. The firm of today has become the focus of social reconstruction because its is the right place in which the equal value of individual know-how, the organizational collectivisms and the products and their markets are brought about (Technologie, Employ, Travail, 1988, p. 8).

Figure 2: Human-centered or Anthropocentric R&D

To sum up, a development has been set into motion which can no longer be ignored. The volume of academic research in this area has remarkably increased, a vanguard of innovative companies has launched initiatives, and regional and international institutions have given the initial impetus to policy development through the introduction of new R&D programmes. The first step has been made: *Anthropocentric R&D should be part of technological and socio-economic and business policy*. This idea is gaining momentan in a number of fairly prominent institutions.

This is an important turning point: It represents the transition from well-intended local pilot projects in the margin of our companies and technological institutions to the build-up of a coherent knowledge base for integral design and diffusion of knowledge and insight in actual practice. This has also led to *a second step in the policy development: the design of effective strategies* which can give shape to this development. The objectives of such policy are as follows:

- *The development of new knowledge* relevant to other actors than the ones involved in local projects. In other words, R&D efforts have to go beyond local problem-solving;
- The policy has to be directed at *the transfer of this knowledge to and use in other design situations* in such a way that it can be used in design practice.

This development is *still in the very early stages*. The R&D activities are still carried out in the margin of technological development. For example, the wellknown project 1199 was the only anthropocentric project among all the 74 CIM projects within ESPRIT. The link between these projects with the hard core of the technology programmes is weak (e.g. in the German programme Fertigungstechnik). Most of the anthropocentric projects from the programmes mentioned here are carried out in the margins of the technological institutions. The majority of the researchers interviewed in the context of this exploration have indicated how isolated they feel within these institutions. *The co-operation between disciplines continues to pose a big problem* for most of them. The initial steps towards the development of a good scientific infrastructure in this area have barely been touched upon. Moreover, interviews with the programme managements have indicated that the diffusion of knowledge and practical experience is still an extremely difficult task, even in countries such as Germany, Norway and Sweden where investments in the improvement of the quality of work has been going on for years.

Consequently, the build-up of an anthropocentric R&D policy and effective R&D strategies is of even higher importance. In doing so, we shall have to learn from past experiences and picture the various development models that were used in the past and that are going to be used in the future. *Attention must be focused on the areas of tension in which R&D has to develop*, for example, those between:

- design experts and users;
- academic research and the business practice;
- the various design disciplines;
- technology push and demand pull.

The ultimate goal of anthropocentric R&D is to integrate within the main flow of technological and organizational innovation. This means that anthropocentric design has to become an outdated concept in the future. At the moment, the concept is proving its value, because it indicates that important elements are lacking in technological development. The logical consequence is, however, that this type of R&D is making itself superfluous. Policy strategies must focus on achieving that stage. This must be considered when giving shape to the areas of tension mentioned above.

Before focusing attention on this issue, we will delve deeper into the diversity of activities that are brought together here under the denominator of anthropocentric R&D.

3.1 Practice

Anthropocentric or integral design of new production systems encompasses a broad category of activities. Projects vary from computer simulations in the laboratories of technical universities to the radical new design of an automobile plant. In order to provide insight into the range of existing projects, we will describe some of them below. These projects have been selected from some 100 projects described in the documentation of national and regional programmes.

Integrated Computer-based Systems for the Newspaper Industry (UTOPIA)

UTOPIA was a Scandinavian R&D project (1981-1985) on trade union based development of and training in computer technology and work organization, especially, text and image processing in graphic industries: This project was a combination of encompassed technological exploration, design workshops with graphic workers, technicians and work researchers, simulations and off-site technical developments and training (The Utopia project, 1985). The aim was *to foster active participation of the workers and the development of a knowledge platform for future work*. The project never led to a new, concrete operating system. The main value had been in raising awareness and workers' expertise in dealing with new technological developments in the industry. UTOPIA can be regarded as one of the pioneering projects in the field (Ehn, 1988).

Flexible Manufacturing Systems

The FMS project was started under the leadership of Rosenbrock (1983) at UMIST (UK) and was directed at the design of a prototype of an FM-System which was a subordinate to the workers instead of the other way around. The concrete design consisted of an NC lathe, a milling machine and a robot. A social scientist was member of the design team. The project is another pioneering study and although it was a small scale experiment, it has been a source of inspiration for the studies that followed.

Integration of Computer-Aided Tools in Mechanical Design

This project is part of the Swedish MDA programme. MDA stands for: People, Computers and Work and is carried out at the University of Lund. The goal of the project *is to link advanced support systems to CAD with a well-functioning manmachine interface.* As a basic resource, an advanced integrated CAD laboratory is being built at the Institute fo Technology in Lund. In this environment the experimental work is being conducted by integrating current CAD systems with commercially available support systems. The experiments are conducted in close co-operation with craftsmen and experts from a large and a smaller company. The results are fed back and tested by a dozen or more companies within the engineering industry (MDA projects 1989).

Shopfloor Involvement in Planning and Design (ESPRIT 1199)

This project is designed within an engineering environment, developed within the framework of the EC's ESPRIT programme and implemented by a conglomeration of European research institutes. A Danish group designed *a kind of electronical sketchbook.* This sketchbook facilitates the involvement of the craftsmen in discussions with designers and other technicians about the design and work preparation in.CAD environments. The objective of the British group was *to design a prototype of a human-centered control system*, based on the principle of *operator control* for a production cell of four workers. In this approach, the dialogue system had to be suitable for the craftsmen as planning people and should be operated by the craftsmen themselves. The prototype had been implemented within a British firm (BICC) which produces radio frequency connectors. The company decided to introduce the concept on a larger scale and other companies have become involved in the programme as well (Brandt, 1990).

The Integrated Design of a New Automobile Assembly Plant (Volvo-Uddevalla)

In August, 1988, Volvo opened a brand new factory in Uddevalla for the production of the 740 on the basis of a revolutionary concept. The cars are assembled as stationary bodies by production teams of seven workers. Each team produces a complete vehicle. The design of this factory can be regarded as an industrial experiment of high significance because it offers *a real example of integrated organizational design.* The layout, logistics, material handling, the training system, the organizational structure, administrative procedures, machinery and so on are designed as integral parts of a coherent master design. Even the product design logic and methods are focused on *the objective of creating a learning organization.* At the same time, the factory is not an experiment but has to operate as a normal business producing 40.000 cars by 1.000 workers in 1992, in a harsh and competitive market (Pornschlegel, 1991).

The Integrated Design of a Computer System for Share Trading

This project is part of the Dutch programme on Technology, Work & Organization (TAO). Researchers from the social faculty of the Rotterdam Erasmus University work together with technical and organizational internal experts and outside consultants from the brokerage department of a large Dutch bank (NMB). *The goal of the project is* twofold: *(1) to design a system in which the quality of work and organization are incorporated; and (2) to develop an active co-operation between the various design experts and managers and stockbrokers* (Ramondt et Al., 1990).

The Humanization of Work in an Engineering Firm

A German producer of electro-motors (Felten & Guillaume) restructured its organization by the creation of production islands and group work in its parts production and assembly. The project has been developed within the framework of the German federal programme on Humanization of Work (HdA, now renamed *Work & Technology*) and took seven years. The final result was *a new* and *integrated design of the whole factory* in Nordenham. The factory is presently composed of 60 to 70 semiautonomous groups. The redesign involved all aspects and parts of the firm: production technology, work organization, logistics, design technology and its interface with production technology (CAD/CAM), the accounting system, the training system, the wage system and so on. Researchers from IFAS in Siegen took care of what is called *accompanying research* (within the HdA programme Klingenberg & Kränzle, 1988).

Humanization in the Garment Industry

In 1977 the German garment industry started an organizational innovation project which was to last seven years (Betriebliche Humanisierungsforschung, 1987). It was set up as a *'Branchenprojekt'* (a branch in this context meaning an industrial sector, such as the garment industry, mining, etc.) within the framework of the HdA programme. Such projects are based on an active co-operation of employers' federations, unions, researchers and professional associations. The project started with an aim set on the basis of an analysis of the range of problems in a specific *'branch'* or line of business. At that time, given the estimation that automation would not offer new opportunities on the short and medium term, the idea to approach this aim by humanization of work and flexibility in organization was reached. The organizational redesign work took place in individual companies which volunteered for the project. The redesign activities focused on broad changes within the firm instead of changes in the individual work places. As the project went on, it became increasingly evident that technological developments could make an important contribution to the objectives of the programme, especially in the areas of sewing, cutting, transport and planning.

It may be obvious that the new construction automobile plant in the Swedish city of Uddevalla is totally different from the development of software design tools and computer simulations in the university laboratory. Projects falling under the wide umbrella of anthropocentric design may show large differences. Below, we will attempt to give *a global characteristic of R&D strategies.* Considering all of these differences, there are two main common dominating features:

- the binding objective is to *influence the development and implementation of technologies;*
- resources are invested in *the development, transfer and diffusion of new knowledge.*

The strategies described can form a mix: small-scale laboratory experiments can, for example, play a role within large-scale, firm-based, redesign programmes.

3.2 In-Company Development Programmes

The main objective here is *to develop realistic alternatives for present production systems.* The new design must operate under normal business conditions and compete with the old ones. Extra investments are made but have to be paid back within reasonable time. The Uddevalla plant and the project in Felten-Guillaume are examples of such programmes. Some of these projects might be referred to 'experiments', because they represent a radical step in the long-term policy of a large company or line of business. The learning experiences must be transferred to other parts of the organization or line of business. In another sense, these projects cannot be characterized as experiments, because the change process cannot be reversed any more. This certainly holds true for companies as Felten-Guillaume: When the 'experiment' fails, the firm will be broken.

Among these in-company projects is a difference in the extent to which they contribute to broader societal objectives, to which (public) external institutions are involved in the design, and to which public means are used. Both of them, the Volvo and the Felten-Guillaume project, had some kind of mixture. The latter project was part of an extensive program initiated by the German government: the Humanization of Work (HdA). The central authorities financed half of the change cost. Thus, the project is serving *both, public and private objectives*:

- *to develop a knowledge base* for the integral design of production systems and to contribute a renewing German business;
- *to contribute to the business development* of the firm in question.

External researchers are involved in conducting accompanying research which leads to public reports, and the experiences are systematically transferred to other firms. The company has accepted a number of rules when deciding to participate in the HdA programme. These rules are concerned with:

- the way in which *employee representatives* are involved in the design process, and
- their *participation in 'accompanying research'* and in activities aimed at the *transfer of experience.*

The Volvo project is concerned with *the company's policy.* The business objective is of prime importance, but it also serves public objectives. The plant was also set up for the development of an economically weak region. Public means also play a role here, maybe in the form of settlement conditions. The company itself has overall control of the transfer of experiences. The Trade Unions also play an important part: it largely determines the scope in which the company is allowed to operate. The Volvo projects are interesting for the *trade unions* because they *take up a pioneering position in Swedish business* in the area of organizational renewal.

3.3 The Academic Workbench

In the first instance, the MDA project is *carried out within the walls of a technical university.* In experimental configurations and on the basis of existing technology, the researchers concerned are working on new configurations and applications which meet the anthropocentric and business-administrative objectives. The experts, mainly academics, have defined the project's aims, issues and working methods. The academic development of knowledge is the prime objective, on the condition that the project places emphasis on a relation with new societal objectives: the development of human centered technology which offers realistic perspectives in a normal business environment. In view of this societal objective, the project is not purely an academic affair. The researchers are well aware of the fact that something can only be proven in actual practice. Therefore, *implementation outside the laboratory is one of the objectives.* In addition, the researchers realize that the contribution of technicians and craftsmen from companies are indispensable in a similar project, both in terms of their knowledge and experience as well as in terms of considering their wishes and priorities. Implementation and participation are two of the conditions set forth in the MDA programme.

Although two main differences catch the eye - *The ESPRIT project* belongs to the same category. Firstly, the project is part of a much more extensive technological programme, in which it is the only project with a clearly anthropocentric objective. Secondly, the project is performed by a co-operative of researchers from various countries and institutions.

3.4 Dialogues

The UTOPIA project did not focus on the design of an integrated system for the newspaper industry, although in the beginning this was being considered. In a purely technological sense, halfway through the project it appeared to be outdated, for example, by the products which were introduced into the market by Apple. Working with make-ups and dummies might seem amateur in a technical sense, but it is innovative in a socio-political sense. The central issue was *to initiate a democratic dialogue between employees, technicians, social scientists and companies*. A meeting place was created where the various groups could be brought together and from where exploratory studies, simulations und minor experiments could be performed. Nevertheless, the continuation of the dialogue in local unions and companies was the prime objective.

3.5 Networking

The German project in the garment industry was also part of the HdA programme. It was a 'Branchenprojekt' in which humanization objectives were integrated with objectives concerning the renewal and survival of the industrial sector as a whole. This project was based on an elaborated consultation structure in which all actors (firms, Branchen, organizations, professional associations, researchers and Trade Unions) are represented. Despite the fact that the local projects formed the hard core, there existed a clear common development.

The Norwegian HABUT and *the Swedish LOM programmes* both belong to this category: Syndergy by jointly operating from the separate responsibilities of participating firms. These Scandinavian programmes *focus more on the development of a diàlogue*. The core of all projects will mean learning the *democratic dialogue* (Gustavsen & Engelstadt, 1986; Gusstavsen, 1988; Engelstadt, 1990). Eventually, this dialogue is expected to result in changes in the work environment. Remarkably, little attention was paid to the design and design methodology (at least in the Englishlanguage publications concerned). Research was particularly focused on the development of a dialogue process, which is expected to lead to knowledge that can be generalized. Design is based on the development of local theories and continues to be generalisable to a limited extent, according to the philosophy of both programmes.

The third form of networking can be found in *the German Production Programme "Fertigungstechnik" (1988)*. The common development of a group of companies with one or more specialized non-profit institutions is the centre of this programme. In these projects the humanization of work is just one of the objectives. The further development and application of new technologies (e.g. robots) for the companies in question is the first matter of importance. The government contributes to their cost. In the first instance, the applications will be available for the participating (and co-financing) companies, which also define the contents and form of the programme. An important part of the activities was performed within the walls of the specialized institute (particularly in the initial phase).

3.6 Accents in Terms of Content

Not only is it the context of the anthropocentric R&D projects that differs, but also the content. Emphasis may be placed on *a variety of* issues, such as:

- *the design of concrete technical systems* as prototype or as a direct application;
- *the design process*, either the co-operation within the innovation chain or the participation of shop floor workers;
- *the development of design tools*;
- *the development of anthropocentric criteria*, for example in terms of qualifications;
- *the evaluation of designs* on the basis of such criteria;
- the development of *training scenarios*;
- *job design* ranging from ergonomic aspects of individual manmachine relations to the design of complete group tasks and
- *the organizational design*, ranging from work cells to the renewal of the organization as a whole.

A broad range of mixtures can also be found here;

4. Considerations

As indicated above, anthropocentric R&D can be weighed on the basis of two *criteria*: the development of new knowledge and skills and its *diffusion*. Of course, a third criterion should be added to that: its *feasibility*, both in terms of cost and in terms of political relations.

We will further elaborate on a number of considerations, or rather dilemmas, below to weigh the feasibility.

4.1 Design Versus Dialogue

Most of the approaches attach *high importance to an active participation* in the design process of those who eventually have to work with the technology. This goes for the majority of the projects, but more particularly for the Scandinavian ones. The key question is 'how' rather than 'who' should participate rether than 'whether or not' Armeone should participate. Both the *HABUT and LOM programmes* have put participation at its centres. General theories and general solutions are rejected. Local theories and local solutions are the key. In other words, these programmes reject the standard solutions brought up by the sociotechnologists in the seventies. Only process-orientated interventions have general value, such as dialogue conferences (Gustavsen & Engelstad, 1986). This express choice for the process side of organizational renewal, however, evokes reactions which emphasize the importance of the structural side. In this respect, De Sitter and Den Hertog (1990) argue that *integral* (or as the Swedes call it: 'holistic') *approaches can be developed only on the basis of generally applicable design knowledge* encompassing a design theory with design rules, *a design sequence and design methodologies*. Local design and local customizing of design models is only possible in this view from a similar common design theory. One might use a civil engineering metaphor here. If one wishes to build a bridge, the design depends on the local circumstances, on the function that the bridge must serve, and on the resources available for that. In this process, nobody will overlook the design knowledge which has been gained within civil engineering. In this vision, it cannot be the intention that a great deal of time is invested in dialogue and that a design is never achieved.

The conclusion that we would like to draw from the above mentioned is: *the link between dialogue and design in the establishment of anthropocentric design deserves prior attention*. In this respect, a number of options are possible. The Dutch sociotechnical systems design, e.g., has chosen an approach in which the members of the organization are trained for self-design through intensive design courses (De Sitter & Den Hertog, 1990).

4.2 Technology Push or Demand Pull?

During the development of the QWL movement, the idea of technology being considered as given from the black box of the technical designers, has been increasingly rejected. *Emphasis is placed on the new social demands which technologies must meet*: learning opportunities, responsibility autonomy and a good ergonomic design. In the last years, this argument was enforced by referring to *the economic relevance of a good integral design* of the organization and technology.

One might say, that *this strategy places emphasis on demand pull*. Internal or external suppliers are faced with new demands and are expected to adopt their designs. During the last ten years, the Scandinavian and German Trade Unions have, to a large extent, performed similar roles. In countries like Sweden, Norway, Germany and The Netherlands, these new demands are further incorporated in the legislation regarding working conditions.

A remarkable turn has occurred in *the development of anthropocentric design*. It has developed from pure technological into *integrated* design. This calls for the use of socioorganizational expertise, which can be provided by both, the social scientists and the workers themselves. In other words, they take the place of the designers. This development implies that *anthropocentric design develops into a new form of technology push*. The inherent risk is that it leads to the same problems as the traditional technology push:

- hobbyism which is far from realistic;
- isolation of the core of the business development;
- vulnerability at a time of turning tides;
- alienation of the workers eventually concerned;
- implementation and diffusion problems.

By separately referring to this technological approach, there is a risk that the new efforts will not be taken seriously and, in turn, end up in the margin of the technology. The history of work structuring pre-eminently illustrates this phenomenon (Van Assen & Den Hertog, 1984). This analysis points to a new question: *What is most effective: the development of a new technological offer, or influencing technological demand?* This issue is concerned with the key question regarding the effectiveness of future anthropocentric R&D. Let's give *two examples* to illustrate this point.

First, it cannot be denied that *the introduction of the Apple computers* have been an important step in anthropocentric design. What anthropocentric designer could have outlined the perspective offered by Apple a year before the introduction of its first computers? The members of the UTOPIA team would not have been able to do so at any rate! This is not only related to the development within the company itself, but also to the efforts of other firms to meet the same market demands (e.g. Windows from Microsoft). The diffusion of computer applications originated in the western world from the private R&D machinery on the basis of latent and manifest market demands. What the UTOPIA team will be able to do, is to *make latent anthropocentric needs manifest*.

Second, *an example from metal working*. For over ten years (Brödner, 1990), it has been clear that two routes can be followed in the development of a metal-removing machine: The development of systems where the programming is done by the work preparation team, and of *systems that are programmed by the craftsmen themselves*. During the last few years, we have observed that the accent is gradually shifting from the former to the latter route (Brödner, 1990). The most important reason for this is that the engineers and managers that are responsible for purchasing the equipment, have discovered that the practical knowledge of the craftsman is indispensable for the programming. Similar considerations have induced the Dutch steel company Hoogovens (Reitsma, 1989) to take serious account of criteria relating to qualifications, man-slash-machine interfaces and training when purchasing equipment for the new continuous casting process.

From the above examples, it appears that *influencing market demand* offers an important perspective towards the realization of anthropocentric objectives. This provides the opportunity to set the enormous technological potential of the producers of systems into motion. One might consider *the development of anthropocentric consumerism*, analogous to the role played by the consumer movement during the last ten years. Trade Unions, purchase departments and managers should indicate that they no longer accept products which people cannot work with effectively, which break down human capital and which lead to frustration, illness and a high level of employer turnover.

4.3 Organizational Theory and Practice

Theory and practice are also a continuous area of tension in anthropocentric design. The *legitimacy of this effort is measured directly in terms of its practical value*. This certainly goes for the projects which are directly financed by the government and companies, which stem from new policy and must be converted into hard cash within a limited period of time. One tends to overlook *the long lead time which applies to every technological innovation*. Inherent in the philosophy of anthropocentric R&D, however, is the accumulation of knowledge. The formation of theory is vital to such accumulation. The least that one can expect is that the current basis for organizational knowledge is incorporated into the integral design.

In the *German and Dutch legislation regarding the working conditions*, one of the demands is concerned with job design, which should be based on the achievements of the current business sciences. When looking at its current stage, the state of the art, we see that this is also happening in specialized and segmented areas, e.g., the man-machine interface. Only limited use is made of more extensive and content-orientated models for integral design. Occasionally, references are made to sociotechnical systems design and organizational learning, particularly to clarify the objectives, but seldom to found the approach. Some recent reports which plead for anthropocentric R&D seem to completely ignore the richness of organizational-theoretical knowledge that was built up after the Second World War. It seems as if technologists who have discovered the social system want to start from scratch (figure 3 illustrates this as follows):

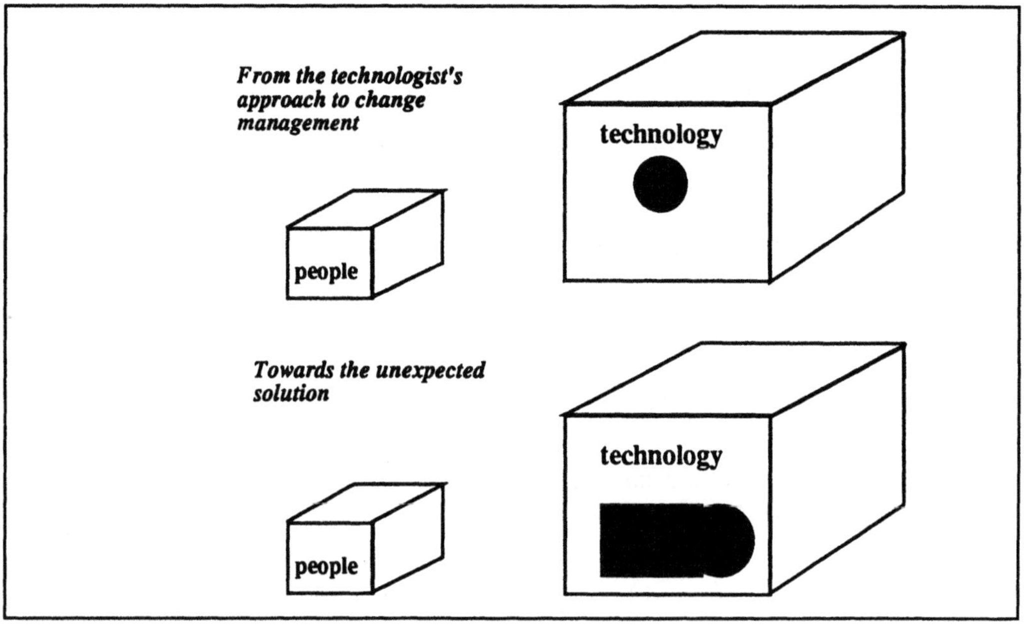

Figure 3: People and Technology (Kidd, 1990, p. 14)

The interface between technologies and human beings is represented in diagram 2 as the relationship between a round bar and a square hole. In order *for anthropocentric design* to be taken seriously, *its advocates must also take the organization and, hence, organizational theory seriously*. The modern insights into organizational theory and design are of great relevance to the 'social shaping of technology'. Technology and work are embedded in an extensive organizational system. If the interaction between them is isolated, there is a risk of 'Kurieren am Symptom' ('curing the symptom'). The Dutch variant of sociotechnical systems design (De Sitter, 1990; De Sitter & Den Hertog, 1990) is illustrative in this respect. Theory and practice of this approach tell us that in *the (re)design of the total system, the key is in redesigning the production structure* with the latter being defined as the grouping and coupling of performance functions. The aim of the redesign is to organize the more or less homogeneous flows where the necessary interfaces between these flows are reduced to the highest possible extent. A process of parallelisation and segmentation allows for the creation of simple and transparent organizations with complex tasks. Subsequently the control structure is built up bottom-up. Everything that can be controlled at the basis, i.e. locally, should be controlled at the basis. Subsequently, all remaining processes that can be controlled independently (interlocally) without the involvement of the hierarchy, nust be controlled amongst themselves at their respective sites (interlocally). Only the tasks that remain after this selection are, the object of control at the next system level within the organization.

The argument is clear: First, *the organization of the work flow* must be strongly simplified and only then can *the control structure* be set up, followed by the design of *the information structure*. Along this line of reasoning, it makes no sense to start with the design of the information structure in an organization which is far too complex. *The temptation* is strong in controlling complex processes *to find solutions in the growing possibilities of information technology*. At the level of government regulations, this tendency is the rule rather than the exception (Den Hertog & Van Spiegel, 1986). An example of this is the application of Artificial Intelligence to deal with the complex rules of social security (v.d. Beselaar, 1991). Policy-makers and politicians tend to ignore the complexity and transparency of such rules as long as the executing administrators assure them that the computer systems are able to manage them. Only when designing these systems and implementing the new issuing of rules, does it become apparent of how complex and inconsistent it is.

Along this line of thought, *the design of anthropocentric information* systems makes sense only after a redesign of the production structure (in the last example: the issuing of rules), from the global to the detailed structure. If the coherences between aspect and subsystems are ignored; there exists a considerable danger of continuing to work in the margin of the organization. Good intentions and technical competence are not sufficient here.

4.4 Technology and Business Development

In the nineties, companies are faced with the task of bringing into the market a growing diversity of high-quality products and services which are being replaced more and more rapidly and with shorter and shorter delivery times. This is possible only when researchers and developers take more account of the implications of their choices in the early stages of the innovation trajectory for those who follow later in the trajectory; process developers, manufacturing designers, producers, maintenance engineers, etc. Those working downstream will have to better anticipate the ideas that are thought of downstream. There exists a considerable risk that in this relay race the baton falls to the ground, particularly in firms with strong functional demarcations and bureaucratic structures. This implies a considerable switch for many companies. It calls for co-operation in multidisciplinary design teams, direct involvement of system designers and system users, open communication. In short, it calls for *integral approaches in which design problems are approaches in their mutual relation* (Cobbenhagen et Al., 1990; Kanter, 1983; Takeuchi & Nonaka, 1986; OECD, 1988). *The anthropocentric approach* fits excellently into this development, because it *places a direct relation between designers and users and interdisciplinary co-operation* at its centre. A precondition is that the anthropocentric approach is incorporated into this renewal process of the organization. If this does not happen, every single attempt in this area will, sooner or later, be rejected by the organization.

5. Policy Implications

In recent years, *the pleas for a balanced technological and social development* have become louder on a societal level. On the level of the firms, awareness is growing that holistic or integrated approaches are needed in order to make technology effective. Research and development focused on the interface between technology, work and organizations, might support the efforts to put intentions into practice. The present investments in this field of (inter)national and regional governments, firms, universities and other institutions show that they take this challenge seriously. Since the end of the seventies we have witnessed relatively small but interesting 'crucial experiments' in this field. Range from small-scale projects in the engineering lab to intensive and time-consuming (re)design programmes of the business level. Two general lessons can be drawn from these efforts: *the integrated design projects in a real business environment take years of intensive involvement of many workers, design experts and managers*. The redesign in Felten/Guillaume, which took years to complete, is no exception. Most companies, which start projects of these kind (Volvo), can already build on years of experience in the field of work reform. The other lesson which is taught: *the 'laboratory experiment'* has proven to be a very difficult task to create a common ground for social and technical experts and representatives form practice. There still remains a large gap between small-scale experiments and full-swing organizational change in practice. The contribution of anthropocentric R&D to the balanced development of technology, work and organization will depend on our ability to embed local efforts into more embracing policies which surpass the level of the individual work place, production department, research group or even firm. The considerations presented in this paper could offer *a preliminary list of basic elements of such anthropocentric R&D policies.*

1. Integration in the Engineering Curriculum

Most of the university-based experimentation is still taking place within the margin of the activities in our technological institutes. They can, and should be, the starting point for a change in the hard core of the education of engineers. Technical universities should be encouraged to foster this development.

2. Linking of Technological and Anthropocentric Government-Sponsored R&D Programmes

Anthropocentric government - sponsored R&D is, and has been, predominantly carried out in projects in the periphery of the main technological development programmes. In future technology programmes, work and organization related criteria should be incorporated as self-evident specifications in the development of new process and product technologies. As a first step, more attention should be given to evaluation and implementation research (or: 'accompanying research') on the shop floor level (cf. Voss, 1988).

3. International Co-operation and Exchange on Program Level

On an international basis, the national and regional programmes on work and technology offer a good starting point for mutual learning. They offer a more or less institutionalized framework to facilitate (and not *organize)* co-operation. Priority should be given to the exchange of information among the people who actually do the work. Up until now, too few publications have been made available in any other language other than the native language in which it was composed and too little of the travelling is done by the people involved in the projects themselves.

4. Activating the Demand Side

Managers, 'Branchen', organizations, procurers, users, works councils and Unions and any other group of actors which makes up the market or demand side, should be stimulated and enabled to express their demands in terms of work and organization. Technological systems should not only be appraised according to technical but also to organizational standards. One may speak in this respect of a new anthropocentric *consumerism.*

5. Use and Development of Design-Oriented Organizational Theory

Integrated design should be based on comprehensive and comprehensible organizational design models. R&D proposals should be weighted against the standards of contemporary organizational science.

6. Learning from Innovation in Real Life Business Environment

Anthropocentric R&D should focus more on the transfer of knowledge and experience gathered in technological and organisational (re)design processes initiated by business management. Research institutions should be stimulated to do so and firms and their "Branchen" organisations should be stimulated to open their doors and to participate in networks for the transfer of learning.

7. Dialogue and Design

The development of expertise within firms should be one of the main objectives of anthropocentric R&D. In other words: the accumulating design know-how should contribute to participative design processes in practice. The search for effective strategies to bring about the integration of dialogue and design should be one of the central issues in future anthropocentric R&D programmes.

The central word in the considerations presented here *is 'integration'.* Integration means that something specific is becoming part of a larger whole and vanishes as an entity existing in its own right. This might be the fate of anthropocentric design. As long as something is missing in the whole, it makes sense to give it a name. The challenge for the wide variety of actors which call humanizing of technology their cause, is to show that in the whole, an essential elements is missing and that *there are effective ways to restore the 'Gestalt of the socio-technical systems' we work in.* By you, that point has not yet been reached. In the meantime, they have to deal with the contradictory task of developing something specific without staying apart.

References

Agurén, S. (1989), *From industrial relations to learning organizations.* Stockholm: SAF.

Assen, A. van (1980), *Organisatie-ontwerp. Een analytisch model voor werkoverleg en werkstrukturering*, in: A. van Assen, J.F. den Hertog, P.L. Koopmann (eds.), Organiseren met de menselijke maat, Alphen aan den Rijn: Samsom.

Assen, A. van, Hertog, J.F. den (1984), *From job rotation to organizational design*, in: P. J.D. Drenth, H. Thierry, P.J. Willems and C.J. de Wolff (eds.), Handbook of work and organizational psychology, London: Wiley.

Beek, H.G. van (1964), *The influence of assembly line organization on output, quality and morale*, Occupational Psychology, vol. 38, No. 3/4, pp. 161 - 171.

Besselaar, P. van de (1991), *Technology assessment van artificiële intelligentie en kennistechnologie (concept)*, Den Haag: NOTA.

Betriebliche Humanisierungsforschung in der Bekleidungsindustrie, (1987), Campus, Frankfurt.

Brandt, D. (1990), *Advanced experiments, European case studies on anthropocentric production systems*, Aachen: KDI/HDZ.

Brödner, P. (1985), *Fabrik 2 000, Alternative Entwicklungspfade in die Zukunft der Fabrik.* Berlin: Sigma.

Brödner, P. (1990), *Technocentric-anthropocentric approaches: Towards skill-based manufacturing*, in: M. Warner, W. Wobbe and P. Brödner (eds.), New technology and manufacturing management, Strategic choices for flexible production systems, London: Wiley.

Buchanan, D.A., McCalman, J. (1988), *High performance work systems: The Digital experience*, London: Routledge.

Buitelaar, W. (1989), *Alternative Automatisierungsoptionen und deren Auswirkungen auf die Hafenarbeit*, Apeldoorn: TBO/STB.

Clegg, CH., Corbett, M. (1987), *Research and development into "humanizing" advanced manufacturing technology*, in: T.D. Wall, Ch. Clegg and N.J. Kemp (eds.), The Human side of advanced manufacturing technology, Chichester: Wiley.

Cobbenhagen, J., Hertog, J.F. den and Philips, G. (1990), *Management of innovation in the processing industry: A theoretical framework*, in: Ch. Freeman, L. Soete (eds.), New explorations in the economics of technological change, London: Francis Pinter.

Cooley, M. (1989), *European competitiveness in the 21st century, Integration of work, culture and technology*, Brussels: EC/FAST.

COR/Philips (1973), *Wat doet Philips in Nederland aan werkstructurering?*, Eindhoven.

Dankbaar, B. (1987), *Social assessment of workplace technology: Some experiences with the German programme "Humanization of work"*, Research Policy, vol. 16, pp. 337 - 352.

Dankbaar, B., Hertog, J.F. den (1990), *Labour process and socio-technical design: Living apart together?* New technology, Work & Employment, vol. 5, no. 2, pp. 122 - 134.

Davis, L.E. (1976), *Developments in job design*, in: P. Warr (ed.), Personal goals and work design, London: Wiley.

Does de Willebois, J.L.J.M. van der (1986), *Werkstructurering als organisatieontwikkeling*, Eindhoven: Philips.

Ehn, P. (1988), *Work-oriented design of computer artifacts*, Stockholm: ALC.

Emery, F., Thorsrud, E. (1969), *Form and content of industrial democracy, Some experiments in Norway and other European countries*, London/Tavistock.

Engelstad, P. (1990), *The evolution of network strategies in action research supported socio-technical redesign programmes in Scandinavia*, Presentation at the 1990 annual National Academy of Management Meeting, August 12/15, 1990, San Francisco.

FAST: *Monitor programme, Draft working programme, 1989 - 1990*. (1989), Brussels: European Commission.

Fertigungstechnik Programm 1988 - 1992, (1988), Bonn: BMFT.

Forschung zur Humanisierung des Arbeitslebens, Dokumentation 1987, Bonn: Bundesministerium für Forschung und Technologie.

Gustavsen, B., Engelstad, P.H. (1986), *The design of conferences and the evolving role of democratic dialogue in changing working life*, Human Relations, vol. 39, no. 3, pp. 101 - 116.

Gustavsen, B. (1988), *Creating broad change in working life, The LOM programme*. Totonto: Ontario QWL Centre.

Hertog, J.F. den (1987), *Arbeitsstrukturierung, Experimente aus Holland*, Huber.

Hertog, J.F. den, Spiegel, E. van (1986), *Overheidsautomatisering als organisatievernieuwing*, in: T. Huppes (ed.), Arbeid en management in de informatiemaatschappij. Leiden: Stenfert Kroese.

Hertog, J.F. den, Schröder, P. (1989), *Social research for technological change*. Maastricht: MERIT (89 - 028).

Jaikumar, R. (1986), *Postindustrial manufacturing*, Harvard Business Review, vol. 64, no. 6, pp. 69 - 76.

Jenkins, D. (1983), *Quality of working life: Trends and directions*, in: H. Kolodny, H. van Beinum (eds.), The quality of working life and the 1980s, New York: Preager.

Kanter, R.M. (1983), *The change masters*, London: Unwin.

Kidd, P.T. (1990), *European research in advanced manufacturing*, in: P.T. Kidd (ed.). Organization, People and Technology in European Manufacturing, Interdisciplinary Research for the 1990s, Macclesfield: Cheshire Henbury.

Klingenberg, H., Kränzle, H.P. (1988), *Humanisierung bringt Gewinn, Modelle aus der Praxis*, Eschborn: RKW Verlag.

Margulies, F. (1980), *Why "HCC:" again?* in: A. Mowshowitz (ed.), Human Choice and Computers 2. Amsterdam: North Holland.

MDA, *An Interdisciplinary Research Programme*, (1986), Stockholm: The Swedish Work Environment Fund, The Swedish National Board for Technological Development.

MDA projects, (1989), Stockholm: MDA-secretariate.

MDA, (1989), *People, Computers and Work, An Interdisciplinary Research Programme in Sweden*, Stockholm.

MODEM Bulletin, vol. 1, no. 1 & 2.

Mumford, E., Weir, M., (1979), *Computer Systems Design, The ETHICS Method*. London: ABP.

Nadler, G., Robinson, G., (1987), *Planning, Designing and Implementation Advanced Manufacturing Technology*, in: T.D. Wall, Ch. Clegg and N.J. Kemp (eds.), The Human Side of advanced Manufacturing Technology. Chichester: Wiley.

OECD, (1988), *New Technologies in the 1990s: A socio-economic strategy*, Paris.

Pornschlegel, H., (1990), *A New Concept of Integrated Manufacturing, The Volvo Company's Automobile Assembly Plant in Uddevalla*. Dortmund: Sozialakademie.

Prakke, F. (ed.), (1989), *Human Factors in System Design: Methodology and Cases in Factory Automation. Report to the Communities DG Employment, Social Affairs and Education*. Apeldoorn: The CIRP Human Factors Action Research Group.

Reitsma, D., (1989), *Personele gevolgen van technologische innovatie*, in: A. Simonse/W. Kerkhoff, A. Rip (eds.), Technology Assessment in ondernemingen. Deventer: Kluwer.

Ramondt, J.J., Dreu, J.H. de Standaart, C. A., (1990), *Integraal ontwerp van een informatiesysteem, ervaringen en leermomenten in een bankproject*. Rotterdam: RISBO.

Research and Development Programme, Work and Technology, (1990), Bonn: BMFT.

Sitter, L.U. de, Hertog, J.F. den, (1990), *Simple organizations and Complex Jobs*, Maastricht: MERIT (90-012).

Smits, R., Leyten, J., (1991), *Technology Assessment: Waakhond of speurhond? Naar een integraal technologiebeleid* (Proefschrift). Zeist: Kerckebosch.

Takeuchi, H., Nonaka, I., (1986), *The new, new product development game*. Harvard Business Review, vol. 64, no. 1, 137 - 146.

Technologie, Arbeid & Organisatie, Uitgangspunten voor een onder zoekstimuleringsprogramma, (1986), Den Haag: TK, 1985 - 1986, 19 542, nrs. 1-2.

Technologie, Emploi, Travail, (1988), Paris: Ministry of Science and Higher Education.

The UTOPIA project, (1986). Graffity, no. 7, p. 2.

Utbold, M., (1988), *Towards a Learning organization*. Stockholm: The Swedish Work Environment Fund.

Voss, C.A. (1988), *Implementation: A key issue in manufacturing technology: The need for a field of study*. Research Policy, vol. 17, pp. 55 - 63.

Walton, R.E. (1972), *How to counter alienation in the plant*. Harvard Business Review, vol. 50, no. 6, pp. 70 - 81.

The LOM Programme
- Swedish Experiences and Assessments

Lars E. Björk, Danderyd, Sweden

1. Some Facts about the Swedish LOM Programme

Between *1985 and 1990* the Swedish Work Environment Fund launched *a five-year research and development programme on Leadership, Organization and Participation* (Swedish abbreviation: LOM). The programme supported research and development both in private and public organizations.

The programme had an annual budget of about 1.7 million US dollars. *Policy decisions* were made by a board consisting of representatives from the central parties on the labour market and the managing director of The Work Environment Fund. Day-to-day decisions were made by a secretariat where one employers' representative was working together with two union representatives, one researcher and the full time programme director and his assistant.

2. Background

Already in the late Sixties and early Seventies Sweden had national programmes on the improvement of productivity and satisfaction-participation. Those programmes looked differently in the private and the public sectors. But they were shop-floor orientated and they rested on the co-operation between the social partners locally and centrally. They were able *to create new organizational solutions in the existing* work. This was often done with a difficult involvement by researchers who sometimes became the spokesmen of the workers.

The results showed that although *autonomous groups were created*, that showed a better productivity and flexibility than the old organization, the *diffusion did not work*. On the company level most of the middle managers kept their old roles and the changes were often incapsulated on the shop-floor. The diffusion dit not work between companies either.

Reviewing these results in the early eighties *the aim was to engage the companies more actively* themselves in the development of work and to diminish the role of the researchers. This is what the LOM programme is about.

During the mid-seventies the traditional co-operation between the central parties on the Swedish labour market ceased and turned into a conflict around the proposed law on co-determination. This situation lasted several years and did not change until the early eighties. The symbol of the Swedish model working again was *the Agreement on efficiency and participation* signed in *1982* by the SAF (the Swedish employers' Confederation), the LO (the Swedish Trade Unions' Confederation) and the PTK (the Federation of Salaried Employees in Industry and Services). The agreement referred to as the Development agreement states some important *common values* between the parties. It is then commonly "agreed that *efficiency, profitability and competitiveness require constant development* in all aspects and at all levels of the company's activities".

"SAF, LO and PTK are agreed that *decentralization and delegation of responsibility* and decision-making create conditions in which decisions can be made as far down in the organization as possible. The parties are also agreed that this increases the chances of achieving the objectives of this agreement".

These commonly expressed values are partly the result of a tougher international competition facing Swedish industry in the beginning of the eighties. The LOM programme *should contribute to the learning-process of new forms of practical co-operation* within private and public organizations.

3. Tasks for the LOM Programme

Three tasks were formulated in the beginning of the programme:

1. *Initiating and supporting development* projects where the local parties co-operate in the change of work.
2. *Supporting long-term research and growth of knowledge* within this field at the local universities, the main universities and other research institutes.
3. *Supporting a fruitful co-operation between companies, public organizations and researchers*. Information about the experiences shall be distributed to stimulate the development within the field.

4. Characteristic Features of the Programme

The programme has had some noteworthy characteristics.

The different projects in companies and public institutions all aim at solving problems that *have been defined locally*. That as opposed to the definition of local problems by a consultant or researcher. The local demands for development will guide the content for change of work not models or ideas from outside experts. These ideas are in line with the agreement on development and worth noticing as an unusual aspect of a national programme.

To start up projects in companies *a starting up conference* is often arranged. Four or five interested companies will be invited. Every company is represented by *a "vertical group"* consisting of about eight people including management, supervisors, employees and union representatives. In this 24-hour-working conference everybody should have the possibility to exchange experiences among equals and also to inspire each other to think more freely on the problems in ones own company. These conferences are repeated during the running of the projects with the task of exchanging views of the change of work and its organization.

In the single company the programme tries *to support a broad participation* among the employees on different levels around the development tasks. This is partly done by the researcher whose role is more process supporting and less that of task expert. The idea is to create forms of co-operation where the employees can use their experiences and knowledge as a base for the change of work. Of course this also implies that the daily work includes some degrees of responsibility, decision making and learning. The change of work should as far as possible be carried out *in the form of real dialogue* between the people involved. That often means the creation of new forms for meetings. It also implies supporting individuals and groups with less experience of expressing their views and ideas.

To establish and support the local development projects the LOM programme co-operates with the already established *development committees on industrial trade level.* Such a committee is made up of employers' and employees' representatives in one industrial trade, e g the paper and pulp industry. Its task is to support local development in the companies of the trade. This infrastructure is of great value and means a real link between the development projects in the companies and the greater network of other companies within the same industrial trade.

5. Description of a Typical Project

To clarify how the practical work is done we will give a short description of a typical LOM project.

- Mutual contacts between the LOM secretariat and the Development Committee of an industrial trade.
- The Committee selects around a dozen interested companies.
- The LOM secretariat contacts the companies and visits the ones that want to proceed. Discussions are carried through with management and the local parties.
- The starting-up conference is carried through together with three or four other groups from different companies. Preliminary areas of development are proposed.
- After the conference a project organisation is formed in the company with the support of a researcher.

- Further analysis and dialogues of the problem and the developement of a projectorganization.
- Working on the problem with all involved. The group/groups report regularly also encouraged by the visits of the researcher.
- Exchange of experiences with the other companies in the group at a conference after 6-12 months. Especially experiences of how to manage and organize the development work.
- Evaluation of the project in the company and discussions of further development among the parties locally.

6. Scope of the LOM Programme

The programme supported about 50 projects. The following *eight industrial trades* were engaged in the programme:

- the engineering industry, hotels and restaurants, trade and distribution, process industry, paper and pulp industry, printing industry, wood processing industry and building components;
- in the public sphere projects are under way in state institutions like the postal service, the police etc, in the county concils and in the local municipals.

A few local universities are still engaged in *creating groups and networks of projects* within their region. These groups will often mix private and public organizations and seem very promising.

Some 50 researchers have been occupied in the programme, usually on half-time basis responsible for a group of projects and assisting in others. Their background is mainly behavioural science but also technicians and economists work with the projects. The researchers worked in many aspects like consultants but they were also responsible for documentation and analysis of the change of processes and its organization.

7. What is the Expected Outcome of the Programme

The many projects within the programme ought *to give good examples of how to manage and organize work* in order to stimulate local development with a broad participation, how to create a self-changing organization with learning and change possibilities for all employees. But the experiences can not be summarized as a cheque list with general rules to be followed by all companies occupied with development work. We believe that the local circumstances often differ so greatly that general rules are of very little value. But we can draw *certain conclusions about which factors that lead to success and which that lead to problems in a change process*. These conclusions can serve as a stimulus in local discussions or in considerations taken by industrial trades. They can be compared with the local experiences based on the company's history, line of business etc.

The development committees of different industrial trades will *learn about the character of chance and development projects* during the programme and so will the local universities. They will hopefully continue to support local development in the nineties.

8. Further Discussions

One of the corner stones of the LOM programme was *the aspiration for broad change*. It involves both the intention of getting a broad participation from the employees and also to inspire a great number of organizations to start changing in new ways. In many respects the programme seemed to come up to the same difficulties in this area as have previous efforts. Even if the LOM programme has succeeded in launching many changes in the processes there seems to be *a limited spread of changes in this direction*. Perhaps our aspirations were not fully adequate. The impact of an R&D programme seems to be *more in the field of long-term dissemination* of fruitful ideas and concepts. That was for example the case with several of the projects run twenty years ago when the concept of work organization emerged and was filled with substance. Many of these projects did not survive in their difficult environments but the concepts put forward by the researchers are widely used today.

This might also be the case with the LOM programme. *Concepts like Democratic Dialogue* might be useful long after the cases are forgotten and even if we did not get the broad change we wanted.

The Learning Organization: Swedish Concepts - Aims and Experiences from two of the Swedish Work Environment Fund's R&D Programmes

Peter Docherty, Stockholm, Sweden

1. Competence and Learning as Worklife Issues in Sweden.

Sweden is facing today the same changes and demands in both industry and the service sector as other European countries. We feel that these are probably more profound and more far reaching than we have experienced in the last fifty years. The radical integration of geographical markets, e.g. in trade blocks, of corporations, e.g. in the multinational cooperations, of technologies, e.g. in information technology, and of products and services, e.g. the technoservice sector, are placing tremendous demands on organizations and individuals to adapt and develop. The demands relate to all the central features of an organization - its business ideas, its structure and work organization, the attitudes, skills and knowledge of its personnel and its technology. Irrespective of which of these areas is deemed by top management to be of highest priority or the starting point for necessary changes in the individual company, *the execution of the required changes will entail adaptation and learning on the part of its workforce.*

The development of the knowledge and skills of workers has thus been *a central feature of the Swedish Work Environment Fund's R&D programmes* since the early Eighties. This paper will mainly be concerned with two of these programmes; the recently completed Development Programme for New Technology, Working Life and Management which was the first of the present suite of such programmes and the *recently started Learning Organization Programme*, which is the latest in the series.

The aim of *the Development Programme for New Technology, Working Life and Management* was to explore new ways of introducing computer-based technology in organizations to further the development and realization of business goals through parallel, integrated efforts to develop the work organization and the work environment. The programme would illustrate practical ways of realizing the ambitions contained in the joint agreement in the private sector regarding efficiency and participation. The programme supported pilot and development projects in business, industry and public administration. A common feature of the projects was *the utilization of computer-based technology in production and administration* to further efficiency and competitiveness. The values forming the point of departure for activities in the projects were that specific goals should be formulated in each project concerning business ideas, personnel, technology and work organization. These were to be balanced against each other during the introduction process. The new technology was to be utilized to maintain and further develop the knowledge and skills of individual workers. Thus the emphasis was on know-how, commitment and determination rather than level of technology. The programme was administered by a board, a steering committee and a secretariat, the majority of whose members were representatives from the labour market organizations. The programme ran from 1982 through 1987, included forty projects and cost somewhat over 70 million Swedish crowns.

Several leading industrial nations have already identified *continual skill formation as an increasingly important competitive factor* for industry and commerce. I use the term *"competence"* instead of skill here as it is given a broader meaning in the Swedish context, covering not only knowledge but also the will and opportunity to apply the knowledge. Similarly competence development is more than such activities as formal training and education, the planned systematic growth of experience through on-the-job learning playing a key - many feel *the* key - role in competence development.

Competence development strategies in organizations in different countries reflect strong cultural elements related amongst other things to educational and industrial relations traditions. The Centre for Education and Innovation at the OECD has identified three strategies which they name:

- the Human Resource Intensive,
- the Polarization and
- the Mobility Strategies.

Only the first aims at developing *all* the members of the organization.

The *Swedish Work Environment Fund's programmes* have actively adopted this strategy. They include, in fact, examples of our lifting the "vision of the possible", e.g. retraining individuals from an assembly line for electrical components to be software programmers in the same organization. What then are our main experiences from the Development Programme?

2. Experiences from the Development Programme for New Technology, Working Life and Management

As stated above *the aim* of the Development Programme for New Technology, Working Life and Management was *to explore new ways of introducing computer-based technology in organizations* to further the development and realization of business goals through parallel, integrated efforts to develop the work organization and the work environment.

An integral part of these efforts was the development of workers' competences and the main avenue open for this is *learning at work or on-the-job learning in a planned and systematic fashion.* The programme aimed at including a variety of projects that would illustrate different means of facilitating and supporting this learning:

- new job design, work organization and supervisory practices,
- new means of technological support, such as local personal computer-based planning systems,
- temporary organizations, such as the project organization of the change project itself, problem-solving groups and action learning projects.

The programme even *organized interest group networks and series joint seminars*, though these did not constitute learning at work to the same extent as the other forms of temporary organization. These project activities were organized within and between work units and/or functions and in some cases together with customers, suppliers and other companies.

The *basic design of projects in the LOM-Programme* required the utilization of start conferences and the exchange of ideas and experiences between small groups of companies in the same business sector. The Development programme utilized even regional groups. The *MDA-Programme* on People, Information Technology and Working Life has a special subprogram devoted to the question of technical support to learning.

2.1 The Programme's Theoretical Roots: Competence and Learning

Competence is the ability to execute tasks and meet external demands in a manner relevant to the situation in hand. It includes not only knowledge but also the will and opportunity to apply the knowledge. This definition of competence is an expression for a view of *the individual as an action-orientated and problem-solving being*. The facilitation of competence development requires organisation for learning. This is realized through the design of the jobs, the work organization, the communication structure, management systems, technical support and leadership. Figure 1 summarizes a number of aspects that must be taken into consideration in achieving a working environment that facilitates learning.

The projects in the Development Programme have drawn on *different experiences and research*. Some of the more influential are those regarding organizational learning, theories of action, activity theory and organizational development.

Organizational learning is currently discussed in terms of concepts such as "rational calculation" and "experiential learning". (March and Olsen, 1975) The concept *"rational calculation"* means that organizations are intentional and goal-orientated: expectations about likely future outcomes are actively utilized as the basis for selecting among possible current courses of action. *"Experiential learning"* assumes that organizations are responsive entities that continually adjust their activities through a consideration of previous experiences in an effort to optimize their competence. (Ryan et al., 1989)

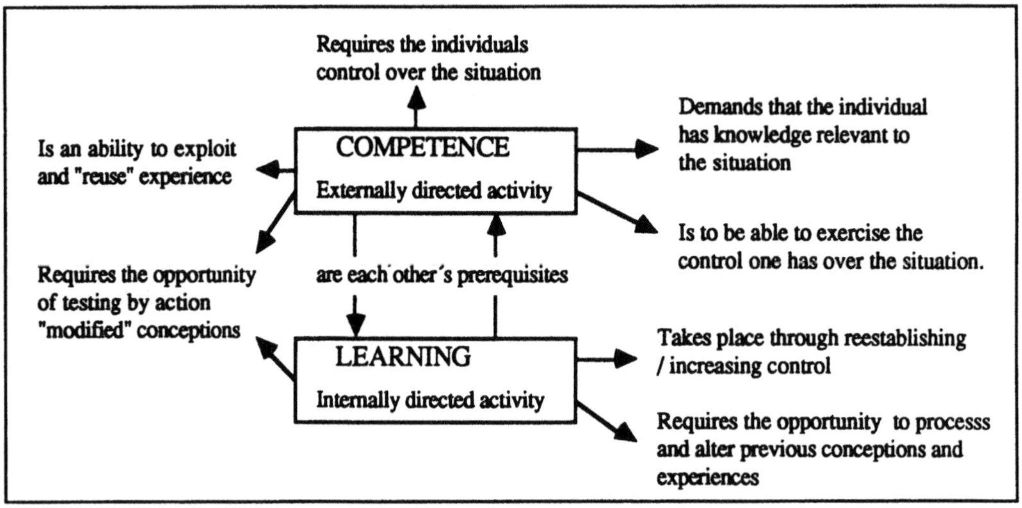

Figure 1: A number of aspects important for a working environment that facilitates learning. (Dilschmann, 1990)

Theories of Action

Argyris and Schön (1985) propose that *organizations develop "theories of action"* through their ongoing learning activities. Such theories represent beliefs that organizations have about the environmental consequences of their actions and enable members to share organizational "frames of reference" which become institutionalized as consensual knowledge. These authors also use the concepts single- and double-looped the learning that have similarities with the concepts of rational calculation and experiential learning.

Morgan (1986) elaborates the *behavioural models of organizational learning* in the context of the distinction made by cybernetic models between single- and double-loop learning. The learning abilities conceivable in single-loop learning are limited in that the system can maintain only the course of action determined by the operating norms and standards guiding it. He points out that this is alright as long as the actions defined by those standards are appropriate for dealing with the changes encountered. But when this is not the case, the "intelligence" of the system breaks down, for the process of negative feedback ends up trying to maintain an inappropriate pattern of behaviour. Double-loop learning adds a second feedback loop which depends on being able to take a "double look" at the situation by questioning the adequacy and relevance of current norms.

Morgan notes that systems that are capable of double-loop learning: "Are often able to detect and correct errors in operating norms and thus influence the standards that guide their detailed operations. It is this kind of self-questioning ability that underpins the activities of systems that are able to learn and self-organize."

The relation between individual and organizational learning is not clear. Pender (1989) underlines the concept of the Learning Organization as a metaphor "which can serve as a guiding star of where the company may want to go. It can help people to think and act together." He sees the learning organization not so much as an entity but as a "generative theory" mobilizing energy to shape the shared perceptions and the will to change as well as leading to concrete action. Argyris (1985) proposes that organizational learning is generated through the actions of individuals acting as "agents" for the organization. The organization does not "learn" unless people in the organization learn. To avoid confusion here the term learning is restricted to individuals for analyzing processes at the individual level.

Activity Theory

The activity theory developed by Russian and German psychologists is a main source of inspiration in current Swedish research regarding learning in work (Leontjev, 1978; Hacker, 1986; Volpert, 1984, 1988; Aronsson, 1990; Aronsson och Friedrich, 1988; Bokander,1990; Leymann, 1987; Docherty et al., 1988). A basic assumption in activity theory is that *thought and action, planning and executing a task are regarded as a single unit* and are not separated into mental processes on the one hand and observable behaviour on the other. Activity theory aims to bridge the gap between cognition (thinking) and action.

Other important concepts in activity theory are *regulation demands and levels*. Hacker (1986) defines three main regulation levels: the sensory motor, the perceptual-conceptual and the intellectual level. The lowest level, the sensory-motor level, is made up of automatic skills triggered by signals from the environment. The highest, intellectual planning, is characterized by consciousness and language. Between these is a level in which flexible activity patterns predominate.

Starting in activity theory Volpert (1988) has drawn up a composite *list of criteria for the design of jobs for the facilitation of personal development.* His research group is now engaged in developing a series of instruments for job analysis which has been given the name "Contrastive Task Analysis" (KABA). Volpert's group has direct contact with several Swedish research groups. One Swedish project on the relation between new technology and learning focuses on the individual's control over his or her work situation. (Docherty et al., 1988) Their operational breakdown of the concept of control is to a large extent based on the KABA criteria, e.g. being able to perceive the task situation as a whole, to have and to perceive degrees of freedom / room for action, to see and to handle obstacles, to develop a personal work style and to be able to communicate the situation to others.

Organizational Development

Much Swedish work on learning at work is carried out within the context of organizational development. Learning often occurs in such contexts as an unplanned result of variations within and around the organization. As in Kolb's (1984) ideas on experiential learning *key concepts are action, feedback and reflection.* A more conscious and planned learning strategy involves the active utilization of such variations by their analysis, handling through conscious action, the outcomes of which are evaluated and fed back to those involved. (Björk et al., 1990). This can be applied, for example, to the steps in organizational change. Recent Swedish experience indicates that the character of such change processes is of similar import for the final functioning of the organization as the actual organizational and technical changes that may be introduced - *"the process gives the solution".*

The concept *"local theories "* has been coined in Scandinavian to describe how organizations are rather the result of mutual understanding and agreements within organizations than of external analyses and general theories. A local theory consists of the common ideas within the organization, and holds for just that organization and the people who formulated the ideas or theory. The ideas become common ideas through discussions, the dialogue, between the employees. Feedback leads to the reinforcement or review of the local theory.

	Everyday Improvements (On-the-job Learning)	**Innovations (Major Step Learning)**
Changes	Gradual and constant	Abrupt and volatile
Timeframe	Continuous and incremental	Intermittent and nonincremental
Pace	Small steps	Big steps
Involvement	Everybody	Select few "champions"
Spark	Conventional know-how and state-of-the-art	Technological breakthroughs, new inventions and new theories
Practical Requirements	Requires little investment but great effort to maintain it	Requires large investment but little effort to maintain it
Focus (Effort Orientation)	People	Technology, Products
Approach	Collectivism, group efforts, systems approach	Rugged indiviualism, individual ideas and efforts

Figure 2: Features of Everyday Improvements (Kaizen) and Innovations (Imai, 1986)

Learning and training often come onto a company's agenda *in connection with major innovations*: changes in the business idea, ownership, new products or models, plant and production technology investments or new organizational structures. These are often dramatic and far-reaching changes requiring radical adjustments in people's way of working. Core competences can be reduced to marginal competences over night. Basic elements of the local theory may need to be reevaluated. An essential complement to this, however, from a Swedish perspective the everyday improvements - the continual, incremental development - the everyday learning, the daily reinforcement of commitment at work. Figure 2 compares several central features of these complementary learning processes in the organization.

The extensive debate concerning work qualifications and its associated research have, of course, not passed Sweden by. Several branch studies have been conducted to ascertain if technical developments have resulted in upgrading, degrading, polarization or regrading. The picture emerging from Swedish studies would appear to be as mixed as in other countries. There is however one topic that has been receiving special attention in Sweden, namely that of "learned helplessness". (Lennerlöf, 1986) Similarly, work at the Centre for Work Life Research in Stockholm has focused on concepts concerning knowledge at work: propositional or theoretical knowledge, practical knowledge or skills, knowledge of familiarity and tacit knowledge. (Göranzon and Josefson, 1988; Göranzon and Florin, 1990). The worklife context for this research has been the "unbridled" and unreflected introduction of information technology into many work situations and the risks arising of negative consequences for workers' competence.

In summary, *work conducted* in the Swedish Work Environment Fund's programmes *has a variety of theoretical roots*, drawing amongst others on organizational learning and theories of action, activity theory and organizational development. The projects within the Development programme have resulted in the formulation of a number of lessons that form a point of departure for the current Learning organization programme. These lessons can be summarized under two headings: Competence requirements and Preconditions for learning at work.

2.2 Tendencies in the Programme Regarding Competence Requirements

The Development programme comprised *forty cases centred on development projects* defined and financed by the companies and agencies participating in the programme. A key element in all the studies was the introduction of new computer-based technology. The actors on the local level, the management, workers and unions, appreciated that the projects would result in new demands being made on the workers in terms of the knowledge and skills required of the workers. Reviewing the cases as a whole it is possible to discern a number of general trends which are shown in figure 3. Several of the changes shown in this figure overlap. Several are interrelated in as much as the one follows from the other. The first trend is away from specialization to a broader integration of skills to enable the individual worker to understand and handle a wider span of tasks, stations in the production process and / or a wider range of tasks connected to a specific task, station in the production process, such as planning, administration, maintenance and quality control and not just the elements of production.

Competence Requirements: Current Shifts

Niche specialization	→	Broad integration
Concrete, sensient base	→	Abstract symbol manipulation
Rule-based performance	→	Knowledge-based performance
Steady-state orientation	→	Disrupted-state orientation
Puzzle-solving orientation	→	Problem-solving orientation
Discontinuous growth	→	Continuous growth
Individualized knowledge	→	Collective knowledge

Figure 3: Current shifts in the competences demanded at work

The latter extension is related to the second point - *the growing abstract character of the work demands*. This is perhaps most clearly seen in the process industry in which for much of the working time the continuous exposure to sensory stimuli of sounds, smells, nuances of light and kinetic sensations are replaced by the abstract symbols on the instruments and displays in control rooms.

The *registration and manipulation of symbols required* is usually steered by underlying models of the production process. In all events the worker is required to think abstractly and inductively - to create a model in his head of what is happening at the workplace. Rasmussen (1986), whose main area of research has been the working environment of process control workers makes an important distinction between rule-based behaviour and knowledge-based behaviour. In the former situation the operator matches the situation demanding action against "stored rules" regulating action. Difficulties arise when the matching process breaks down because, for example, the current situation cannot be recognized - the problem has not arisen before. In the higher level knowledge-based behaviour the operator has goals and a theoretical frame of reference which facilitate the construction of plans of action when required - even if the situation has not arisen before.

Rule-based behaviour may also be described as puzzle-solving behaviour. Puzzle-solving entails limited demands on the creativity and initiative of the solver - there is usually only one correct solution to the puzzle and this is contained within clearly delimited frame of the puzzle, e.g. the pieces available. *Problem-solving* is more analogous to knowledge-based behaviour. Heavier demands are made on the creativity and initiative of the solver than in the previous case. The solver must often identify and define the problem himself and the limits to its solution are less often dependent on the control or limits imposed on the worker by the organization, but more often by his competence.

A similar analysis has been presented by Koike (1989) regarding the skills demanded of Japanese production workers. He classifies production operations as "usual" (routine, repetitive and monotonous) and "unusual". This latter category is subdivided into "changes" (in products, production, methods and labour mix) and "problem-solving" (detection, diagnosis and rectification). These *unusual operations* require "intellectual skills" regarding such topics as the structure of machines and the logic of production processes. This Japanese approach lies behind the shift from a steady-state to a *disrupted-state orientation*. The rationalization tradition of most Western countries has geared the job demands of production workers to handle solely the execution of production tasks in the steady-state continuous flow situation. When contingencies arise - machine breakdowns, materials shortages, emergency orders - they are handled by other specialist groups, maintenance engineers, supervisors, production engineers. It is becoming more common to extend the skills of workers to be able to handle what Koike called "unusual operations" when such disruptions occur.

Again the reasoning above on knowledge-based performance can be related to Davydov's (1977) discussion of *"empirical" and "theoretical" ways of thinking*. He proposes that the empirical and the theoretical knowledge represent two qualitatively different levels of reflecting reality.The "world of experience" comprises our individual experiences - a world we have experienced independently through our senses. This corresponds to Kolb's *experiential learning*. The "world of knowledge" is a world presented to us as theoretical knowledge - a world that transcends our practical experiences and is mediated predominantly through language (Vygotsky, 1978). Empirical knowledge or the individual's world of experience is not developed through formal generalization and has limited scope for inference. Theoretical knowledge provides mediated knowledge expressed in abstract models via symbolic relations. Empirical knowledge is what I also refer to here as "individualized knowledge". *"Collective knowledge"* is attained through a dialogue between the employees (cf. "local theory"). At best this collective knowledge will also include the relevant theoretical knowledge. (The distinctions here are made clear by the example from the Finess Långa papermil, Rehnshow and Utbult, 1988)

Finally the continual occurrence of "unusual operations" as a result of the high rate of change and turbulence in organizations and their environments *places continuous demands on workers to improve their competence* in the spirit of Kaizen noted above. What then are our experiences of the preconditions necessary to facilitate such continuous competence growth?

2.3 Examples in the Programme of Preconditions for Continuous Growth in Competence

The projects in the Development Programme varied considerably in their aims, strategies and contexts. They appear also to have varied considerably in the extent to which they managed to handle the *issue of competence growth*. Attention has undoubtedly focused however on a number of preconditions that appear to have supported and facilitated learning in those cases which were more successful in this area. These preconditions are shown in Figure 4.

Preconditions for Competence Growth: Current Shifts

Organization

Redundancy of parts → Redundancy of functions

Centralization → Decentralization
* Goalsetting
* Planning
* Problem-solving
* Feedback (including the reward system)

Uniform organization → Polyform (Multiprinciple)

Single-loop conditions → Double-loop conditions

Technological support

Automating → Informating

Development Process

Expert relationships → Collaborative relationships

Individual focus → Social, group focus.

Figure 4: Current shifts in preconditions for competence growth

The first group of preconditions concern the design of jobs and the work organization. The first listed in Figure 4 is a principle long recommended by the sociotechnical school, namely *the redundancy of functions*. This entails the company investing in the worker's competence so that he or she has the ability to execute more functions than he or she is called upon to do on an average "steady-state" day. The traditional approach has been to have narrowly defined jobs to which nonskilled workers can be quickly trained in order to reduce vulnerability to personnel turnover.

The second point concerns *decentralization*. The first aspect here is *goalsetting*. In the Development Programme this has mainly concerned the reformulation of company goals in terms which are directly related to the production group's own situation, in particular the degrees of freedom over which it can exercise control. That the workers understand, accept, can relate to and see how they influence the fulfilment of the business idea and goals of the company entails a learning process probably resulting in the workers sharing essential elements of management's values. (Agurén, 1990; Björkmann et al., 1987; Ranhagen, 1986; Rehnström och Utbult, 1987).

The second decentralization point concerns the delegation of *planning*, e.g. for the scheduling of production in terms of determining the order in which different customer orders are handled, determining the work in progress to attain high capital rationalization, planning materials supply and machine maintenance. Successful examples of all these cases were found in the Development programme. Other variations of delegated planning were allowing workers to build their own plant (an automated welding line) at the suppliers and allowing papermill operators visit overseas customers to determine their quality needs. Naturally this delegated planning has required both *special formal training* in the technical subjects in question, e.g. capital rationalization, and the support of new supervisory roles, i. e. the changes for the production workers also entail development for the salaried personnel.

The third aspect of decentralization is that of *feedback* - a basic requirement for learning. Naturally it is important for the workers to ascertain how successful their planning has been and it is not particularly difficult for companies to arrange this - though government agencies have reported difficulties in defining key indicators of efficiency in their business. Another difficulty by no means confined to public sector is to achieve congruence between the different feedback systems in the organization. One very important set of such systems is the reward systems. These may lag behind and come out of phase with new planning systems. For example workers responsible for capital rationalization in a production plant may still be paid using a straight piecerate system, i. e. in the worst case the more efficient they are at rationalizing capital, the less they are paid.

The forth aspect concerns the allocation of *problem-solving*. Problem-solving is also regarded by many as an essential prerequisite for learning at work. The need for this may be regarded in this context as a natural corollary from the preceding discussion.

Many organizations have stressed the need for uniform organization, i. e. exactly the same work organization in all units, irrespective of regional or local customer and labour markets. These demands are sometimes related to conceptualization's regarding the company's market image or are seen as a means of guaranteeing service and product quality. On the other hand more and more organizations seem to be *moving towards a polyform organization* where different offices or production groups are organized according to different principles in order to offer the best services possible to the local market and to utilize and develop as best as possible the competence and experience of the local personnel. These *variations* may entail differences regarding:

- their *degree of job specialization* (some groups being highly specialized, others displaying an extensive task integration in individual jobs)
- *the range of tasks and responsibilities* within individual groups (displaying different mixes of production, quality control, planning, administration and maintenance).
- *The degree of autonomy in the work group* (ranging from groups with traditional supervisors to others which are quite autonomous).

In the Development programme's project in two Labour Employment offices, the project groups chose radically different organizational solutions to meet local needs.

Two double-loop learning conditions would seem to be of special importance in an organizational context. The first is the provision of formal organizational mechanism for reflection - *the analysis of feedback and the studied evaluation of necessary review or action.* Many studies report that the pressure of circumstances lock the worker in an act-react loop, excluding the possibility of reflection (Ramnarayan and Mohan Reddy, 1989). The learning organization at Teli producing software for digital public telephone exchanges organized a series of different meetings or organizational arenas to foster such reflection (Utbult, 1989). The second feature of double-loop learning prerequisites is the power to change "the rules of the game" or at least some obstructing practices or policies. Again in the case of the Labour Employment offices the personnel were given the brief that "the Law was the Limit" to the changes they could introduce, i. e. they could "setside" the thick regulation handbooks produced by their own agency.

New computer-based technology is a feature of most decision support systems, e.g. planning systems. Gutek et al.(1984) report that the issues of work design and the development of professional competence have been accorded little weight when introducing computer-based systems. On the contrary the main goal with the introduction of such systems in the sixties and seventies was the rationalization of decision making. The introduction of decision making algorithms was usually regarded by the users as a form of job impoverishment. Others researchers report that competences have "withered away" in such circumstances through non-use (Karlsen och Oppen, 1988). Another reported phenomenon is "bureaucratic rationalism" where systems may give rise, amongst other things, to goal displacement and puzzle-solving.

Zuboff (1982, 1988) has coined a new term which clarifies the crossroads we face when introducing decision support systems today: *Informate or automate?* Informating means providing the decision-maker with better information as a basis for his or her decisions. This in fact is very much in line with the stance of the Development programme which was: "A worker's primary role is not only to push buttons to control the process but also to use the information generated to push the business - to redefine process variables, to improve quality and to reduce costs" (Quote from G. Schuch, Organizational Dynamics, 1985).

Regarding the development process we note a change in the stance of the facilitators or support resources. The postulates of the traditional consultants, "Do this", and trainers, "It's like this", are being gradually replaced by *a questioning approach of the leader or coach*, "How can we collaborate in this venture?". The individual focus reflected in aiming at imparting new job skills to individuals, usually through formal training, is gradually giving way to a group focus. This has several different aspects. Firstly, the realization of the role of local theories, collective knowledge and the establishment of dialogues between different groups and different levels in the organization for the efficient functioning of organizations. Secondly, the realization of the need to include social and political components in qualification research (cf. Gulowsen, 1987). The need for coordination follows naturally from the increasing level of integration in production processes (Wood, 1987).

3. Current Effort: The Learning Organization Programme.

The issues of competence development and the learning organization grew out of the Development programme. They became so central that they were the subject of one of the programme's thematic reports summarizing the experiences from several cases, *"Towards a Learning Organization"* (1988). Key topics in that report are mobilization and learning to participate in change processes, "shared learning" between persons from different functions between and within organizations and skill formation through continual on-the-job learning.

The Swedish Work Environment Fund has now launched a *new R&D programme* that will build on the various experiences form recently completed and ongoing programmes that have broached and broach learning at work in complementary ways. The new programme, *the Learning Organization Programme*, is a five-year programme with a budget of 50 million Swedish crowns. The programme reports directly to the board of the Swedish Work Environment Fund and it has a management committee and secretariat in which the majority of members are representatives for the social partners in the labour market. Its main tasks are to initiate and finance research and local development projects and to disseminate the results continually through the course of the programme by creating networks, organizing seminars and workshops and distributing reports.

The programme will generate new knowledge on and *support the development of learning at work through involvement* in

- innovation projects,
- local projects testing practical combinations of different measures such as job design, team-work, new forms of technical support, such as embedded knowledge-based systems and local planning systems, new work organizations and new supervisory practices.
- studies of the preconditions for positive and continuous learning and growth.

The projects will be broad in that they will cover personnel from different levels and different categories and sectors. The role of formal training as an important complement to on-the-job learning in overall competence development programmes will be studied. It is planned that projects within the programme will have the *following characteristics*:

- They will constitute joint ventures between the social partners in the labour market at company and branch level.
- They will be initiated by the local stakeholders who will have a strong commitment to the values behind the programme.
- Both long-term and short-term efforts will be represented
- Studies of research questions will conducted in field sites.

The latter will cover such questions as:

- The positive driving forces for and the forces obstructing learning processes,
- The roles of work organization, management systems and job design in learning,
- The roles of organizational culture and climate in learning,
- The ways in which communications and other technology can facilitate and support learning,
- The extent and nature of the utilization of workers' knowledge, skills and experience in the workplace.

References

Agurén, S. (1990), Att organisera för lärande i högteknologiska produktionssystem, in: Edgren, J. (Ed.), *Lära i jobbet.* Stockholm: SAF.

Argyris, C. and Schön, D. (1978), *Organizational Learning: A Theory of Action Perspective.* Reading, MA: Addison-Wesley.

Aronsson, G. (1990), Organisation, teknik och kvalifikation, in: Lennerlöf, L. (1990), *Människan i Arbetslivet: Beteendevetenskaplig Arbetsmiljöforskning.* Stockholm: Almänna förlaget.

Aronsson, G. and Friedrich, P. (1988), VERA- ett instrument för analys av lärande i arbetet *Arbete-människa-miljö.* 2, 88, 97-108.

Björk, L., Docherty, P., Forslin, J. and Stjernberg, T. (1990), *Att Behärska Föränderligheten.* Stockholm: Arbetsmiljöfonden.

Björkman, T., Glansén, C.and Gurnell, I. (1987), *Med rätt att utveckla egna idéer.* Stockholm: Arbetsmiljöfonden Utvecklingsprogrammet.

Bokander, I. (1990), *Arbetsuppgifter i belysning av handlingsregleringsteori.* Lund: Lundsuniversitet, Psykologiska institutionen working paper.

Davydov, V.V. (1977), *Arten der Verallgemeinerung im Unterricht.* Beiträge zur Pädagogik Nr 8. Volk und Wissen.

Development Programme (1988), *Towards a Learning Organization.* Stockholm: Arbetsmiljöfonden.

Dilschmann, A. (1990), *Kompetens och lärande - Kan vi mäta det?* Stockholm: Statskontoret och Kato gruppen MDA-project working paper.

Docherty, P., Elling, M. and Leymann, H. (1988), *Erfarenhets bildning, Kunskaps-baserad teknik och Arbetskvalifikationsutveckling i förvaltningen.* Stockholm: Statskontoret.

Gulowsen, J. (1987), *Kvalifikationer og arbeidermakt.* Tromsö: Universitetsforlaget.

Gutek, B., Bikson, T. och Mankin, T. (1984), Individual and Organisational Consequences of Computer-based Office-information Systems. in: Oscamp, S. (E), *Applied Social Psychology Annual Vol.5 s. 231-254.* Beverley Hills,CA: Sage.

Göranzon, B. and Josefsson, I. (Eds.), (1988), *Knowledge, Skill and Artificial Intelligence.* Berlin: Springer-Verlag.

Göranzon, B. och Florin, M. (1990), *Artificial Intelligence, Culture and Language: On Education and Work.* Berlin: Springer-Verlag.

Hacker, W. (1986), User-centered System Design: Design of Mental Tasks, In: Docherty, P. et al. (Ed.), *System Design for Human Development and Productivity: Participation and Beyond,* Amsterdam: North-Holland.

Imai, M. (1986), *Kaizen: The Key to Japanese Competeitive Success.* New York: Random House.

Karlsen, T. och Oppen, M. (1988), Professional Knowledge and the Limits of Automation in Administrations, in: Göranzon, B. and Josefsson, I. (Eds.), *Knowledge, Skill and Artificial Intelligence.*Berlin: Springer-Verlag.

Koike, K. (1989), *Human Resource Development on the Shop Floor in Contempory Japan.* Stockholm: Arbetslivscentrum, working paper.

Kolb, D.A. (1984), *Experiential Learning: Experience as a Source of Learning and Development.* Englewood Cliffs,NJ: Prentice Hall.

Lennerlöf, L. (1986), *Kompetens eller hjälplöshet? Om lärande i arbete. En forsknings översikt.* Stockholm: Arbetarskyddsstyrelsen Undersökningsrapport 1986:32.

Leontjev, A. N. (1977), *Problemer i det psykiskes udvikling.* Köbenhavn: Rhodes.

Leymann, H. (1987), *Teknik, arbetsvillkor och lärande.* Stockholm: Arbetsmiljöfonden and STU, MDA-report 1987:14.

March, J. and Olsen, (1975), The Uncertainty of the Past: Organizational learning under ambiguity *European Journal of Political Research,* vol 3, p. 141-171.

Morgan, G. (1986), *Images of Organization.* London: Sage.

Pender, M. (1989), *The Need for Self-Learning Organizations as a Part of an Overall Business Management Strategy.* Paper to the Eurotechnet work shop "Developing People's Ability to Learn Continuously- Goals and Methods", Maastricht, 2 2-24 November, 1989.

Ramnarayan, S. and Mohan Reddy, N. (1989), Institutional Learning: The Essence of Strategic Management *Vikalpa.* vol. 14, no. 2, p. 21 - 33.

Ranhagen, U. (1989), *Man and Technology in Interaction.* Stockholm: Arbetsmiljöfonden Utvecklingsprogrammet.

Rasmussen, J. (1986), *Information Processing and Human-Machine Interaction.* Amsterdam: North-Holland.

Rehnström, P. and Utbult, M. (1988), *Kompetens för Kvalitet.* Stockholm: Arbetsmiljöfonden Utvecklingsprogrammet.

Ryan, G., Aboulafia, A. och Björn-Andersen, N. (1989), *Intelligent Support Tools Organizational Learning about IT Uptake: The case of the IT-USE Project.* Köbenhavn: Handelshöjskolen i Köbenhavn.

Utbult, M. (1989), *Factory Worker into Software Worker.* Stockholm: Arbetsmiljöfonden Utvecklingsprogrammet.

Volpert, W., Oesterreich, R. Gablenz-Kolakovic, S., Krogoll, T. and Resch, M. (1983), *Verfahren zur Ermittlung von Regulationserfordernissen in der Arbeitstätigkeit (VERA), Handbuch und Manual.* Köln: Verlag TÜV, Rhineland.

Volpert, W. (1988), *What Working and Learning Conditions are Conducive to Human Development*? Stockholm: Arbetsmiljöfonden Swedish-German Workshop on the Humanization of Working Life, Dec., 1988.

Vygotsky, L. S. (1978), *Mind in Society: The Development of Higher Psychological Processes.* Cambridge, MA: Harvard University Press.

Wood,S. (1987), The deskilling debate, new technology and work organization *Acta Sociologica,* vol.30, no. 1, p. 3 -24.

Zuboff, S. (1982), Statements of Concern. *Office: Technology and People,* vol. 1, s. 66-70.

Zuboff, S. (1988), *In the Age of the Smart Machine: The Future of Work and Power.* Oxford: Heinemann.

Working Structure and Production Planning in Accordance with the Employees' Participation - An Integral Approach[1]-

Ekkehart Frieling, Kassel, Germany

1. Problem

The new technology of *information processing* and its use in the integrated control of whole fields of an enterprise are hitherto basically limited to *large companies*.

Therefore ergonomical studies on the effects of the new technologies on the structure of work and on the employees' qualification are most often related to those companies (Kern/Schuman, 1986; Brödner, 1986). Though in the nearer future CIM technologies - as they are called - will more and more enter medium and small business. The need for discussion and information about the problems caused by the new technologies are in any case existing, as the German congress "Produktivität und Zukunft der Arbeit" in October 1986 in Munich and the 2nd "TechnologieZentrumNord-Kongreß: Die CIM-fähige Fabrik" in January 1989 in Celle (Bullinger/Warschat, 1989) as well as different anthologies on the concerned subject verify (Baitsch/Ulich, 1988; Kam, 1989).

Within the recent discussion it is more and more accepted that the *introduction of new techniques* has always to be looked at *in connection with the organization of work and the qualification of employees.*

Either it is maintained that the new technologies require a new organization of work, e g flexible production systems, or that they are linked to extending possibilities for shaping the internal organization of the company. However, these opportunities are not sufficiently used (Sydow, 1989). Besides it is in particular the trade unions to point out *the risks of dequalification or qualification polarization.* The employers on the other hand complain about the lack of qualified employees. Therefore an extension of the employees' interoffice qualification is widely demanded for.

While the interrelation between technology, organization and qualification already is widely accepted, the process of change and *the concept of employees' involvement or participation* in these changes are given relatively little attention.

Now the question appears *how to include* the four formal concepts of technology, qualification, work structuring and participation *in innovation processes and investment decisions in medium-sized firms* and how to view and shape them in their interrelation.

---

1) This article was written in context of the HdA-project "Neugestaltung der Arbeits- und Organisationsstrukturen mit neuen Techniken in der Büromöbelfertigung zum Abbau von Belastungen - Hauptphase" (code no. 01 HH 038). The project is supported by the "Bundesministerium für Forschung und Technologie (BMFT)".

2. Methods

With a project carried out at a company it should be investigated into the question mentioned above. In accordance with the process of change within this company it should not only be looked into but also given advice for changes by the persons who are in charge of the project. The project mentioned is supported by the upholders of the programme: "Humanisierung des Arbeitslebens (HdA)" and is scientifically accompanied by the authors of this article.

Within the framework of an analysis of the actual situation to determine "flaws", the investigators basically use standardized methods of investigation to gain knowledge of a wide range of workplaces and to include as many employees as possible in their examination. Besides the conditions of noise, climax and lighting and organizational requirements are recorded.

A *questionnaire* investigating in *the state of health and further questionnaires on the subjective analysis of working conditions* (Udris/Alioth, 1980) complete the reflection on the question mentioned above.

Numerous analyses of work stations and talks with single employees or with groups of employees lead to *quantitative results* concerning the problems in view. Talks with executives, discussions with teams of employees, interviews with and observations on employees as well as analyses of material flow and information flow shaped an idea of the company and the interoffice processes of change.

3. Results

3.1 Preconditions

The company looked at produces office furniture and it occupies 900 employees. The whole process of innovation is caused by the management decision *to replace the existing electronic data processing system by a new one*. This new system makes it possible to process in one single system all data that is relevant for production control and financial management (from the registration of orders up to the balancing). The main reason to get the new integrated system is to improve the allocation of material, an area that is not EDP-directed until now.

By introducing the new system the company takes *a step towards Computer Integrated Manufacturing (CIM)*, that means to integrate CAD, CIM machines and PPC systems to get through and to control all areas of the company with the help of information technology.

An external software-house was hired to carry out the CIM project. After having analysed the existing structures of organization and information flow, it developed *an "overall concept"* that did fit with the changes planned. Basically that concept dealt with aspects of information technology and only in parts with question of organization. The problem of *the future users' qualification and their participation* in the shaping of the system *was not taken into account*.

At the same time the project was designed - in co-operations with the Gesamthochschule Kassel - for *the realization of a human and social introduction of new techniques*. It was applied to the HdA project-upholder, now called "Arbeit und Technik (AUT)". The mentioned AuT project includes all of the *four principles of shaping*, as they are listed in the beginning:

1. technical shaping of machines and information systems;
2. qualification of employees;
3. form principles in organization and production;
4. employees' participation.

This project developed into a work-oriented CIM conception designed by *scientists* who have specialized in ergonomics and by the technical director of the company analysed. The AuT project now has reached its most important phase that is planned to go on until the end of 1991.

3.2 Project Organization

Three project teams were set up to deal with the first three formal principles. These teams consist of executives and qualified employees of the company, of factory committee members and of the accompanying scientists.

The project committee consists of the project representative, the heads of the three project teams, factory committee members and scientists. It is used for the exchange of information between the project teams and the project-upholder.

3.3 EDP Technique

The first team deals with the shaping of *screen layout, printing and interactive structures* as well as with the *input-output-stations* and the *interfaces* between system components.

In the beginning fundamental hard- and software components had already been implemented. Because essential operational data was missing to a large degree the system still remained in a *testing phase*. The team received written results of work analyses in the fields of material disposition and assembly and results from interviews with employees.

Among other things *the present shaping of internal forms* does not meet the requirements of the German industrial standard (DIN A4). Besides they contain superfluous and redundant information and do not stress important information. An alternative planning was worked out by the scientists to emphasise the shortcomings mentioned before. On the basis of this critique an internal team was instructed to rework the forms and the affected employees were demanded to formulate requirements. Meanwhile an alternative plan was drawn up by an assembly employee. His plan does not comprise irrelevant information and important information is stressed by heavy print.

3.4 Qualification

The basic principle of qualification is *to build up an internal education system* including employees. To perform this concept consequently within the framework of the CIM conception, it is necessary to qualify employees of the computer centre to enable them to teach their colleagues how to handle an EDP system. In comparison with external qualification methods the advantage of this internal concept can be seen in the fact that teachers can complete changes within the system as well. Because of this fact, *a structure of social intercourse emerges* between employees located in the computer centre and employees of other departments. Furthermore this concept can be useful for a better coping with the daily work and in addition it is helpful for an ongoing improvement in technical aid and further education.

This principle has already been used to conduct CAD, CNC and WORD training programmes but has not yet been transferred to EDP elementary training and PPC training. Training programmes in this field only have been taken over by external experts because of the lack of qualified personnel in the internal computer centre.

The whole concept of qualification is not limited to the field of EDP but includes all employees by using concepts of qualification that are already existing in the sales department.

3.5 Work Organization

The third team started its work by *investigation in the field of final assembly*. Its main purpose was to reorganize the structure of work and internal organization.

From this point of view it would have been essential to think about alternative concepts of material flow and intermediate storing, job controlling and feedback and of further education and payment. Taking into consideration the question of reliability it also would have been necessary to discuss the problems of material and information interfaces.

Investigations in job controlling in the field of assembly revealed that time and again internal orders get lost on their way from the central printer located in the computer centre to the foreman's' office in the assembly shop. Therefore it would be more sensible to print the papers wherever and whenever they are needed. So far they are requested telephonic, they are printed in the computer centre and they are available only the following day.

However, observations on this problem would exceed the area of competence of the team members because dealing with this problem means to touch the interactions between other subject areas. Therefore the plannings concentrate on technical and constructive solutions to avoid flaws in assembly systems.

The reduction of physical strain caused by inconvenient posture as well as an improvement in the conditions of lighting, climax and noise are the main objectives of these plannings.

3.6 Participation

Up to now technical innovation was only planned by the executives being responsible. As a rule the persons being affected were included only after all decisions were made. Considering this, the planned changes in assembly - worked out by the third team - appear to be *an important step towards employees' participation.* At least the factory committee members and the lower executives are included in the process of planning right from the beginning.

The shaping of the internal papers was fulfilled by the hired software-house without giving the affected employees the possibility to comment the suggestions or to formulate requirements themselves. Only after the scientists had criticized the proposals and an internal discussion among executives had taken place, the employees were advised to write down their concrete demands on internal papers.

4. Discussion

The depicted problems show that *medium-sized firms* broadly prefer technology-oriented concepts that widely disregard personnel and organizational guide-lines.

Due to the ergonomical studies within the framework of the AuT-project, the one-sided orientation on technology could be broken down in favour of the company in general. Furthermore ergonomical studies stress *the necessity to take into consideration more and more social and psychological aspects.*

Concerning the described processes, the question brought up in the beginning is to be answered as follows: *The starting point of any development in business* is the aim to keep to the latest developments in technical research by transacting investments. As the new technology is more complicated than the old one, it logically brings about a need for higher qualification. Technology and qualification are perceived to be dependent on each other and to be significant for the capacity of development. Aspects of organization and participation still do not influence the perceptions and decisions of medium-sized managements very much. Nevertheless these aspects have to be considered and shaped more intensely to develop into solutions of concrete problems in production.

References

Baitsch, Ch., Ulich, E. (ed.), (1988), *Zukunft der Arbeit*, München: Psychologie Verlags-Union.

Brödner, P., (1986), *Fabrik 2 000 - Alternative Entwicklungspfade in die Zukunft der Fabrik*, 3. Auflage, Berlin: Edition Sigma Bohn.

Bullinger, H.-J., Warschat, F. (ed.), (1988), *Die CIM-fähige Fabrik*, Unterlüß: Technologiezentrum Nord.

Duell, W., Frei, F. (ed.), (1986), *Arbeit gestalten - Mitarbeiter beteiligen*. Humanisierung des Arbeitslebens, Band 77, Frankfurt, Campus.

Frieling, E., Facaoaru, C., Bendix, J., Pfaus, H. und Sonntag, K. (1990), *Tätigkeits-Analyse-Inventar* (TAI), Kassel: GhK-Institut für Arbeitswissenschaft (im Druck).

Geitner, U. W. (ed.), (1987), *CIM-Handbuch - Wirtschaftlichkeit durch Integration*. Braunschweig: Vieweg & Sohn.

Kamp, L. (ed.), (1989), *Arbeit in der Fabrik der Zukunft*, Marburg: SP-Verlag.

Kern, H., Schumann, M., (1986), *Das Ende der Arbeitsteilung?*, 3. Auflage, München: Beck.

RKW - Rationalisierungskuratorium der deutschen Wirtschaft e.V. (ed.), (1986), *Produktivität und Zukunft der Arbeit*, Eschborn: RKW.

Sydow, J., (1989), *Zur Wahrnehmung organisatorischer Gestaltungsspielräume beim Einsatz neuer Bürotechnologien*, in: Jansen, J. D., Schwitalla, U. und Wicke, W. (ed.), Beteiligungsorientierte Systementwicklung, Sozialverträgliche Technikgestaltung, Materialien und Berichte, Band 5, Opladen: Westdeutscher Verlag, 17 - 36.

Udris, I., Alioth, A., (1980), *Fragebogen zur "Subjektiven Arbeitsanalyse" (SAA)*, in: Martin, E., Udris, I., Ackermann, U. und Oegerli, K., Monotonie in der Industrie, Bern: Huber, 61 - 68 und 204 - 207.

Materials Flow Analysis, Sociotechnology and Naturally Grouped Assembly Work for Automobiles and Trucks

Tomas Engström, Lars Medbo, Gothenburg, Sweden

This paper presents an *overview of Swedish research* concerning some principles and experience in *production system design*, namely the creation of meaningful work tasks, technical autonomy and *functional learning*.

From the point of view of final assembly, automobiles and trucks are all surveyable - even when including the infinite number of variants - if they are described and the workplace is organized according to certain principles. However, the prevalent technical and administrative routines have deformed this logic to unrecognizability.

Reorganization of the technical and administrative *structure based on the "inner logic" of the product* makes it possible to *extend the cycle time*. This reorganization is the basis for reforming the assembly line in, for example, the new Volvo Uddevalla assembly plant.

The *results are based on experiments*, industrial implementations, empirical data and action research during a five year period.

This paper reports the results from the assembly of automobiles and trucks in four problem areas:

- The relation between the research done and the results that have led to industrial implementation during the period up to 1985 (Chapter 1-2).
- A description of the difference between the traditional and the new way of characterizing assembly work (Chapter 3).
- A description of an example of the effects of this *new characterization on the technical system*. In this case, how production planning, production control and external logistics can take advantage of the possibilities that now become available (Chapter 4)[1].
- A look ahead at *future consequences* for companies and research of the reorganization of the technical and administrative conditions (Chapter 5-6).

---

1) **This chapter has been written together with my colleague Lars Medbo M. Sc. Department of Transportation and logistics at Chalmers University of Technology.**

1. Swedish Research into Final Assembly

Younger scientific disciplines such as materials handling within the scope of transport engineering has its practical origin in operation analysis and is connected in its traditional application in serial production in manufacturing industries to the following problem areas:

- Work place design and loading ergonomics.
- Buffers and serial flow theory.
- Group technology.
- Construction and design.
- Line balancing models.
- Layouts and picking method analysis.

The research I have pursued over the *past thirteen years began in 1976* by using picking *method analysis* to seek *rational handling chains* for typical cases of equipment and goods combinations; an optimizing problem applied to an existing reality.

The Thesis Work in Relation to the Insight of Today

The problem areas I have treated have changed over the years. This is, of course, natural also for the arsenal of methods I have had at my disposal - a relationship I shall describe below.

The research became really interesting when *work sociology* was *added to the traditional transportation engineering*, so that with this as a starting point alternative systems to lines could be sought. The initial competence in the problem area was provided by a then qualified work-sociologist with more than four years'practical experience from the shop floor at Saab-Scania's factory in Trollhättan and who, moreover, was from the beginning a qualified engineer (ref. 8).

Work sociology with comprehensive empirical elements is a generalizing science of experience, and proved in the chosen problem area to be a *powerful complement to the technology*. The research began with a *sociotechnical approach*, with the result that researchers and practitioners initially framed the problem area for my thesis research:

- "Flow parallelization is the most attractive alternative production system for final assembly of large products. Parallelization in final assembly leads to materials supply becoming a restriction. How do we solve this restriction?"

Today I understand that this phrasing was missleading. But the knowledge sought could nonetheless be crystallized out.

In theory and in a number of case studies, it proved to be both practical and economically viable to supply materials even to large objects, a situation illustrated in my thesis (ref. 10).

The phrasing of the problem, originating in dialogues with people active in the field, was misleading insofar as it presupposed that these people's descriptive categories were axioms, declared as conditions for the problem phrasing.

My thesis demonstrated possible materials flow structures, the product being analysed partly by means of the so-called "design analysis" (a method of assessing how easy or difficult it is to assemble i.e. "degree of design") and partly by means of the so called "form classification" (a method of assessing the influence of the design on a given production system).

The final assessment of the reliability of the production systems was made with the help of a so-called "zero system calculation", a method of calculation which is a development of loss analyses (ref. 5).

The approach enabled, through its raised scale level (a reference system in the form of a Utopian loss-free system was introduced) together with the possibility of linking product and production system, a relation to be determined between "degree of process" (the capacity of the production system to adapt to disruptions) and the "degree of design". This meant that it was possible to determine when it was worthwhile to allow an operator to adjust an "unsatisfactory product" or when it was more suitable to try to raise the "degree of design". It proved to be the case, for example, that with *the low "degree of design" that Swedish automobiles and trucks had, that raising the "degree of process" was always to be preferred.*

The result generated during the thesis research and the larger STU-project[2] which financed the work, was besides methods of analysis, several practical changes in manufacturing industry (ref. 10) for large products, principally Saab-Scania's miniline system in Trollhättan. Complete parallelization was here thought too risky, because of the need for development of control systems and too expensive with the need for mechanized sorting equipment demanded by the large synchronous pre-assemblies. My method was practically applied to its fullest extent in two factories for Zanussi in Italy. Characteristic for my work during the thesis *research was a constant swinging between simple and extreme products* (from sailing YACHTS to vacuum cleaners and back), finally returning to the classic line products automobiles and trucks. The complete work is based on comprehensive empirics (the basic material for the thesis was in excess of 6.000 pages).

The reasoning on degree of design was put to practical test and linked to adjustment in sailing YACHTS assembly. A complete *"zero system calculation"* was made for two separate yards, Najad on Orust and Comfort in Arvika. The sailing YACHTS were thus subjected to the most extensive analysis, insofar as all operations were treated, the reason being that here the researcher could obtain comprehensive data. Automobiles and trucks

---

2) Styrelsen för Teknisk Uteckling, Stockholm.

were at this time too difficult to obtain an overview of. Today I understand that we lacked something. It is clear today that the existing description structures lead both analyst and practitioner to false conclusions. This explains today why the efforts of researchers, despite comprehensive empirics and theoretical analysis did not achieve a massive penetration in the really large industrial systems.

In the situation of today with the labour shortage, industries are more prepared to listen. At the same time I realize that the earlier question was also one of personal legitimacy. How could individual researchers argue against the prevailing administrative rationality deeply rooted in routines and administrative systems?

With the help of my collaborators I obtained during the period 1985-1987 unrestricted access to the practioners'theatre of change. *Access was achieved by initially formulating the problem area as a work-place and ergonomics question.* Industry successively placed resources in the form of premises, products and expertise at my disposal (ref. 15, 16 and 17). After exhaustive exploratory studies these resources gave a compact empirical operationalization of the sociotechnology. Operationalization ranges today from an extremely low level up to querying of total factory structures.

2. Streamlining and Development of the Original Problem Area

Over the last few years my research has once again *converged on industrial assembly*, where for the final assembly of large products it was necessary to develop pioneering method instruments, which generated new knowledge of the final assembly process on the basis of existing preconditions in the form of product and human. Choice of layout, materials supply and administrative systems were a consequence of basic processual knowledge, *formalized in the so-called "assembly-geographical atlas"* - a considerable departure from the analytic model crystallized from the original problem phrasing in my thesis. I have chosen the term "assembly-geographical atlas" since one or more operators, depending on conditions, choose their route on the "maps" that illustrate the product. Alternative names are "assembly anatomy", "final assembly topographical maps" or "spatial assembly anatomy".

The difference, in a few words, is that the methods earlier applied had their origin in a given design (a base object), with the operator density and cycle time that proved empirically possible, while today principles for naturally grouped work allow quite different work content and work organisation through the grouping/formalising of the assembly work so that functional learning is achieved (rather than additive learning). The work is described in maps designed on characteristics always found in all products and production systems.

The *materials handling system in the assembly plant* consists then of *combined supply principles*, from completely centralized batching to decentralized material areas with floating borders with the assembly areas. Characteristic is that it is small components (screws, nuts, clips, washers etc) that have to be treated separately. We started an experiment in this respect by batching these small components in plastic bags ("kitting") in a prototype. Today Volvo has further developed this idea in Uddevalla, the reason being that these small components are in the majority, it is impossible to keep the stock balance updated for several parallel material addresses and small components show the largest variation and therefore visualize the assembly work and the product much more precisely than the larger components, which the assemblers moreover learn more quickly.

Solutions with decentralized material areas and combined materials supply systems were penetrated in my thesis, but did not become industrially feasible or mentally tangible until they had been proved by the descriptive structures of the people in the field and through the Uddevalla project being successively forced, with the help of the researcher's results and own experience, to follow principles for naturally grouped work.[3] The work that I and my colleagues did on the layout during the summer of 1987 was the final breakthrough. The present situation in Uddevalla is that functional grouping of assembly work has been tested, introduced and found superior to earlier assembly principles. The groupings have been technically and administratively formalized through assembly instructions with support systems, layouts in final assembly and kitting fixtures ("kit racks") developed by myself and my colleagues. It is naturally satisfying for the researcher that practitioners have accepted knowledge, a knowledge that has influenced the design and establishment of several final assembly plants and related projects. To my assistance i have had excellent colleagues welded into a knowledge compact group, and the support of senior research competence in occupational learnings.

Summary of the Research up to and Including 1985

The sociotechnical school does not presuppose that humans are lazy and uninterested. The reason for lack of interest in the work lies rather in the technology, organizational structure and the bosses' expectations of their subordinates.

An important ingredient in work *design in industrial systems is to design the activities as rewarding or tension reducing in themselves* (ref. 6). The work itself must consist of *meaningful unities* and *demand skill and mental capacity*. The human must himself be able

---

3) Naturally grouped work presupposes that the traditional disintegration is broken and professional skills created - skills involve a number of tasks being combined in work functions. There are generally four characteristics in a skill:
- Natural rythm surveyable over time.
- Wholistic views of one's work in relation to the product.
- Functional grouping.
- Result orientation, where the central element is not time but "what one is doing" (ref 15).

to choose the way in which to perform the work and must continually be supplied with information on his performance (conditions discussed in ref. 1, 2, 3 and 7).

The *technological structure* which influences the *attitude to responsibility* has a decisive influence, so that most accept *responsibility as the price paid for technical autonomy* (the possibility of working-up) and *administrative autonomy* (the elimination of supervision and detailed control). Working-up is moreover, in serial production systems, a precondition for administrative autonomy and presumably the strongest reason for most to accept responsibility (ref. 8).

Preconditions for functional learning must be created, where the work can be learned in the working environment (ref. 9).[4] If meaningful unities, technical autonomy and functional learning do not exist, then it was, for example, absurd to learn more than 20 minutes' work content in automobile assembly. Buses on the other hand gave a limit of up to two hours. The reason was found principally to be *differences in product* design (the bus functioned as a reversed cassette, while the automobile consisted for the most part of hidden fitting-in movements).

Today I realize from my latest research that the difference was *mainly dependent on the design of the materials flow pattern* and the *way of describing the products*. Buses, automobiles and trucks are all surveyable - even including "the infinite number of variants" - if they are described and the work/work place are organized according to certain principles. But the previously *prevalent technical and administrative rationality has deformed this logic to unrecognizability*. For automobiles this was exacerbated by the product being transported at this time along a long line. The buses on the other hand were assembled in a threestage dock, where the chassis was moved sideways, which enabled the assemblers themselves to find structures of their own.

3. Knowledge-wise Breakthrough in Relation to the Difference between the Traditional and the New Assembly Work

The development in the final assembly of *automobiles and trucks* is in Sweden *moving from production systems with short repetitive tasks*, which are controlled outside the work process, *towards more connected, complicated tasks*. In these new tasks the worker *makes important decisions together* with his co-workers on, for example, variation in pace and how exactly the task is performed. To make this possible in the Uddevalla plant, a *radical renewal of various technical* and administrative sub-systems was necessary.
Some of the preconditions for this renewal were to be found in the characteristics of the *production system that has dominated* until today:

---

4) I would like to point out the difference between functional learning which is an established concept within occupational learnings, while final assembly functional grouping here refers to the practical groupings in final assembly I have developed for automobiles and trucks.

- To achieve successful production in the traditional line system it is not important for the individual operations to be meaningful for the worker performing the assembly. Each part has a low information value for the worker. A minimum of language and concepts is therefore required. It is sufficient for the worker to know where the component is to be fitted, it is sufficient for the materials handler to supply the correct quantity of components to the right address.[5]
- This situation means that the *formation of concepts* (language) on the production shop floor, *can be kept to a machine-friendly syntax, i.e. suitable for computer-based storage and communication*. Any necessary adaptation to the production organization can be made after a short training session and with a limited social system. Characteristic of this adaptation are a low level of communication, and a minimal amount of planning in the separate work tasks, which leads to a one-sided utilization of work potential and not of intelligence.
- The control lies in the fact that each worker has only a limited number of possible combinations. One of these is to teach him the correct assembly sequence, an assembly order consisting of a number of meaningless part numbers in assembly sequence (a knowledge of little value or permanence).

In such a production system it is important that the preconditions at the level of the *individual are prescribed in detail*, i.e. that tools, materials and *work methods* are trimmed together and defined beforehand.

The *possibility of humanizing* the work lies *outside production itself* and must be introduced through *some form of distribution mechanism* (in the general meaning of "The Swedish Model" this has been accomplished through active union participation via mutually agreed systems of rules).

In a traditional production system, production is thus planned in advance down to component level by others, not by the workers. The *forming of concepts around production theory* is thereby divorced from the workers' concepts. Humanization then becomes a question of replacing production with other activities.

---

5) On a line, work content will continually move between work stations for reasons, among others, of balancing. Components that are identical or have a connection with characteristics in the automobile or truck will seldom be assembled at the same station. If, on the other hand, the object stands still during the whole assembly, direct adjustment can be made, i.e. the worker can correct faults directly, immediately after the assembly is completed.

This is illustrated by the Arendal project within Volvo Truck Corp. at the beginning of the 1970's. When the worker had mastered the whole chains of tolerance for the assembly work, the components were first pre-fixed. When they were correctly related to each other and the chassis, they were then tightened. This procedure is no possible in short cycle line work, since the worker deals only with a few components. Components which are not obviously part of the same chain of tolerance. In the final assembly, of an automobile of the "degree of design" of Swedish automobiles, there is not time on a paced line with no intermediate buffers for direct adjustment (ref. 10 and 11).

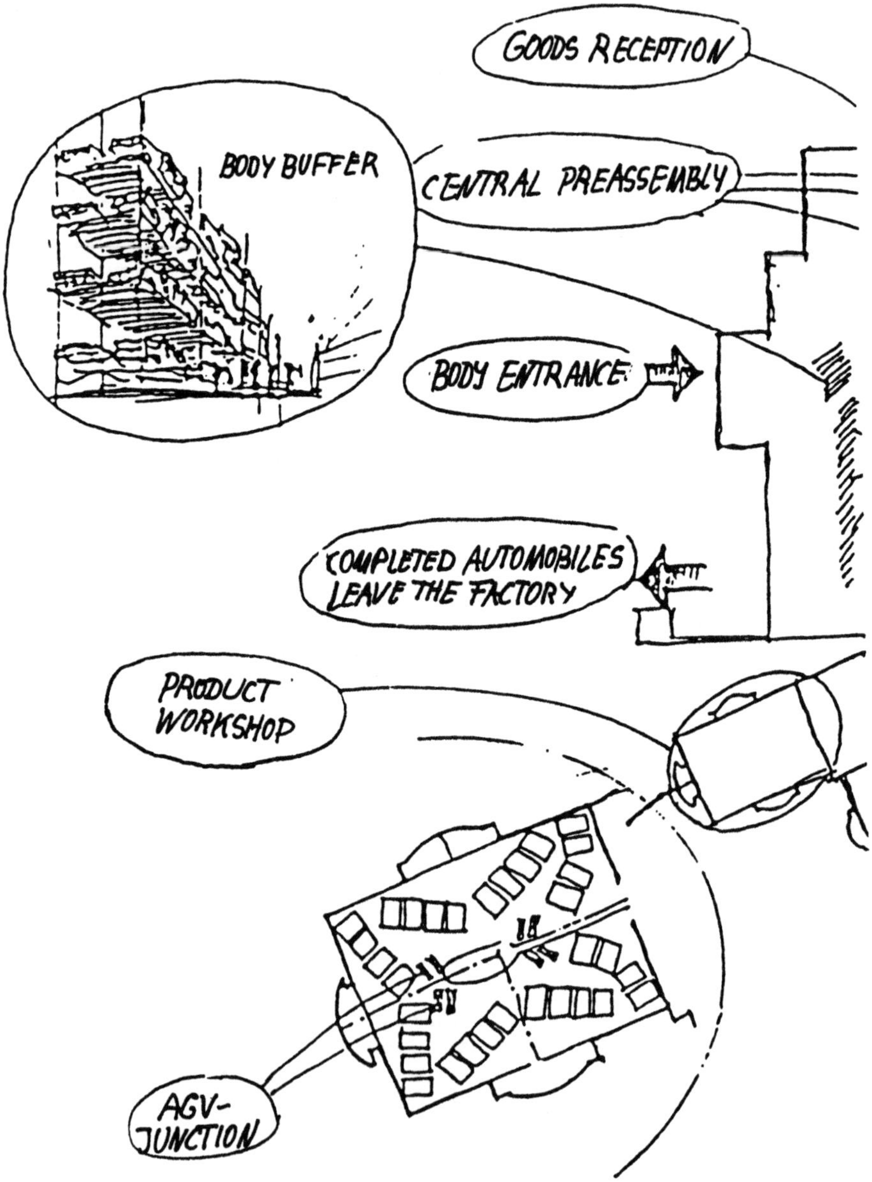
GOODS RECEPTION
BODY BUFFER
CENTRAL PREASSEMBLY
BODY ENTRANCE
COMPLETED AUTOMOBILES LEAVE THE FACTORY
PRODUCT WORKSHOP
AGV-JUNCTION

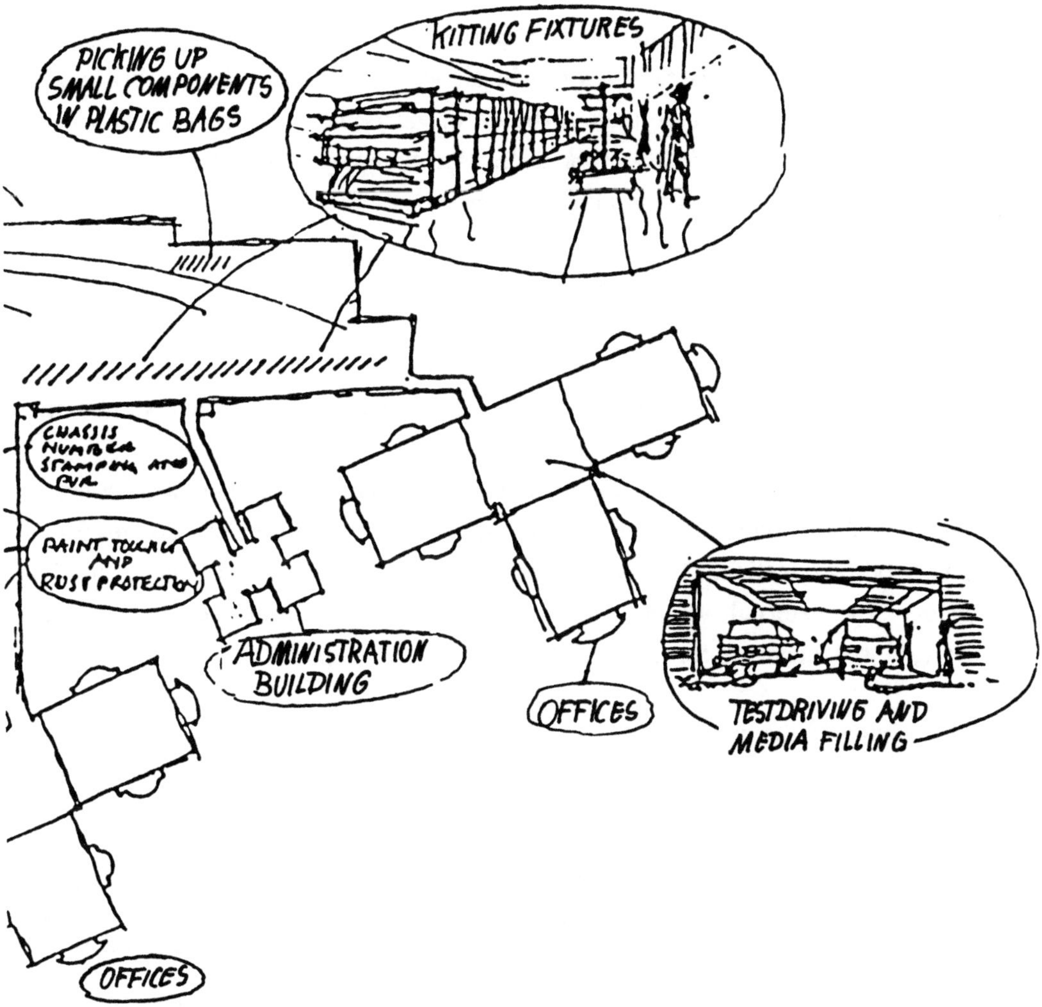

Figure 1: *The Uddevalla plant with its six parallel product workshops. These product workshops are grouped around two test workshops where media are added and the automobiles are test driven. The materials workshop prepare and supply in kitting fixtures to the final assembly. These fixtures serve both as a work instruction and as a "clock" to give the work teams an overview of the total assembly time.*

Characteristic for *naturally grouped assembly work* in, for example, the new Volvo assembly plant in *Uddevalla* are:

- A *nuanced language in the assembly work*, developed from classification systems which allow a richly developed use of language. A professional grammar related to the work, containing amoung other things a *syntax which permits the individual worker* to perform assembly processes which previously were regarded as impossibly long.
- An advanced administrative system (based on *relational databases*) for *distributing assembly sequences within the work team*. An economy system which feeds back experience to the workers and allows autonomous correction.

A comparison of the Uddevalla plant illustrates the differences:

- Increased complexity in materials handling with combinations of high and low mechanization. The body itself is moved few times while the components to be fitted have several simultaneous supply methods.[6]
- An *administrative system* which can *break down each single component* with its correct name to each individual automobile, no matter what structure week applies for the automobile in question. This system functions as a "protective membrane" around the factory. The external administrative structures at team level are compared and changed to achieve agreement with the assembly principles.[7]

One of the bases in the principles behind the Uddevalla plant is thus the *reclassification and regrouping of the assembly work*. This regrouping allows a *deeper understanding of the*

---

6) The body is moved to its assembly position. It stands still during the final assembly itself, while the materials that are to be fitted to the body are transported to it in kits for each individual automobile (ref. 12). The flow pattern is organic (comparable to a tree with an ever narrowing branch system). Characteristic is a successively diminishing degree of mechanization with a maintained sorting capacity between the object and the materials kits for each individual automobile. In such a flow pattern, there are points where mechanized equipment is applicable, depending on the degree of mechanization of the equipment (set-up time and operation time in relation to the total final assembly time for the object, and/or the total time at the disposal of the work team, determine where in the organic flow pattern each piece of equipment will be placed). The result is that the production design as a whole is not determined by the degree of mechanization at a given time of one piece of equipment. But the position in the complete flow is determined by how far mechanization has been driven in the direction of manual or machine work. This is not possible on a line or in more traditional layouts with mechanistic flow patterns (ref. 15).

7) Where one of the basic principles is that there must be concordance between carrying out the work, displaying the materials and describing the work. It must be possible to visualize the work in advance, but also to determine if the wrong materials or the wrong description have been supplied, or if one is uncertain of how to carry out the work - otherwise the worker is forced to test-fit the materials, with the consequential risk of being forced to dismantle.

relationship between *production sequences, component grouping*, the sequence restrictions of the components and variant affinity, etc.[8]

The consequences of this deeper understanding has implications in two directions. In the one, a considerable potential is made available for the design of new administrative systems and automobile models, with consequences for the degrees of ease of production built into future designs and also gives new guidelines for future logistical and distribution systems at the plant level. In the other, the new classification allows *the opportunity to develop the training pedagogics for the professional automobile builders* of the future.

The application of occupational pedagogical principles in *long cycle work means* that, unlike in short cycle work, a *system of concepts and a language have been developed*, which facilitate *communication*, the building of *experience*, and *orientation in the reality of assembly*. This language allows the transfer of production engineering and pre-production activities to a more qualified personnel than that of today, namely to the spearhead competence on the factory floor (such spearhead competence is in use abroad, see ref. ref. 14).

This new language forms a firm base for the mental insight into context and structure that is required to enable the new production system to function generatively and yield the efficiency desired.

By linking the components to names which connect them to function and sub-system, a foundation is laid for communication between planning, assembly and materials handling work.

The most important contribution of this new use of language is that it makes it possible to *communicate* with regard *to materials and parts and the interrelation of the components*, their spatial position and their sequence in the assembly. It becomes possible *to create natural grouped work.*

---

8) In order to operationalize occupational pedagogical theory and to convince practioners of, for example, the possibility to supply materials, the Uddevalla project employed researchers (ref. 8, 9 and 15). These were given access to an experimental workshop ("The Red Shed") and the company's products (including structural information from the administrative systems). The researchers worked together with assembly workers and production engineers in the education workshop that was set up before the Uddevalla plant was completed. Both established and newly developed knowledge could in this way be articulated as "nuts and bolts". It was necessary to depart from an evaluating "label research", whereby researchers try only to describe and classify that which already exists. This type of research does not lead to any breakthrough as regards knowledge.

In order to influence large industrial systems, research established on thorough empirical data is required, where researchers and practitioners work together in action research projects. Industrial changes occur so quickly that action research is a precondition for my aims to transfer and, together with industry, develop new knowledge and in order to influence the ongoing projections in industry.

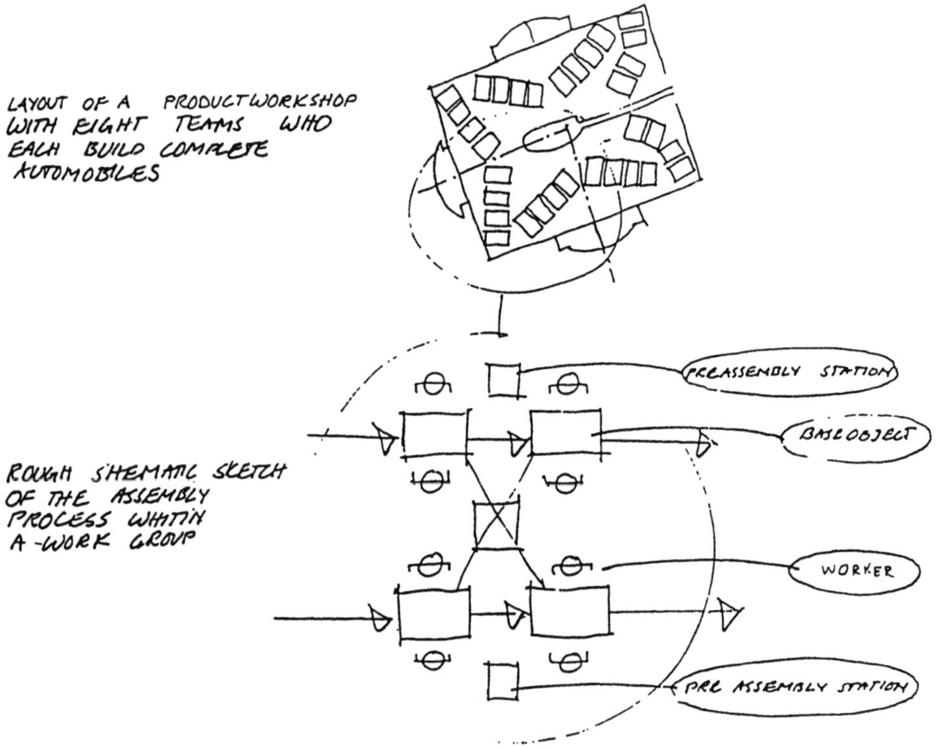

Figure 2: *Schematic function of final assembly within a team zone where one work team assembles a complete automobile. The automobile is assembled in two phases, with a sideways transfer within the team zone. The labour intensive pre-assemblies (principally doors, engine and dashboard) are integrated in the team. The aim is to increase the internal assembly-active buffer and to allow variant-resistant work patterns on the team level.*

Through the components being given a meaning by linking them to a language, it also becomes possible to regard assembly work as a meaningful occupation. It then becomes *a job* to which it is possible *to have an intellectual relationship*, thereby enabling the learning of an ever larger work content.

There is thus an intimate relationship between reality (in the form of components and their construction), and the linguistic and visual representation of this reality.[9]

9) Both a verbal and a visual network have been developed for the Uddevalla plant, which make it possible, without exception, to relate individual parts to different wholes. This network of texts, words and pictures is named "the assembly-geographical atlas". This atlas is based on characteristics always present in all vehicles (there are certain generic characteristics which I have used to obtain a description that is resistant over time and between different products irrespective of whether these vehicles are automobiles or trucks). My colleagues and I have carried out a similar analysis for Volvo Truck Corp. and the results are today in the process of being applied.

4. Assembly Variants and Production Control

For final assembly of automobiles and trucks in highly parallelized processes with long cycle times to be successful, the products must be able to be described in a, comprehensible way, from the point of view of final assembly. This is especially important, in order to be able to plan and control activities both within and between units in production systems, such as, for example, materials kitting, centralized preassembly workshops and parallel team zones assembling the final product.

Existing descriptive structures consist of market-oriented product numbers and of design-oriented material control codes. There are no descriptions relevant to the total final assembly in the Swedish motor industries of today for either complete vehicles, or parts of them, based on assembly similarity.

The present descriptive structures lead to erroneous conclusions when they are applied to the new production systems. They are not able to discriminate in the way required to support and create wholes as holistic maps in assembly and materials handling work.

As discussed earlier, one of the bases for Volvo's Uddevalla plant was a *reclassification of the products' components*. It was hereby possible *to revise the perception of the product*. This reclassification allowed the description of these wholes, in respect of understanding of the product itself.

The *assembly work content* was then able *to be radically changed*. This changed work means, among other things, *that the materials display* and the *administrative description are normally in agreement with* the work itself. In this context assembly variants are an administrative description of the products' similarities and differences.

Assembly variants involve access to a more nuanced language, a broader application of the "assembly-geographical atlas", a language that from the point of view of final assembly, identifies the products. A language suited to the purpose is a precondition for effective communication. The identification has its origin *in the categorization of the characteristics* of the products. The basic idea of the categorization is that different examples of the same assembly variant will be interchangeable. This means that there are three principal categorization criteria:

- *Competence requirements*.
- Assembly time.
- Tool and equipment needs.

There may be further demands, related to specific production systems (for example, low frequency variants), or if the concept is applied outside the chosen production system (for example, external logistics). To obtain *assembly variants*, the *characteristics of the products* are grouped in levels:

Level 1: The *same assembly work*, but with *different components*, for example panels of different colours.

Level 2: The assembly work is *marginally different*. Different components but relatively "obvious" regarding assembly; a gearstick boot, for example.

Level 3: Assembly tasks that have been decided that *all workers must be able* to manage, both in respect of knowledge and equipment. For example, all the characteristics of the doors can be placed here. The doors can then be used as balancing pre-assembly in a team work pattern.

Level 4: *Assembly differences dependent on competence*, assembly time or equipment. It is characteristics on this level that make assembly difference and thereby define the majority of the different assembly variants. *Low frequency variants on this level must* in practice be treated separately.

In order to describe automobiles with assembly variants, we have applied these to all variants of one model range for three principal markets.

All characteristics with some form of variation were described in 140 areas of characteristics, so-called variant families (with respect to the current descriptive structures used by the design department).

Of these variant families, some 100 were administratively "over defined" (the description had a greater variation than reality). If the areas of characteristics are classified according to level, the following numbers of areas of characteristics are obtained:

Level 1:	23	different characteristics
Level 2:	8	different characteristics
Level 3:	4	different characteristics
Level 4:	8	different characteristics

Thus, on level 4 there remained eight different areas of characteristics which describe the automobiles, *seen from the point of view of assembly*. These areas were doors, engine, gearbox, climate control system, braking system, market and extra fuel tank; the extra fuel tank was a very low frequency variant and was therefore treated separately. One specific assembly variant could then be described as follows:

- 4-door saloon.
- Manually operated sun roof.
- B230FS engine.
- Manual gearbox.
- Air conditioning.
- Anti-lock braking.
- Swedish market.[10]

---

10) The Swedish market is a collective term for characteristics which in their turn are unambiguously determined by the market concept. Note that the characteristics have here been ordered according to generality. It is clear that the most general characteristics had the lowest assembly relevance.

Since one or more team zones build only certain types of these variants, these areas of characteristics lack relevance there.

When a selection of products in a given production program are categorized into assembly variants, the individual products will be positioned somewhere on a scale ranging from one product to all products.

The number of different assembly variants is considerably smaller than indicated by both the present design-oriented codes and the material. When the frequency of the different assembly variants is known, it is possible to optimize the production system.

In order to test this new categorization, we selected one month's production of 10.000 automobiles, corresponding to 1.900 *different so-called product numbers* (the present system of describing the complete product).

Categorization resulted in *125 different assembly variants*, a small number of which represented the majority of the produced automobiles; 20% of the assembly variants represented all of 88% of the production volume. This is illustrated in the figure below.

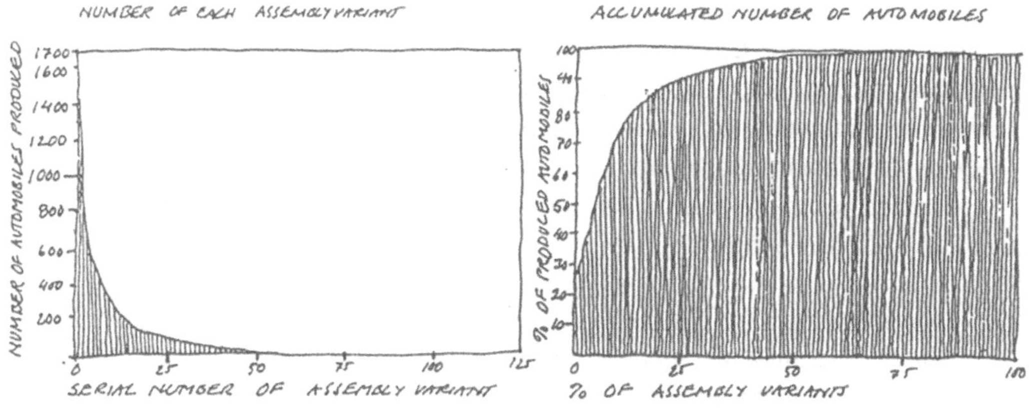

***Figure 3:** Absolute and percentage distribution of automobiles with regard to assembly variant, from high to low frequency. The diagrams are based on one month's production of 10.000 automobiles.*

Consequences of production planning and control by means of help of assembly variants have been studied through simulation. Regarding production in highly parallelized final assembly with centralized materials kitting, planning and control on an individual automobile and on an assembly variant basis have been compared.

The results of these simulations indicate unambiguously that assembly variants improve precision in time for both assembly start and supply precision. This is true even in cases of wide distribution in throughput time per team zone. There is less need for replanning and buffer volumes in the material and assembly workshops are reduced, see figure below.

It is important to use a description suited to its purpose at any given time. Different descriptive methods do not exclude each other but have a different relevance in the separate stages of the manufacturing process.

The effects of using assembly variant as an identity concept are summarized below. They give the products discriminating identities, from the point of view of final assembly, resulting in:

- A tool for personnel to be able to plan production in an appropriate way for them.
- Greater technical and administrative autonomy between different stages in the production chain.
- Simplification of reordering and interchanging of individual products between and within units.[11]
- Easier handling of sequence demands. An example of a sequence demand is that automobiles with assembly similarities should be grouped in twos (resulting in a time-gain as regards "mental set-up time" for the second automobile).

There is less need for detailed centralized production planning in highly parallelized production systems. The number of individuals in the plan fell in our case by 80% when assembly variants were used instead of product numbers.

Greater freedom results through the assembly variants being used for product identity in production planning without lessening the precision demands on the centralized planning. This means that:

- There is less need for replanning in the material and assembly workshops.
- Materials consumption sequence is in better agreement with the planned sequence, leading to less buffering and better JIT efficiency.
- A centralized store is not affected negatively if parallel team zones do not follow in detail a previously determined time plan.
- Delivery precision of completed automobiles is in better agreement with the plan.

5. Future Consequences

Consequences of the Reorganization of Final Assembly Work.

The changed perception of the product itself and the thereby consequential assembly work has brought to the fore such questions as:

---

11) High frequency products with assembly similarity will automatically be interchangeable. This argument is true for both the whole production system as well as parts of it.

- How will it be possible to exploit efficiently the precise work of the designer, so that it will be an asset in the various processes in final assembly, at the same time as experience attained locally is formalized so that it can be used to advantage in design and pre-production?
- How can the central pre-production work be made to support all local final assembly processes?

The answer can be found in the way of describing the products. I see a need for new product and process structures, or, to be more specific, the automobile industry of the future presupposes completely different descriptive structures to those of today, if the line is to be replaced.[12]

It is therefore necessary to change the technical and administrative preconditions so that these are not an obstacle to surveyability. This surveyability is described, and related to both individual (person, department, production stage, etc) and, as previously, centrally formalized knowledge. It is then that the product description becomes and integral part of the individual's personal professional competence. The products contain an inner logic that it is possible to utilize. This is important, not least because the products are predetermined in detail right from the design work itself, in respect both of correct names and that component functions and systems are known. This is knowledge that interrelates the assembly components to each other and to quality demands.[13]

---

12) There are at least three different languages applicable in final assembly. These are: 1 A language related to the fact that components have a name (brake cylinders have some connection with brakes). 2 A language related to functions in the sense of how things work (a difference must be made here between functions in general and a mastery of Swedish). 3 A language that is a consequence of the new groupings. The last type of language is today under development and we must allow that certain terms and concepts are temporary, not established and not always well chosen. They are working names; when "the diagrams of the new assembly work" have been fully developed in the future, it will be easy to duplicate them in many plants.

Besides these three languages, there exists a visual language to enable understanding of the necessary co-existence between time and space. If these new languages are established the introduction of future assembly processes will be considerably simplified. In the Uddevalla plant, final assembly functional groups are referred to - these must not be confused with the functional groups 1000 - 9000 that Volvo today uses to describe its vehicles. Note that the term functional is used to mean suitable for a purpose. The final assembly functional groups are suited to the purpose of final assembly.

13) Early in the research my colleagues and I realized that this inner logic in the final assembly of automobiles did not become obvious until 1/4 of an automobile could be surveyed. It is much simpler to interpolate separate work tasks or sequences of these, based on general views known and stipulated from the outset. These views are both verbal and visual (I make a difference between final assembly functional groups and variant tracks). It is only when this perception of reality helps to provide orientation in reality that the new groupings are correct - not the reverse. It follows from this that assembly work as it is carried out is a consequence of work content and its characterization.

In order, in the future, to be able to create and maintain a system of concepts related to final assembly it is necessary to create new product and process structures. These will be a precondition, partly because of the need to reform the factory floor work itself, but also necessitated by the fact that different assembly plants today have specifically different production systems (as long as line assembly was all-dominating, the central pre-production work was more accessible to the local processes).

Today line assembly is being replaced, for reasons of efficiency, demands for humanization of the work, and a wider range of variants which, in combination with a larger number of direct deliveries of complete pre-assembled components, lead to irrational solutions for materials supply to the line. The differences between the various production systems become so great, that the designer's way of describing the product in combination with the procedures of central pre-production means that the amounts of information to be transferred between design, central pre-production and local process must be reorganized on the basis of formalized knowledge on the factory floor if final assembly with considerably greater work content is to be achieved. A reorganization based on the product itself and its variation.

The basis for this future work is the "assembly-geographical atlas", a classification used today in the Uddevalla plant. A verbal and visual network must also be created, extending over the height, depth and breadth of all the plant's organization. I have illustrated this in the figure above.

Figure 4: *An example demonstrating the difference between using top and not using assembly bottom variants; the consequence is that buffer volumes are either exploited for reasons of commonality (pooled buffers) or develop their own separate queues. The figure shows a simplified example with four different assembly variants with a frequency distribution according to the previous figure. The difference in function between the buffers is shown schematically in both cases. The large upper triangle indicates that the grouping of the assembly variants gives the automobiles differing degrees of interchangeability depending on the frequency of each assembly variant (shown by the smaller, inserted triangles). The highest frequency ones will automatically be interchangeable and thus form a common buffer for all queues to parallel work teams concerned. The significance of this is that there will be degrees of exploitation of commonality ranging from the highest frequency to the extremely low frequency assembly variants. Such low frequency variants must always be treated separately.*

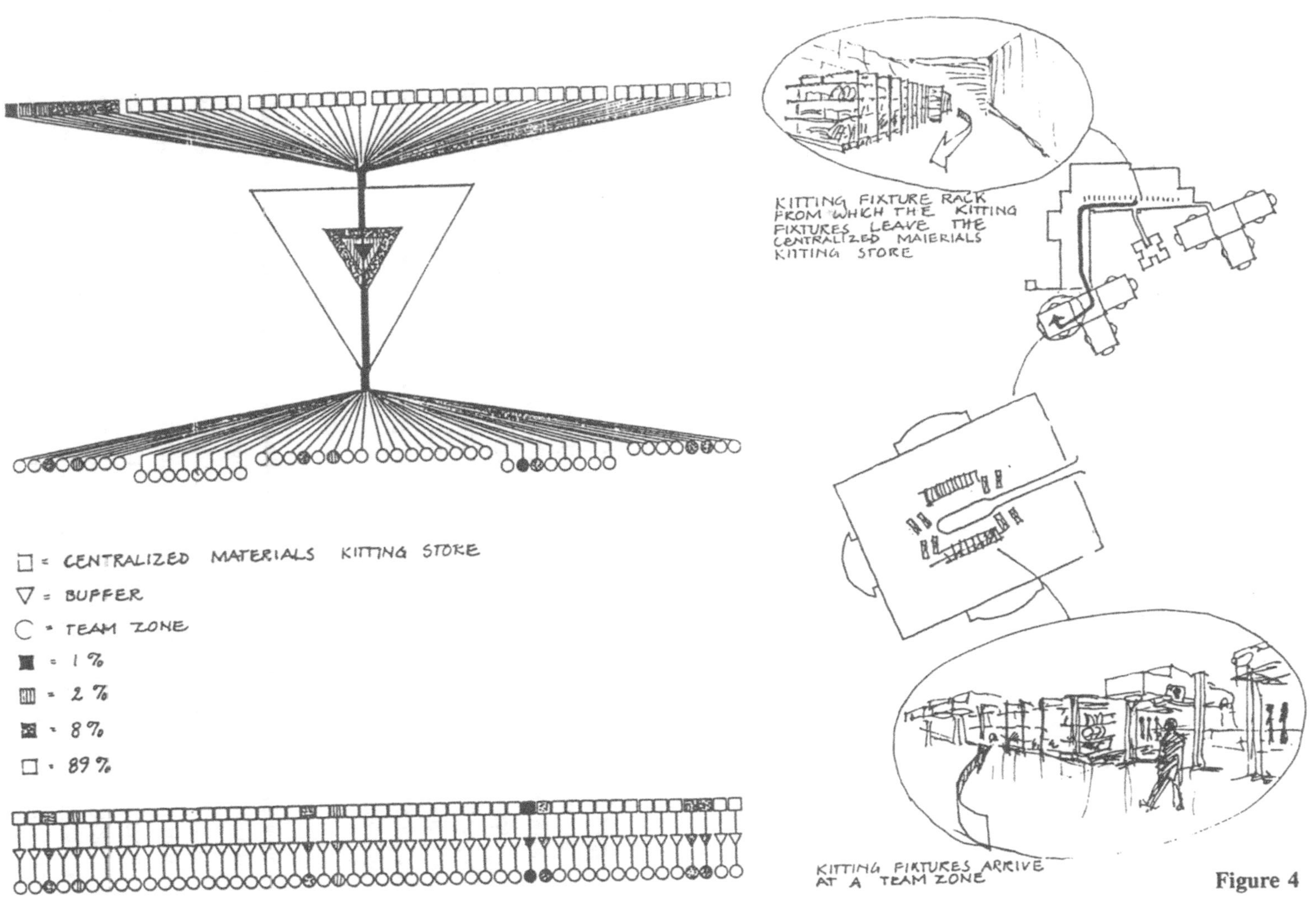

KITTING FIXTURE RACK FROM WHICH THE KITTING FIXTURES LEAVE THE CENTRALIZED MAIERIALS KITTING STORE
KITTING FIXTURES ARRIVE AT A TEAM ZONE
□ = CENTRALIZED MATERIALS KITTING STORE
▽ = BUFFER
◯ = TEAM ZONE
■ = 1 %
▥ = 2 %
▩ = 8 %
□ = 89 %

Figure 4

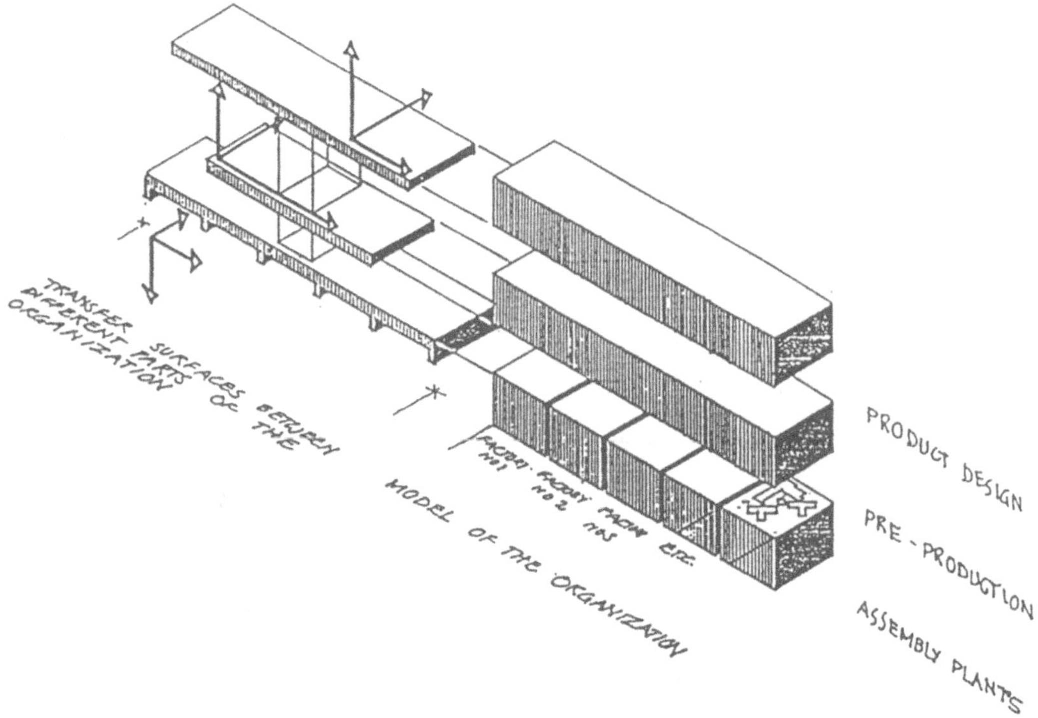

Figure 5: *Three 3-dimensional model of the verbal and visual network. The different parts of the organization separated from the transfer surfaces.The arrows in the figure show how the product is described in different parts of the organization, i.e. how the product description changes according to who is using it. Suitable units must be found for the axes in the transfer surfaces between the different parts of the organization. It is not least important to find common surfaces, lines and points of reference for the whole organization, in order to allow the various users to be able to move between product and process structures without losing their mental orientation.*

The greatest obstacle within the automobile and truck industries to the work of tomorrow (either as an assembly worker or as a designer) does not lie in the work itself, but in the current description of the work and the consequential perception of one's own work in relation to the performance of others. Since it is so, that when my conceptions help me to handle reality, then they are right, not the reverse. However, these conceptions can initially be derived from qualities existing in reality, and later assembled into logically resistant patterns. Rational thought patterns of this kind have previously not been needed in final assembly, and therefore they have never been identified and formalized.

Future Research and Development Work

My conclusion is that the assembly work of tomorrow presupposes a qualitatively new research direction, in which the point of departure will be how the technical, economic and organizational system will be able to create and take advantage of human competence. To achieve this, in the engineering industry, for example, work and materials will have to be grouped according to completely new principles. New forms of job description will also, therefore, be required.

This must be based on knowledge of human learning as well as of technology. Some of this knowledge has today been empirically tested and scientifically proven. But it has not in its entirety, or even in detail, been finally reported in the academic world.

References

1) Dahlström, E., (1966), *Teknisk förändring och arbetsanpassning*, Prisma, Stockholm.
2) Dahlström, E., (1969), *Fördjupad företagsdemokrati*, Lund, Prisma.
3) Lawler, E., (1969), *Job Design and Employee Motivation*, in: Personal Psychology, 22.
4) Thorsrud, E., Emery, F-E. SAF, (1969), *Medinflytande och engagemang i arbetet*, Stockholm.
5) Wild, R., (1975), *On the selection of mass production systems*, International Journal Production Res. No 5.
6) Friberg, M., (1975/76), *Är lönen det enda som sporrar oss att arbeta?*, Sociologisk Forskning no 4, 1975, no 1, 1976.
7) Dahlström, E., (1977), *Effiency, Satisfaction and Democracy in Work*, Acta Sociologica.
8) Karlsson, U., (1979), *Alternativa produktionssystem till lineproduktion*, Sociologiska Institutionen, Göteborgs Universitet (doktorsavhandling). Göteborg.
9) Nilsson, L., (1981), *Yrkesutbildning i nutidshistorikt perspektiv*, Pedagogiska Institutionen, Göteborgs Universitet (doktorsavhandling), Göteborg.
10) Engström, T., Karlsson, U., (1982), *Alternativ montering*, Institute for Management of Innovation and Technology, Chalmers Tekniska Högskola, Göteborg.
11) Engström, T., (1983), *Materialflödessystem och serieproduktion*, Institutionen för Transportteknik, Chalmers Tekniska Högskola (doktorsavhandling), Göteborg.
12) Asterhall, H., Åkesson, S-Å., (1986), *Rum för arbete*, Industriplanering, Chalmers Tekniska Högskola, Göteborg.
13) Koike, K., (1987), *Human resource development and labor-management*, The political economy of Japan, vol 1, Stanford University Press.
14) Ellegård, K., Engström, T. och Nilsson, L., (1989), *Principer och realiteter, projekteringen av Volvos bilfabrik i Uddevalla*, Arbetsmiljöfonden, Stockholm.

15) Ellegård, K., (1989), *Akrobatik i tidens väv - En dokumentation av projekteringen av Volvos bilfabrik i Uddevalla*, Kulturgeografiska Institutionen. Göteborgs Universitet, Göteborg.

16) Ellegård, K., Engström, T., och Nilsson, L., (1990), *Principles and Realities in the Reform of Industrial Work - the planning of Volvo's car assembly plant in Uddevalla*, The Swedish Work Environment Fund (in press), Stockholm.

17) Johansson, M., (1989), *Product design and materials handling in mixed-model assembly*, Department of Transportation and Logistics, Chalmers University of Technology (doctoral thesis), Gothenburg.

The Need for Holistic and Innovative Research: Approaches in Studying Work

Flemming Norrgren, Gothenburg, Sweden

In contemporary research publications on work life issues much emphasize is given to *the topic of change or transformation within the organizations or between organizations and their environment*. Be it technological, behavioural or economic research all introductions pay their visits to the gospel of change rate, knowledge obsolescence, internationalization or magnitude of change as rationales for accumulating more systematic knowledge.

In the setting where I am presenting this very paper we have a session which title reflects some of the issues involved when dealing with the change phenomena - *holism and innovation*. One of the rationales of addressing holistic strategies and calling out for innovativeness at work is that organizations increasingly have *to cope with combining rapidly growing knowledge within various knowledge* areas and at the same time integrate specialized knowledge into a meaningful whole. Another task is *to respond innovatively to the increased output of new competitors* that launch new products more frequently than in earlier times. At times one might wonder whether the scenarios put forth about increased change and complexity reflects what actually happens in all branches of work life or if what we observe is the result of researchers postulating the same themes at so many conferences that it becomes true only by passing the "conference test of validity".

If we however accept the importance of addressing the change factor and the importance for work organizations to deal with holistic and innovative approaches I would like to bring up the issue *in what way we as researchers match our way of organizing and conducting our own work*. Have we adjusted our *research designs and tools* in order to be able to understand and catch what is actually happening out there in the jungles?

In short I would say that we need *holism* not in the limited sense that we are equipped with *a broad set of perspectives*. We should also have *a readiness to cope with unpredicted events* occurring in the field settings we are studying. As pointed to by Gustavsen and Hart in a forthcoming article in Human Relations we should *accept the relevance of interactive logics in addition to linear logics* (1). Iterating between different theoretical models and observations of what happens in state of art could enhance the innovativeness of for example sociotechnical approaches.

In most academic settings we are trained and guided by research handbooks to stress careful planning, rigorism in selecting and developing *methods of data collection and following a step-by-step procedure*. *Problem formulation* is in many cases derived solely from theory and/or from existing knowledge bases within narrowly defined fields. Our merit and promotion systems reward and underline the ability we have to adhere to these procedures and great attention is paid to the relevance that the individual researchers work has within his or hers specialized discipline. *A safe way of getting promotion or positions* is to maximize the number of specific articles published within specific areas by using a specific set of data repeatedly for different publications. There is accordingly more to gain in academic career strivings by *carrying out micro-oriented, piecemeal research* and staying within the field of a specialized area such as cognitive ergonomics, leadership etc. Hence it is better to stay away from trying to study the often disorderly mechanisms of innovativeness or complex matters of holistic happenings.

My position is that we to some extent *have to reconsider our own processes and academic structures* if we want to adequately understand, describe and predict what is happening in our study object in work life. If we shall collect valid data about holistic and innovative processes *we must make it legitimate to design and conduct research* that captures some of these characterisation:

1. *Iterative and interactive problem formulation* where a dialogue between representatives from state of science and art contribute.
2. *Interdisciplinary and team based research* to get a comprehensive picture at the same time that specialists' contributions are maintained.
3. *Faster and parallel designs of research* to avoid presenting results long after they may be of high relevance.
4. *A readiness to use explorative and empathic data collection methods* that encourage our human study subjects to continuously contribute to the validity of our observations, and an awareness of researchers on *how* their *methods affect validity* (3).

To give *one anecdote* to illustrate the opposite hallmarks of much research I would like to give you an example reported by one of my appreciated colleagues Dr. Hörte (2) who attended a conference on ergonomics. The main discussion theme envolved around the issue of mismatch between scientists and practicioners. The science community complained that practicioners complained about that their carefully collected and rigourously analyzed data were found to deal with problems not relevant or already solved at work places. That is to say that *reality progresses too fast in comparison to our pace of presenting theoretically sound models*. One might wonder if we should ask practitioners to slow down to fit our processes or if we should contemplate on our way of working.

Let me close by giving you *one example* of how we at the *Gothenburg Centre for Work Science* try to tackle the problems. Through initiatives taken both by the Work Environment Fund and scientists at Chalmers University of Technology and University of Gothenburg our interdisciplinary research centre was set up in 1988. The centre was set up and designed through a collaborative effort involving parties of the labour market, top officials at the Work Environment Fund and scientists in which *structure, goals and time perspectives* were evolved in a process characterized by mutual adjustment. A few but central common denominators were laid down such as focus on short lead-times in research, joint evaluation of research areas, preoccupation on field research in industry and a commitment to add both to academic knowledge and action assistance to state of art representatives. The centre involves six disciplines, in total approx. 24 employees of whom eight are professors and 16 are doctoral students from various disciplines. *Our research* has mainly *been organized in programs*

- one on Product development
- a second on blue collar work in the Nineties and
- the third on models for Development and Change.

Both the Centre itself and the programs run on sharp time limits but at the same time are given good resources and considerable autonomy to ensure innovativeness. The programs involve both technicians and social scientists - at least three different perspectives in each programme.

The program I head is the one that has been running longest, starting in 1988, and is labelled *Product Development in the Nineties*. It involves 11 researchers who jointly share in the formulation, presentation and external contacts with companies, managerial networks, unions and the international scientific network engaged in similar work. Within the program *ten different projects* are carried out dealing with management and leadership, organization and change of product development in industry. About 20 Swedish corporations are studied in detail and complemented by more extensive data collection methods. Every year we meet a network of top management representatives to discuss our mutual interests and also a network consisting of middle management and union representatives from the engineers unions.

In addition to traditional research reports and dissertations we get *coverage in media* and are now committed to *the creation of an international scientific network within the area*. Our joint analysis of the pooled data started this month and will be presented in a joint book in 1992.

After almost two years of work we can conclude that we so far have accomplished *both scientific and pragmatic goals*. This is
not only due to our positive resource situation but also dependent on the conscious building of internal and external relations and not at least to good group climate.

Let me close by saying that I am not proposing to turn all research in the direction we have chosen. I am merely pointing to the fact that *we need complementary structures and processes* in order to cope with our changing environment. If we shall be able to meet the relevance criteria we must value and support several alternatives in organizing and conducting research.

References

1) Argyris, C., (1970), *Some unintended consequences of rigourous research*. Psychological Bulletin, 70, (pp 185).

2) Hörte, S.Å, *Personal communication.*

3) Gustavsen, B., Hart, H. and Hofmaier, B., (1990), *From linear to interactive logics: characteristics of workplace development as illustrated by projects in large mail centres*. In: Human Relations.

Assessing Work Organisation Needs in the Light of New Technology

David Grayson, London, United Kingdom

1. Introduction

It is a truism that the development and growth of an organisation depends upon the extent to which it is prepared to become involved in innovation. The most pervasive form of innovation in recent times has undoubtedly been the introduction of what have been called the new technologies. It is difficult to draw clear demarcation lines between new production technology and new technology concerned with the automation of administrative activities. However what unites *all types of new technology* and sets them apart from previous technologies is *their ability to analyse, transmit, integrate and respond to information*. What is clear is that they have influenced and altered not only the work processes in many organisations, but also products and services and the kind of work that people do. The all embracing effect of new technology can be expressed diagrammatically (Figure 1) and it is to some of the elements identified that this paper is addressed.

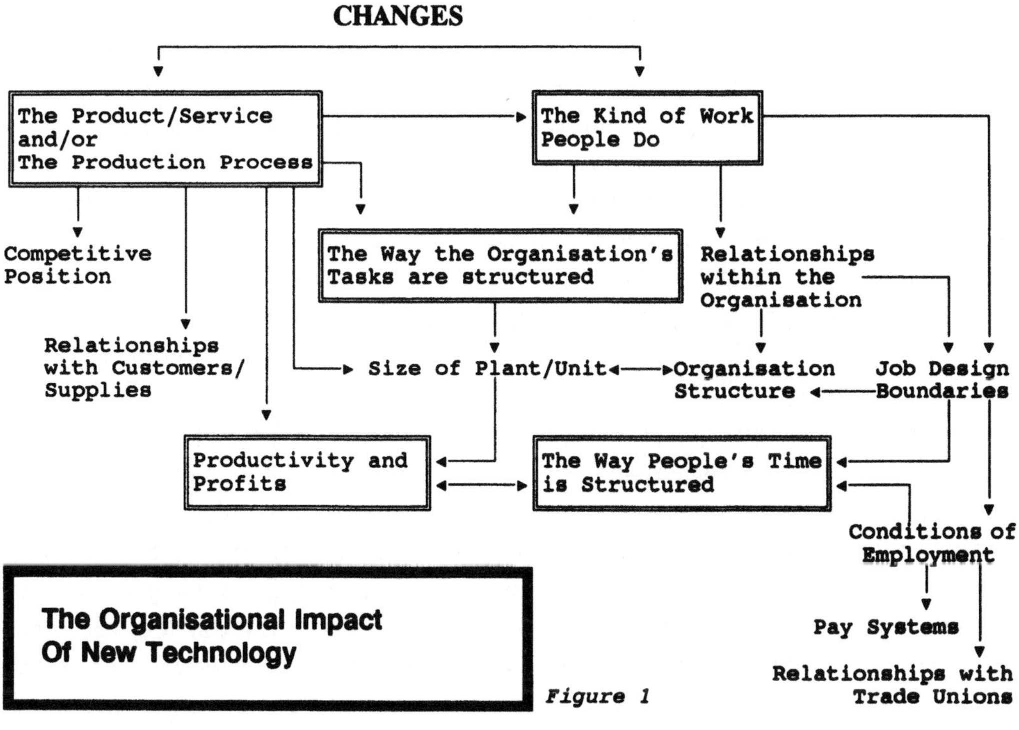

Figure 1

2. Technological Determinism

A particular view of the relationship between technology and work organisation is that the type and level of technology in use has a determining influence on the organisation of work. This view of what has been called *technological determinism has its roots in the scientific management movement* which advocated that an important outcome of technological and organisational change was to reduce the dependence of employers on workers; that by changing production systems, control over the quantity and quality of work could increasingly be given to managers who could determine the most efficient methods. There is little doubt that scientific management is still a common management practice.

3. Management Values

However, the proposition of technological determinism has been increasingly challenged particularly as a result of more widespread experience about the introduction of new technology within the manufacturing, commercial and service sectors. The challenge is based on the argued interdependence of technology and people and the *different characteristics which the new technologies exhibit* compared with the past such as compactness, relatively low cost and, in particular their flexibility and adaptability. The conclusion drawn is that *new technology is much less deterministic of work organisation* than the earlier impact of automation. A particular justification is that the impact of the introduction of new technology on work organisation or jobs may lead to *either positive or negative outcomes for employees dependent upon the values, beliefs and knowledge of those implementing it*. This is what has been termed the *"strategic choice approach"* which emphasises the role of choice in explaining variations in organisation and behaviour at work. Thus using new technology, management can still organise work on scientific management lines to give a high degree of central control and standardisation. Alternatively management can use it to organise work in, for example, self regulating work groups based on flexible work roles with non-standardised tasks. In the design of jobs, new technology can be seen as a vehicle for more simplified, deskilled, less autonomous jobs. It can equally be seen as a means to create better jobs in which the individual is able to exercise expanded skills and control. At the organisation structure level traditional understanding about the concept of centralisation (tight control, minimalist delegation and authority) and decentralisation (greater local autonomy, freedom of action) also become less meaningful because of the *interplay between technology and management philosophy*. The use of information technology can mean increased central control even within a decentralised structure. It can therefore be argued that the introduction of new technology has no uniform impact on work organisation; that the form of work organisation ultimately adopted reflects the outcomes of choices which are not governed solely by the nature of technology adopted. In such choices non-technical factors are equally significant. Management and human relations matters play as important a role as technical requirements.

4. The Total Systems Approach

Organisations have technical, manufacturing, and administration systems, which consist of equipment, layouts, and the principles, techniques, and rules of their operation. Allied to these technical systems is a *people system of values, roles and relationships with further support systems* to maintain them. *Emphasis is frequently placed on the technical system* because the rewards of getting it right are more likely to be reflected in short term results and in financial statements. Investment of time and expertise in getting the human aspects of the system right may not show a return in the short run. It tends to be more difficult to support with clear cut immediate evidence of return and is often wrongly seen as an operational rather than an investment cost.

It was recently quoted that *when introducing advanced manufacturing technologies (AMT)* it has been estimated that organisations spend more than 90 per cent of resources such as finance time and effort on the technical issues while less than 10 per cent is focussed on human and organisational details. Yet surveys reveal more than 75 per cent of production managers are not happy with AMT performance. It appears that this perceived failure stems to a considerable degree from *misunderstanding and neglecting human reactions to the technology*. People are expected to be adaptable enough to fit into the technical system where they are needed if they are given appropriate training, together with financial and other inducements. Additionally, organisational effectiveness and the well being of individual employees are often seen as independent objectives separately pursued and sometimes conflicting. There is therefore considerable value in *planning the work system as a whole*, reviewing the tasks that people will have to do, management philosophies and styles, personnel policies as well as the production matters and appropriate organisational structures and procedures. The broad aim of such an *holistic strategy* is the optimum "fit" of technology and people. Development of the strategy will involve acknowledgement of the need for replanning and changed approaches in the light of experience and may well benefit from a *policy of "minimalist initial specification" to allow for maximum flexibility*.

5. Supportive People Strategies

The principal effects arising from new technology are likely to be technical changes both to the product or service and to production processes. However as previously suggested in its successful introduction non-technical factors, particularly *human relation factors, play as equally an important role as the technical requirements*. This means that to manage and operate the new equipment effectively organisations need to examine the continuing relevance of existing people strategies and to consider whether different ones might be needed to support new directions being taken. Such examination and evaluation is likely to take as long as the commissioning of the new technology itself, a factor which needs to be recognised and allowed for in the overall planning process. Some people *strategy issues* have already been touched upon, for example, *the changes to job content, job boundaries*, and the likely alteration to the way that *tasks within the organisation* will in future be structured.

In addition the technical changes are likely to affect any one or all of the following:

- *Relationships within* the organisation (vertical, horizontal and diagonal) and therefore the organisation structure; likely moves away from functional hierarchical organisations towards flatter more flexible models.
- The *size of plant and unit* with the associated manpower implications.
- *Management style.*
- *Skills and knowledge* required, reskilling, multi-skilling.
- *Selection and Training.*
- *Communications.*
- *Conditions of employment*, harmonisation.
- *Pay systems*; the need for skill related pay.
- *Relationships within and between trade unions.*
- *Patterns of working time.*
- *Quality of working life*, job satisfaction.

The association of scientific management with technological determinism was earlier noted. Equally recognised is its close association with control forms of management style. New technology however provides both the potential and the opportunity for embracing individual worker discretion and responsibility. It will not work effectively unless it is *linked to a culture evolving from a style of management which is responsive and adaptive*, which sees people as agents, capable of handling a broad range of responsibilities rather than objects, or extensions to the technology. A choice therefore exists between management styles based on imposing control and those based on obtaining commitment. The control style of management is characterised by, for example, autocracy, a hierarchy of specialised roles, the top down allocation of authority, the general wish to establish order and exercise control over the workforce. The commitment style of management, reflecting current and changing expectations about work among employees, is characterised by the creation of jobs involving greater flexibility and more responsibility, relatively flat management hierarchies, removal of differences in status and the provision of what have been called *"stretch objectives" emphasising the principle of continuous improvement both in terms of the individual and the organisation.* Such commitment philosophies are often enshrined in the publication of organisation *"mission statements"* which set out the key values and principles which an organisation wishes to maintain and through which greater employee trust and commitment might be achieved.

6. The Change Process

Much evidence indicates that in the process of introducing technical change human factors are either ignored or given low priority. Yet, the nature of the process used, particularly where work organisation change is involved, can be critical in determining commitment to the change and whether the change is seen as improving or reducing the quality of work. *The process of change thus becomes equally as important as the nature of the change itself.* Failure to adopt the right approach is likely to mean that the new equipment, whatever the form of the work organisation, will not be operated to best effect.

There are a *number of options available for the process of introducing the technical change.* It can be introduced by management fiat using specialised staff with minimum participation by employees trade unions or operating managers. This is likely to produce an immediate conflict situation, some of the longer term results of which will be implementation later than planned, or operation at below planned efficiency because the implementation is negatively modified by employee reaction. Another approach is to develop an outline plan at management level but to start consultation with employees and their representatives before finally deciding and announcing the plan. Implementation can then be in consultation with employees, their representatives and operating managers. This approach is likely to improve implementation, and whilst there may be some initial resistance will likely give longer term co-operation. The most effective approach is *to develop and implement the plan through joint problem solving* involving engineers, systems design experts, operating managers, employees and their representatives. This approach may well require the *development of new participative mechanisms* (e.g problem solving groups, task forces) and likely shifts in the attitudes of both management and trade unions.

7. The Trade Union Perspective

By and large *trade unions* and trades union officials *have accepted the need to change work organisation in line with technological change*. This has been shown to be more so than with straight organisational change - that is change involving substantial changes in working practice or work organisation but not involving new plant machinery or equipment. The acceptance has generally been seen in the broader context of necessary investment to secure future employment opportunities. Nevertheless it is important to recognise that despite this positive reaction *trade union concerns will still remain*. Three examples can be given.

- First, there will likely be *a fear* that technical and workorganisation innovation will be used by management to *circumvent or reduce general trade union influence*.
- Second, concerns may well arise within and between individual trade unions as work organisation, skills and skills boundaries change and *each trade union seeks to retain its distinct identity, status and influence*.
- Third, there is likely to be a belief that *change will lead to increased control by management* and, for example, adversely affect payment systems, job satisfaction and patterns of working hours.

Concerns such as these will *need to be recognised by management* if the changes to technology and work organisation are to be successfully introduced and sustained in the longer term. If not addressed conflict is a likely outcome, if not also a general deterioration in the industrial relations climate, individual morale and motivation and ultimately overall company performance. A key factor in creating confidence and reducing suspicions will be the opportunity given to all those affected by the changes, including trade unions, to identify and explore at an early stage the organisational and human relations issues. A further factor will be the *necessary understanding* among managers involved in the change, particularly specialist engineers and systems designers, *that work organisation is not solely determined by the form of the technology;* that it depends equally not only on the relative bargaining strengths of managers and trade unions but also on their intentions. That is in the final count the succesful implementation of technology and work organisation change still depends heavily on the responses from those to whom it will apply.

Sociocultural Change and Unity of Purpose in Policies Concerning the Organisation of Work

Andreas Drinkuth, Frankfurt/Main, Germany

1. A New Cultural Model?

A new zeitgeist is abroad: "Corporate Identity". Managers are discovering "ethics" and are becoming receptive to values beyond turnover and profit. The times when one could lead a company to growth and future prosperity without regard for its surroundings, the environment and one's fellow citizens seem to be gone for ever (Göring 1989).

Goeudevert, Managing Director of Volkswagen, defines *a company as a "socially productive system".* The consequences of economic growth and technical development have to be appraised. "Today's manager must learn to think holistically" (1989).

Frau Prof. Höhler, the immensely expressive main speaker at company conferences, interprets *the needs of the "new top performers"* as being that working hours should be used as "profitable" time. "Lifetime as a precious possession, daily portions of which one is compelled to sacrifice to the working world, is conceived in terms of its very preciousness. People know that they have a right to time spent meaningfully at work. They suffer failure when the purpose of their work cannot be made clear to them" (Höhler 1989).

The production factor "human being", "fuelled" with "education", is being rediscovered as the engine for the supposed production factor "technological progress". "The maximum achievers in any company are almost always those who feel at home in their jobs and who are able to identify themselves with the company's aims and products" (Ortner 1989).

Zoll, as one of those social scientists with trade union sympathies, speaks of the "hypothesis of a new cultural model", of a new political culture. People are trying to redetermine their lifetimes in relation to the routine of gainful employment. They are looking for an individual identity which is embedded in a communicative world in which they live and in its culture. Communication free from "power" is required for this. However, according to Zoll, they are not only concerned about more individual leisure. An "ethic of new involvement" can be observed: new company-internal as well as external groups, networks or everyday solidarity (Zoll 1989).

The trade unions also sense that value-oriented questions of a meaningful nature are gaining in significance: personal dignity, recognition, justice, autonomy or knowledge and awareness, as well as requirements concerning the content of work, alongside the rather more classical trade-union fields of organisation such as income, performance limits or job security (IG Medien 1990).

2. Systemic Rationalisation as an Open Quest for Solutions

Not only radical cultural changes but also changing market conditions (the shift from a seller's to a buyer's market) alter the basic framework for company as well as trade-union action.

Traditional patterns of rationalisation and implementation strategies are undergoing change. Works councils and workforces are being confronted with measures concerning the company in its entirety and sometimes involving more than one company instead of rationalisation measures referring to individual jobs or to specific areas. Entire companies are being reorganised or fully restructured ("systemic rationalisation"). In the process, cause and effect frequently become separated over time and space.

Work is allocated in a new manner: quantitatively - within businesses and companies or among them; qualitatively - among workshops, production planning and design departments, among commercial departments and external service departments, through decentralisation or relocation or by means of profit and cost centres.

Technical and cultural models are increasingly influencing company decisions: the company as a "socially productive system" or "CIM".

Being determined by such model concepts, investment plans often only assume concrete form during the process of organisation. Thus, social consequences cannot be clearly predicted, nor causes and effects related to one another nor blame apportioned.

Systemic rationalisation strategies are incorporated into concepts of "corporate identity". These include, for example, some increased-flexibility models which are definitely very attractive to certain groups of employees. Forms of participation such as inter--disciplinary quality assurance, educational facilities or team-work will serve to heighten the employees' motivation.

Of course, there are contradictions. The basic pattern of capitalism continues to have an effect:

Company closures and redundancies leave little room for ethical principles. Conservation or the avoidance of working media which are dangerous to health are paid lip service when costs argue against them. Rheinhausen does not know any corporate identity as propagated by Prof. Höhler and others.

Should the employees' participation as initiated by the company remain effectively non-influential as is the case, for example, with some inter-disciplinary quality assurance concepts, then the employees will recognise this very quickly and the desired motivating effect will be lost.

"Faults" in these management concepts which are inherent to the system can be recognised, such as,

- technical, organisational and/or personnel-related difficulties in realising the solution being strived for;
- runaway costs;

- the sale of the business, whereupon the new owner follows a different concept; or
- quite simply, an incompetent management.

Examples of this can be seen in the countless investment ruins during the introduction of centralised data processing or the increasing number of CIM crashes.

The insight is only now gradually asserting itself that communication is a matter between people and not between machines and that therefore CIM is not to be understood as a technical but as an organisational problem (Lay 1989; Hirsch-Kreinsen 1990).

The companies are trying, with the concept of "cooperative project management" (Hildebrandt 1988), to balance out the sometimes contradictory positions of interest and ideas between various departments in a company as well as the planning uncertainties which arise during the process of carrying out investment and rationalisation measures.

"Company managements are turning increasingly to the development of system conceptions in close contact with the factory and introducing them with the participation of representatives from the company sectors concerned (cooperative project management). This frequently occurs in project teams which are, as a rule, also accessible to the works council. The management has created a specific new company-organisational field with these teams (consultative decision-making) in which technology and work organisation, and thus also their social consequences, can be influenced to a considerable extent" (Hildebrandt 1988).

3. Development of Tangible Utopias for Future Working

Despite many inconsistencies and contradictions, trade union policy is making heavy weather of finding adequate answers.

The basic contradiction between capital and work as a means of orientation for trade union activities is indeed understandable. However, it can only be identified by means of complex analysis of actual situations and is then often eclipsed by the localised interests of individual groups and protagonists. The content of a company's organisational measures is determined more by the various activities which generally take place within the company (Birke/Schwarz 1989) than by rationally (and often also economically) comprehensible decision criteria.

When formulating their ideological concepts (We are, in fact, realists when it comes to practical matters!), workforces, representatives, works councils and trade unions assume, on the other hand, that decision-making processes in companies follow economic or technical necessities. "It turns out, however, that this is hardly ever the case when the ideas take concrete form" (Ortmann 1889). More important are interests relating to assets, resources, income, careers, power and influence of colleagues, departments and levels of hierarchy which attempt, often from greatly differing viewpoints to influence decisions and their implementation (Idem).

It is not sufficient to criticise management decisions concerning systemic rationalisation measures in view of their supposed consequences, e.g. for employees, or to perhaps reject them altogether (mostly ineffectively).

The effects of systemic rationalisation concepts are uncertain at the planning stage, and likewise their possible later consequences. Various groups of employees can be very differently affected in the case of measures which exceed demarcations. Some of these groups have very different interests.

When participation really does take place, such as at Bertelsmann, VW, Bosch or Hoesch (Prekuhl 1990, Workshop Hans-Böckler-Stiftung 1990), trade-union policy finds itself in a difficult position because it cannot counter with any substantial concepts of its own except for a strategy of "unmasking".

How can trade-union policy keep pace with changing attitudes in human society? Has it any conclusive answers to altered company strategies? Can it influence these at all, or does it have to restrict itself?

After all, it is management which determines the subject and the timing of investment and rationalisation decisions. The formal opportunities for participation (supervisory board of directors, workplace labour relations law, wage agreements and so on) are limited. Power structures, differentiated areas of interest, even among the employees themselves, and the employees' own as well as the company' misjudgements all serve to make the workforce's own political actions more difficult.

"How people live, their nationality and religion and how they work: these are not mere footnotes in the revolutionary process. It is these factors which determine the credibility and feasibility of utopias within industrial society." (Negt 1989).

Trade union policy gains the power to clearly define its aims by offensively taking up people's wishes for different ways of living and working, by formulating its own ideas and models for future working and by following these up in the process of determination of objectives and organisation (Ortmann 1989).

This is only possible in an active process of participation with those involved. The subject matter of the models needs to be developed together with those involved and turned into tangible experience. They will then understand the management's models as well as the managements's often contradictory policies. They will be able to properly appreciate the substance of the management's models and policies against the background of their own interests and moral concepts. The often identical terms (e.g. greater flexibility, team-work, participation) can be analysed and imbued with opposing ideas.

Workers' own utopias are thus the touchstone for company concepts; they are the lever with which pressure for authentication of content can be generated, with which the initially existing formal powerlessness can be overcome. The power of trade unions to oppose would unfold through participation and through competence for work organisation in the company. The conflicts over company models, workers' concepts of "corporate identity" as well as of systemic rationalisation strategies could be given political meaning through the workers' own visions of future working life. This would represent a shift of politics back to where it takes place, to where collective and individual perspectives of life are still decided for a great many people.

This concept of unity of purpose in policies concerning the organisation of work requires courage on the part of the works councils as well as of the trade union officials because they have to become involved in an organisational process where the management defines the goals for the time being, where the results can only be predicted with difficulty and success is uncertain. The ability to syntheses competing employees' interests into a central theme and to balance these out jointly is required. This can only be successful if trade-union policy is organised with those concerned and not simply for them.

However, the future can only be gained through policies which possess a democratically developed and justified content. "To go backwards into the future is impossible" (Glotz 1989).

4. Unity of Purpose in Policies Concerning Organisation of Work. Its Implementation within Trade-Union Company Policy

If a trade-union company policy understands itself to be a "protectionist policy on legal foundations" (Birke/Schwarz 1989), in which it attempts, for example, to protect employees against negative social consequences of rationalisation measures by means of company agreements, then it will achieve nothing unless it is dynamically linked with an organisational policy of substantial content which accompanies the process.

If a trade-union collective bargaining policy restricts itself to minimum conditions which can be arranged independently of the company, such as levels of remuneration and differentiation thereof, working hours or procedural regulations with declarations of intent concerning humane organisation of work, then it will hardly be able to exercise an influence on differentiated systemic rationalisation processes.

Trade-union policy therefore has to bring an influence to bear on company decisions concerning general principles, on works planning and introductory processes. It must attempt to actively follow up people's changing attitudes in their relationship towards changing values on the company's part. This requires structural efforts - see IG Metall in this case. Such a fundamental requirement as regards policies and content does not fulfil itself in a self-perpetuating manner.

The following salient points mark the implementation of unity of purpose in policies concerning the organisation of work:

a) Unity of purpose in policies concerning the organisation of work must be closely aligned with the company itself, i.e. where company decisions are experienced directly by people. It will, however, be orientated both towards the company in its entirety and towards areas involving more than one company. It has to open up a viewpoint which "moves from a job-related or sector-related mode of thinking to a company-holistic, systemic mode of thinking, to a mode of thinking in long-term and interlinked processes instead of in static and directly cause-and-effect related contexts" (Hildebrandt 1989).

Such unity of purpose extends trade-union fields of action beyond the traditional organisational fields such as employment security, grading of employees or maintenance of industrial health and safety and is orientated towards the organisation of products, technical and organisational work processes, qualification of employees, use of staff and so on.

Understood in this way, unity of purpose in policies concerning the organisation of work is not one of the many new subject areas with which trade-union company policy would be additionally burdened and overtaxed. In this holistic or systemic view it would serve to integrate the "classical" subject areas.

b) Unity of purpose in policies concerning the organisation of work has, above all, a chance of success when it concerns itself with situations of radical change in the factory (e.g. restructuring of production into production islands, lowering of the depth of production, setting-up of cost centres). However, it must not restrict itself to criticism of such measures. The policy must develop its own ideas concerning the company's concepts. Such ideas have to contrast the contradictions of the employers' policies with people's viewpoints on life and work. Only then can the power to define one's aims and thus the power to oppose be gained.

c) Unity of purpose in policies concerning the organisation of work is based on the representation of their own interests by the employees themselves. Without the specialist and everyday knowledge of the employees, without them being able to articulate and introduce their own individual interests; in other words, without a democratisation of trade-union company policy, neither the power to clearly define aims nor the power to oppose in the factory can be developed. It concerns the shift of politics back to where political processes really take place and can consequently be influenced. "The trade unions have to start working at the grass roots, with the employees themselves, if they want to retain their ability to mobilise" (IG Medien 1990). Trade- union policy at the various levels can then confine itself to its respective essential tasks. The scope for self-regulation of conflicts could be progressively extended in downward direction.

d) Unity of purpose in policies concerning the organisation of work must be able to recognise, summarise and impart the different positions, interests, assets and intentions of the various employee groups and allow them to become capable of consensus, since systemic rationalisation measures have an all-embracing effect on jobs and sectors and thereby affect the various employee groups in completely different ways.

e) Unity of purpose in policies concerning the organisation of work requires harmonised, dynamically linked and supportive structures:

- It must facilitate and organise participation by employees and allow it to become a tangible experience. Participation is a cornerstone of unity of purpose in policies concerning the organisation of work. Without it, organisational alternatives of meaningful content can neither be developed nor accomplished.
- Works council policy has to set the points of emphasis. Unity of purpose in policies concerning the organisation of work can only become feasible if a strategic project is recognised as "strategic" and pursued as such.
- Local trade-union work has to organise the inter-company and regional exchanges of experience. It has to attempt to organise stable networks with subject or area--related points of emphasis, over defined periods of time in each case.
- Also necessary are decentralised advisory infrastructures, e.g. in the form of trade-union technology advice centres or advisory facilities with close trade-union contacts . These can feed additional experience into the organisational process, answer specific questions or provide experts, serve as clearing centres and set qualifications in working methods. However, it would be incorrect to expect them to deputies for the factory or trade-union protagonists.
- Trade-union collective bargaining policy in the Nineties has to support an organisation-of-work policy which is closely connected with the factory. It must open up possibilities for individual solutions for different factories (as it has opened up, for example, the possibility of regular "leisure units" in the case of reduction in working hours) without neglecting its protective functions in the process.

f) Unity of purpose in policies concerning the organisation of work must be supplemented by and dynamically linked with a governmental research and technology policy which is able to supply preventive measures, operates on a democratic footing and which would have as its core the procedural extension of participation (Fricke 1989) as well as the "constructive estimation of the consequences of technology" (Idem /Naschold 1987).

g) Unity of purpose in policies concerning the organisation of work is a trade-union concept of the power to oppose, which counters the organisational power of capital by means of organisational alternatives which possess substantial content, which have been democratically developed and which have been negotiated on a basis of solidarity. It thereby departs from the previous self-restrictiveness of its protective policies and its mostly ineffective "unmasking" approach. It is thereby forcing its way into the taboo areas of <u>how</u> production processes shall function and <u>what</u> form they shall take.

References

Birke, M. Schwarz, M., (1989), *Neue Techniken - neue Arbeitspolitik?* (Neuansätze betrieblicher Interessenvertretung bei der Gestaltung von Arbeit und Technik), Frankfurt/New York.

Fricke, W., (1989), *Forschungen zu Arbeit und Technik*, ed. Friedrich-Ebert-Stiftung, Bonn.

Goeudevert, (1989), quoted according to Göring.

Göring, R.W., (1989), *Manager entdecken die Ethik - Zukunft eines Unternehmens hängt von der Einstellung seiner Mitarbeiter ab*, in VDI-Nachrichten No. 36 dated September 8th.

Glotz, P., (1989), *Modernisierung als Schlüsselbegriff der Linken*, in: Krämer H.L., Legewie, C. (ed.): Wege ins Reich der Freiheit (Andre Gorz on the occasion of his 65th birthday), Berlin.

Hans-Böckler-Stiftung, (1990), *Workshop: Ansätze betrieblicher Arbeitsgestaltung und Qualifikationspolitik bei unternehmensbestimmten Beteiligunsaktivitäten*, June 20th to 22nd 1990 in Büdingen.

Hildebrandt, E., (1988), *Normsetzung und Kooperative Gestaltung - für eine tarifvertraglich flankierte Betriebspolitik*, in: Forschungsprojekt Rahmentarifpolitik im Strukturwandel, Volume 2 - Gewerkschaftliche Tarif- und Betriebspolitik im Strukturwandel. (Seminar papers and discussions at a conference held on September 27th and 28th 1988).

Hirsch-Kreinsen, H., (1990), *Organisatorisch und personelle Probleme bei CIM*, in VDI--Zeitung No. 5.

Höhler, G., (1989), *Von der Askese zur Entfaltung*, Fortieth Anniversary Lecture at the Annual Conference of the Frauenhofergesellschaft.

IG Medien Hessen,(1990), 1st forum on April 20th 1990: *Neue Wege gewerkschaftlicher Betriebspolitik.*

IG Metall, (1989), Resolution 9: *Technologie- und Humanisierungspolitik*, 16th Ord. Trade Union Conference.

Lay, G., (1990), *Die Verbreitung von CIM in der Bundesrepublik Deutschland*, Fraunhofer--Institut für Systemtechnik und Innovationsforschung, Karlsruhe).

Naschold, F., (1987), *Technologieabschätzung und -bewertung - Entwicklungen, Kontroversen, Perspektiven*, in: Fricke, W. et al (ed.), Jahrbuch Arbeit und Technik in Nordrhein-Westfalen, Bonn.

Negt, O., (1989), *Aus produktiver Phantasie - ein Revolutionärer Realist unserer Zeit*, in: Krämer H. L., Legewie, C. (ed.): Wege ins Reich der Freiheit (Andre Gorz on the occasion of his 65th birthday), Berlin, page 63.

Ortmann, G., (1989), *Management und Betriebsrat: Mikropolitik bei der Einführung von EDV-Systemen*, in: Umkämpftes Terrain, Opladen.

Ortner, G.E., (1989), *Unternehmenskultur - Ein Begriff setzt sich fest*, in: Lernfeld Betrieb

5/89.

Pekruhl, U., (1989), Contribution to the 1st forum of the IG Metall Hessen, *Neue Wege gewerkschaftlicher Betriebspolitik*, Springen April 20th 1990; cf also Bundesmann--Janssen, J., Fox, K., Pekruhl, U., 1989: Aktive auf Zeit - neue Beweglichkeit in der gewerkschaftlichen Betriebsarbeit, in: Mitbestimmung No. 4/89.

Zoll, R., (1989), *Neuer Individualismus und Alltagssolidarität - Bemerkung zur Krise der individuellen Identität*, in: Krämer, H. L., Leggewie, C. (ed.): Wege ins Reich der Freiheit - (Andre Gorz on the occasion of his 65th birthday), Berlin.

Integrated Work and Technology Concept in Craft Trade - Initial Regional Experience During the Establishment of a Bremen State Program "Work and Technology"

Ludger Deitmer, Bremen, Germany

1. Aim of the Program

The Bremen State Program of Work and Technology is intended regionally *to promote a social innovation strategy* that enhances economic profitability for the companies concerned while at the same time taking into account the social needs of the latter's employees.

In contrast to a primarily technology-centered strategy of rationalization, this "*social rationalization*" aims at finding new ways of harmonizing technical and organizational options with social needs and at implementing them exemplarily so as to shape technology in a socially acceptable manner.

If creative freedom is to be utilized, *the introduction of new technologies requires cooperation* between company management, employees as well as researchers and consultants; only in this way can organizational problems be solved. Apart from reorganization, employees' motivation plays a key-role in the successful implementation of innovation. *Workers' demands with respect to job content* are increasing, especially where new technologies are applied - these demands may clash with profitability considerations of company management.

The social innovation strategy pursued by the program aims at a combination of social and technical innovations (Programmheft 1990). In order to achieve this aim, the program applies the *integrated work & technology approach* (Sachverständigenkommission Arbeit und Technik 1988), which is being implemented in *regional project networks* comprised of small and medium-sized enterprises and local R&D institutions. The idea is to support this approach by networking the regional innovation potential on both a horizontal (universities, companies and vocational training institutions) and a vertical (organization development, qualification development & technology shaping) level.

The social innovation matrix (fig. 1) shows the complexity the project networks have to deal with. This holistic concept, which is operationally described in provisional guide-lines (Programmheft 1990), is proposed as a supplement to known technology transfer concepts.

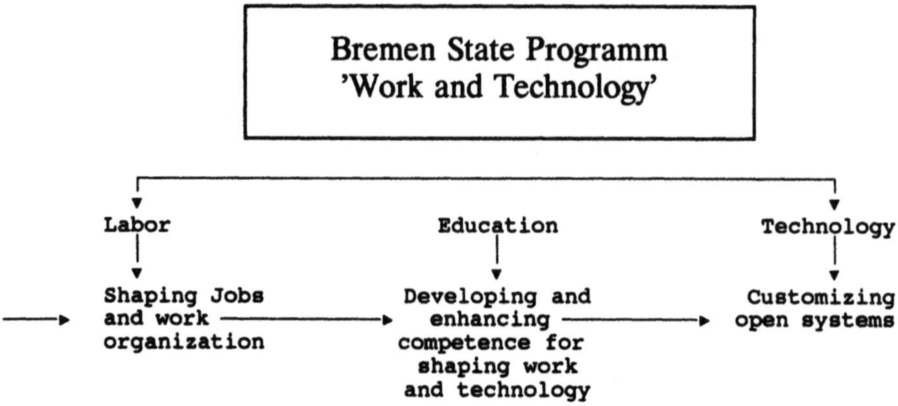

Enterprises	innovation circle - analyzing work - decision-making model - Human Resource Managment	company vocational training - analyzing competence - running courses	rolling development (Corbett et al 1990) - specification - evaluation
Institutions of further training and education	organization development and councelling simulation of work	qualification development and counselling - course developm. - course managem.	laboratory traini ing of work and technology in factory operations
Science	models & theories: industrial science management science work science	models & theories: research on vocational training	technology training and development: - software technology - production technology

Figure 1: Social Innovation Matrix: Linking the regional innovation factores

The program seeks to avoid the hitherto prevailing failure of R&D results being transferred to everyday practice. Previous R&D programs required extra transfer activities for the diffusion of their R&D results. In the long term, our project networks are intended to kick off research and development activities which would be continued even after financial support by the program has ceased. The program is intended as *a catalyst in developing Bremen's regional potential.*

2. Regional Situation of the Craft Trades

In its initial phase, the program concentrate, on setting up *two regional pilot project networks* in the plumbing and electrician trades.
Export-oriented enterprises are confronted with a shift in the terms of trade from an economy of scale towards an economy of scope:

- Rising consumer requirements
- Shortening of product life cycles
- Demographic change
- Less pronounced social differences
- Changes in qualification
- Changes in work culture

For smaller enterprises, here the regional craft trades, there are indirect effects. As suppliers servicing larger companies as well as receiving and buying specific products (e.g. new supply technology; intelligent building control systems), they have to cope with the following shortcomings:

2.1 Contradiction Between Integrated Technologies and Craft Trade Work Organisation

The specific character of the new technologies (modular design; integration of control and execution; network characteristics which combine centralized and decentralized activities) leads to an integration of old and new technologies as instanced by building management systems (BMS) and building control systems (BCS). Heating, air conditioning and light systems are combined with electronic control technologies. The classical work boundaries between the craft trades concerned - here plumber and electrician - impede the trend towards technical integration. These trades do not dispose of know-how which transcends trade boundaries and could thus form the basis for cooperative partnerships. The plumbing/heating trade lacks competence in electronic building control systems. The electricians know too little about the heating, lighting and air-conditioning systems they support with control systems (Pangalos 1989).

2.3 Loss of Planning Competence

Craft trade studies (Ehinger 1990, Nagel 1989) point to the changes in planning and design work in the craft trades. They are losing this traditional planning competence to the service departments of the national vendors. The "American production model" that has been adopted in Europe by some companies in the field rules out the cooperation of trades in the installation and running of building control systems.

2.4 Management Deficiencies in the Craft Trades

The management style in these small enterprises is closely interwoven with the personnel situation. Typically, the master (Meister) plans the work, canvasses customers, makes offers, orders material etc.. He spends most of the day performing operative management tasks such as price quotation, order acquisition, sales, etc.. There is increased market pressure on companies to find new customers and to offer new kinds of services. Knowledge about new products is necessary. Craft trade management has much less time for strategic or human-resource management (Staehle 1989) than medium-sized companies. Decision making is very weak because of trades' limited innovation capacity (Zündorf 1989).

2.5 Lack of Craft-specific Processing and Planning Tools

Craft trade enterprises have specific technological requirements. Available computer software is not well adapted to their specific needs. CAD systems, for example, have been geared towards bigger mechanical engineering companies with their specific organization structures and personnel competence. Many industrial system developers and vendors like to offer these systems to the craft trades as an easy way of making profit. The users, here the craft trades, encounter structural problems, e.g. utilization of capacity, financial commitments, organisational and personnel problems (lack of competence), problems in their companies (lower flexibility) when they try out these systems and attempt to implement these systems (Ehinger 1990).

2.6 Lack of Cooperative Partnership

The craft trades do not make use of partnerships to compensate for their strategic shortcomings in innovation, to enhance the quality of work, competence, etc.

The development of regional networks comprising

- craft trades
- research & development institutions
- consultants
- vendors
- further education and training institutions

is intended to strengthen the regional innovation process.

The northern region of the FRG lacks networks which can successfully tackle the specific shortcomings in work organisation, technology and personnel competence. Most of the companies require help from outside, such as is offered by consultancies or research institutes (Zündorf 1989).

2.7 Employment in the Regional Craft Trades

Economic promotion policy (Wirtschaftsförderungspolitik) overwhelmingly tends to concentrate its support on export-oriented firms. This strategy has hardly any effects on employment. First of all, the export sector is unstable because large-scale business cycles involve higher risks. International conflicts (wars; oil crisis) can affect the large-scale economic cycles and have regional repercussions (June 1985). Secondly, only a minority of the regional employees works in the export sector. Most of the people work in rather more national or regional sectors: the service sector, craft trades, public services, and the professions, such as lawyers, doctors, solicitors, counsellors etc. Figures for Bremen show that the craft trades represent a significant proportion in the overall employment structure of this region. The examples of both Bremen and Hamburg (Caspari, Dörhage, Mattfeldt 1989) indicate that it is wrong to believe that more jobs will be created merely by expansive industrial and export trade companies.

Nevertheless, craft trades play is an important rule in the development of bigger export-related firms. Social innovation through specific project networks should make work in the craft trades more attractive. Even in the North the craft trade enterprises have problems in recruiting personnel. Particularly young people increasingly appear to prefer other jobs which have a better reputation.

3. The Integrated Work and Technology Approach in the Craft Trades

In short, *the craft trades need outside support to compensate for their specific innovation deficit*. The assessment of the classical transfer approach, with its unidirectional research transfer (from research to practice) shows that there is a need for more bi-directional or dialogue oriented concepts (from research to practice and vice versa) to promote a social innovation strategy. Bi-directional means that research should also be oriented towards the practical problems encountered by small enterprises. Through cooperation among experts from R&D, craft trades, vendors and vocational training institutions in a project network, an integrated work and technology approach can be achieved. The network members have to form "value-added" partnerships in the shape of project networks, where the division of both tasks and responsibilities takes place in accordance with the problems to be tackled. Networks of this kind require initiation and moderation by according to specific criteria a project sponsoring institution (J. Ranka 1990). This is an important precondition for fruitful cooperation between the partners as well as for the long-term growth of knowledge required by an integrated work & technology approach. The experience gained by other European programs shows that there is a great need for network strategies and interdisciplinary cooperation. (F.d.Hertog 1989)

3.1 Network Criteria

During the establishment of this new program, *several project networks have been set up* by the project sponsoring institution. Initiation and preparation have been guided by the following network criteria:

Size

Research networks are often far too big. As a result, there is a broad understanding of the problems among all the partners, but the output is quite limited. Hampered by its size, the network is not able to achieve good results.(Dirk Schröder 1990)
To keep networks manageable, *partners should not come from more than five different institutions*. The network should be coordinated by one institution, otherwise the dialogue process cannot be steered towards the preparation of a proposal.

Interdisciplinary Cooperation of Project Partners

Co-operation between researchers and practitioners from industry facilitates their coping with the complex effects of technology application in companies and administrative authorities. Through work organization, technical and qualification-related measures, technical and economic as well as social risks can be better assessed. Experts from the technical and social sciences are involved according to the type of problem concerned. R&D is taking an active role, i.e. it is fully integrated in project responsibility; it departs from the traditional role of supporting practice. R&D and practice have equal status, viewing themselves as project partners. A prime contractor, who additionally takes on project management and coordinating, is selected from among the partners.

User-manufacturer Cooperation

Industrial users (e.g. craft trade enterprises), producers (e.g. software companies, manufacturers of machines and tools) and consultancies are integrated in the network projects so as to ensure that the new products or services developed take into account the viewpoint of both users and producers. The project results are thus not only supposed to benefit "one" user, other interested parties and customers should be able to profit from them, too.

Long-term Outlook

The projects are planned so that continuation of the development activities is ensured even after financing by the program has ceased. All projects foster new cooperation between R&D and practice in the Bremen region on a long-term basis.

Education and Qualification

In all projects education and qualification are regarded as central elements. Organizational and technical changes in a company require without a coordinated concept of prospective vocational training and continuing education aimed at specific company needs as well as the general requirements of industry.

Involvement Approach

The "involvement approach" aims at involving the direct industrial users (e.g. master craftspeople, skilled workers), as well as company management and works or staff councils, in solving problems (product, process or tool level). A new, shaping-oriented field of activity is opened up to management and personnel. Work science provides tried and tested development and participation concepts (see Corbett, Rasmussen, Rauner 1990).

3.2 Work and Technology Approach in the Regional Craft Trades - Initial Experience with Establishing Project Networks in a State Program.

Project A: New technologies in the electrical trade - development, implementation and operation of a building control system laboratory.

By using this laboratory, small and medium-sized craft trade enterprises are to be introduced to building control technology and their personnel is to receive training. In addition to this advisory function, initial training in this field is to be improved. The network project aims primarily to strengthen the regional craft trades as a link between users and various manufacturers of building control systems. The university, a local vocational training school, the district craft chamber, various trades and institutions of continuing education cooperate in the project (Verbundvorhaben No. 2 1989).

As the coordinators in this project, we have learned that setting up project networks is an important prerequisite for developing long-term cooperation between partners who normally do not cooperate. Nevertheless, this process involves learning for all participants; therefore, the setting up of project networks has to be organized accordingly. But holding several workshops is very time-consuming. The whole process took nearly 1 1/2 years from initial meetings up to the final discussion of the project plan. We constantly had to adjust the desired project profile. It is important, therefore, to maintain intensive coordinaton of the network activities through regular meetings in which the partners have to put forward their specific views on the further development of the project.

Project B: CAD in the plumbing and installation trade - development, qualification and organizational development.

Small and medium-sized enterprises in these trades require computer-aided planning instruments so as to be able to plan their work better. The few CAD systems on the market are not tailored to the needs of these enterprises. With the development of a CAD system oriented to the craft trades, the aim is to relieve craftspeople from routine work and support them in their planning and consulting tasks. The involvement of various software firms and of the university will support the craft trades in carrying out the project tasks. (Verbundvorhaben No. 3 1990)

Developing this project network required analyzing the planning and drawing work in the plumbing and installation trade. Only interviews with company owners and personnel yielded information on the problems of the enterprises. In addition, the software offered by the production companies/software houses had to be checked for relevance to practical work in this trade. Helpful advice was given by up-to-date studies (Ehinger 1990; Nagel 1989) in which craft labour has been analyzed especially in respect to new technologies.

Bringing user companies and software producers together was an important task for the further development of the project network. The state of the art of CAD systems for the trades could be highlighted. The early involvement of the Craft Chamber was helpful because the project results can now be distributed much more effectively.

3.3 External Management of Project Networks - Initiation, Implementation and Evaluation

The following steps in setting up a project network have proved useful. The experience is based on the development of several project networks by the project sponsoring institution for the work and technology program:

- *User companies state their problems*. In many cases, these contacts are provided by consultancies. Close contact to these institutions on the part of the program-sponsoring institution has been helpful. *Result*: initial identification of problems in a company sector
- The *project-sponsoring institution* contacts regional user companies and specifies their qualificational, organizational and technological structure; the institution informs about the innovation shortcomings of the user companies and gives advice on the possible R&D support which may be furnished. *Result*: Formulating the task of the project network project.
- *Contacting other users and vendors* in the project field; selection of appropriate partners for a project in close cooperation with the initiating user company. *Result*: Drawing up a list of partners for the project network.
- Organizing and preparing a *project workshop* and defining the project task to be undertaken by the several partners. Definition of the specific outline of the project network on the following levels: technology development, optimization and transfer; qualification, organisation. At this stage, it has proved useful to involve experts in the network.
- *Definition of role of the specific tasks the several partners* have to undertake to achieve the project aim. Discussions in workshops are very useful and productive in stimulating the forming of the project network and in specifying the detailed work plan. *Result*: Work plan for the proposed project
- It is useful to select a network coordinator for the project network. She can later act as prime contractor.

Finally, I would like to caution against any belief that there might exist standardized rules of the "how to set up a work & technology project" type for company reorganization. Conditions obviously vary with branches and even companies, as has been shown abundantly by German industrial sociology. Only after the program has been in operation for some time will we be able to identify key-factors for success. Initial experiences, however, demonstrate that the program has been used by some to legitimize changes in their companies. The project sponsoring institution 'work & technology' expects to continue supporting democratization in companies along these lines.

References

artec, (1990), Forschungszentrum Arbeit und Technik -Forschungsprogramm -*Gestaltung von Informationstechnik und Arbeit*, Universität Bremen.

Caspari, A., Dörhage, W. und Mattfeldt, H., (1990), *Überblick über die Beschäftigungslage in Hamburg*, Hamburg.

Deitmer, L., Grützmann, A., Oehlke, P., und Rauner, F., (1991), *Das Bremer Programm "Arbeit und Technik"*, in: Jahrbuch Arbeit und Technik, Bonn.

Ehinger, Gerd, u.a., (1990), *Felderkundung zur menschengerechten Anwendung von CAD/-CAP-Systemen in Installationshandwerken*; System Consult, Berlin.

Hertog, J.F. den, Schröder, P., (1990), *Social research for technological change: Lessons from national programmes in Europe and North America*, Maastricht.

Junne, G., (1985), Reregionalisierung: *Chancen regionaler Reintegration von Produktion und Konsum als Folge der Entwicklung neuer Technologien*, Bonn.

Landesprogramm "Arbeit und Technik", (1989), *Programm- und Durchführungs- konzept der Geschäftsstelle zur Einrichtung des Bremer Landesprogramms "Arbeit und Technik"*, Bremen.

Nemitz, K., (1989), *Neue Chancen für den Norden? Aktuelle Fragen der Wirtschaftsentwicklung*, in: Dieter Leuthold (Hrsg.), Deutsche Wirtschaftsentwicklung im regionalen Vergleich, Bremen.

Sachverständigenkommission Arbeit und Technik, (1986), *Forschungsperspektiven zum Problemfeld Arbeit und Technik*, Bonn.

Sachverständigenkommission Arbeit und Technik, (1988), *Arbeit und Technik - Ein Forschungs- und Entwicklungsprogramm*, Bonn.

Schröder, D., (1990), *Die Optimierung des Innovationsprozesses in der Freien Hansestadt Bremen*, in: Bremer Zeitschrift für Wirtschaftspolitik (S.57-87), Bremen.

Senator für Arbeit, (1990); *Antragsrichtlinien zum Landesprogramm Arbeit und Technik*, Bremen.

Senator für Arbeit, (1990), Programmheft zum Landesprogramm Arbeit und Technik: *Arbeit und Technik-Forschung in Bremen*, Bremen.

Universität Bremen und Sachverständigenkommission "Arbeit und Technik", (1987), *Arbeit und Technik*-Tagungsband zum 2. Bremer Symposium "Arbeit und Technik" vom 10.-13.6.1987.

Universität Bremen, (1983), *Analyse von Entwicklungen der Technik und Chancen in der Gestaltung von Arbeit*; Tagungsband zum 1. Bremer Symposium "Arbeit und Technik" vom 21.-23.9.1983.

Verbundvorhaben No. 2 im Landesprogramm Arbeit und Technik, (1989), *Neue Technologien im Elektrohandwerk - Entwicklung und Implementation eines Gebäudeleittechnik-labors*, Bremen.

Verbundvorhaben No. 3 im Landesprogramm Arbeit und Technik, (1990), *CAD im Sanitär- und Installationshandwerk-Entwicklung, Qualifizierung und Organisationsentwicklung*, Bremen.

Zündorf, L., (1989), *Die Nutzung außerbetrieblicher Problemlösungenspotentiale als Managementaufgabe*, in: 1. Regionale Technologie-Konferenz Bremen-Niedersachsen BremTec, Bremen.

D

Strategies to Transfer and to Disseminate Knowledge and Know How in the Field of Work and Technology

Better Knowledge Transfer to Support Changes in Working Life

Carl A. Asklöf, Stockholm, Sweden

1. A Difficult Task

This paper deals with *strategies for the transfer of knowledge from R&D to the work place*, directly or through such agents outside the work place as can influence conditions within it.

These considerations are based on the *activities of the Swedish Work Environment Fund*. The Fund supports R&D, training and information and other measures intended to improve working conditions. Through its board, the Fund is governed by the partners on the labour market.

In no research field is there *a straightforward method of putting findings into practice*. The chain from research to development to application is a weak one. Some even argue that there is no such simple progression.

The application of the results of research is difficult, even if there is close contact between the researcher and the site of application. What chance, then, do results from completely different parts of the world have of being used?

What can be done to overcome these difficulties? Before we turn to this question, let us examine how the future may look.

2. Important Trends

Knowledge in the form of R&D into working life is *generated in great variety at an ever faster rate* in an increasing number of countries. The development of the information media means that this knowledge can become accessible very rapidly in every corner of the world. *Networks and other contacts* between people are becoming more frequent, both locally and nationally. This workshop is a good illustration of the way in which this is also true at an international level.

In modern working life there is an almost *continuous need for change* - to adjust to external and internal forces. Full integration of technology at the workplace, the working environment and work organisation at the workplace are needed to stay competitive in the long run. To master such complex processes of change, *all employees at the work place need new knowledge*. In cases where this knowledge comes from outside it *must be adapted to local conditions*. However, external solutions cannot be used indiscriminately. The winners will be those who are best and fastest at moulding together their own experience with the knowledge of others.

In short, a lot of knowledge is produced and a lot is needed. But does the supply fit the demand?

3. Structures for Knowledge Transfer

Swedish working life is highly organised. With the exception of small companies, most work places have links with trade unions and employers' organisations. Within this organisational system there is both a sectoral and a regional structure. Each partner on the labour market has its own structure with various information channels, and there are also joint organisations. Many of these are permanent, others are created for specific tasks.

This system may be intricate and sometimes slow, but it is still an *important means of communication between R&D* into working life *and the work place* itself. This network is used by the Work Environment Fund in many ways.

Recent programmes undertaken by the Fund, some of which are being dealt with separately in other sessions of this workshop, have involved the following *joint organisations*:

- permanent sectoral committees for development questions;
- permanent sectoral committees for work environment questions;
- temporary regional organisations covering the whole of Sweden;
- temporary regional organisations in three particular regions in which programme activities are concentrated.

The formal organisational *structure* of programmes is often either *regional or sectoral*. In practice, however, the division between these two forms is not so clear. The *regional perspective* is tending *to grow in importance*.

In addition to support from the partners on the labour market, *management and labour at each work place* must, of course, also take joint responsibility for development work at their own work place.

Researchers and consultants participate in secretariat functions, and above all *in projects* at the work place. They are also engaged in *programme evaluation*.

By involving researchers and the partners in the labour market in this way, the *conditions are created for the transfer of knowledge* between researchers, practitioners at the work place and the regional or sectoral representatives of the partners in the labour market. Programmes are also encouraged *to develop international contacts*. Contacts and networks created in programmes will ideally continue to be used after the programmes themselves are concluded.

4. Small Companies

It is generally agreed that knowledge transfer has not been discussed from quite a different viewpoint for small companies and for large organisations.

The Fund has run *two programmes* dealing with the theme of *small companies*. In these programmes, many well-known truths were found to be applicable. In small companies, written material is thrown straight into the waste-paper basket. Owners of small companies prefer to communicate directly with people they know and trust.

It is advisable to *use the ordinary information channels* of these companies as well as their *local networks*, as much as possible.

You must integrate the knowledge that you wish to transfer to small companies into their own reality, in which finances and production are important topics. You must look at the development of the company as a whole.

The resources available in Sweden to help small companies tackle their work environment problems consist of two to three thousand people in the *occupational health service system*, over one thousand *regional safety delegates*, and some four hundred *labour inspectors*. A governmental investigation has recently proposed that the number of labour inspectors be increased, and that regional knowledge centres be created. Small companies may receive advice on improvement of the working environment from three specialized research institutes. This service is subsidized by the Fund.

5. Other Channels of Knowledge Transfer

Suppliers of products, systems and services influence working life to a high degree. They set restrictions that sometimes cannot be eliminated from the work place, even with the best of intentions. This applies especially to small companies, which have few resources of their own to adapt purchased products and systems to their own conditions. Some efforts have been made by the Fund *to influence suppliers directly* through supporting development work. Within the framework of German-Swedish cooperation, there is also a project within which Swedish and German researchers work together with a German manufacturer of woodworking machines.

The *results have been rather meagre*. Another way, albeit a slow one, of influencing manufacturers is by means of the *setting of international standards*. Consequently, it is important to transfer knowledge gained via R&D to the European and international standard-setting organisations.

A complementary strategy, especially applicable to suppliers of systems and services, for whom standards do not in theory play a major role, must be to *integrate knowledge of working life into the training and information* that the *suppliers' personnel* receive.

It is important that results from R&D into working life are used in *the education of key people*, such as production engineers and designers. Here, we have a *dual problem*. Firstly *research has to have a more solid basis at universities* and secondly, *research has to influence education programmes*. It is difficult for researchers to maintain good external contacts with working life and at the same time defend their position internally at the university. But it is important that they succeed in this. Only then can they act as the agents of knowledge transfer to their students.

6. In Summary

The sectoral and regional structure of the partners in the labour market has to be used. Temporary organisations should be created when needed.

Networks should be formed regionally and nationally between the partners on the labour market, researchers, consultants and the people at the work places concerned. International contacts have to be stimulated.

For small companies, local networks and normal information channels should be used. The development of the company as a whole has to be taken into consideration.

Reaching suppliers is difficult, but may be achieved through standard-setting and influencing the training and information of key personnel.

Knowledge gained through research into working life has to be incorporated into education programmes within the university system.

The Swedish Commission for the Development of Competence and Skill in Working Life

Christer Marking, Stockholm, Sweden

1. The Commission

In September 1990 the *Swedish government appointed a commission* with the task *to propose means to stimulate the development of competence and skill in working life*. The commission is composed of representatives from the political parties in the Swedish parliament and from the social partners in the labour market, i.e. representatives from the groups that are ultimately the main actors in forming the future policies. Setting up temporary commissions with specific tasks is a frequently used way for the Swedish government to provide the basis for subsequent policy formation. Each ministry of the government can appoint its own commissions in urgent areas. The commission for the development of competence and skill working life is appointed by the Ministry of Labour.

Broadly speaking, the commission should address *two main problems*, on the one hand *the welfare problem in relation to competence and skill* in the labour force and on the other hand *the problem of distribution of learning opportunities* in working life. These problems are considered to be closely interrelated.

2. The Welfare Problem

The welfare problem, from the point of view of the commission, is in general terms *about economizing with scarce human resources*. In Sweden, as in all welfare states today, most people get a good basic education. There are also good opportunities for further education. Taken as a whole, the labour force today is better educated than ever before. That is also supported by the full employment policy. Having a job creates better opportunities for learning to be competent than not having a job.

Even so there is almost *a chronic lack of skilled labour*. That shortage is quite clearly a hindrance to industrial growth and higher productivity. The industry sector in Sweden is considered to be too small, so increased productivity and growth is much desired.

At the same time many people with 11-12 years education are today doing *"one minutes-jobs"*, indicating that a part of the labour force could be termed "overqualified" but is in fact "underemployed". Clearly, this misuse of human resources is improductive in the long run, not least through eventually producing competence barriers to the introduction of technical change, but also in the short run contributing to a high rate of sick leave and labour turnover. Production systems that do not make the best use of available human resources constitute a hindrance to growth.

Which problem is the more serious: *the underemployment of an overqualified workforce or the overemployment of an unqualified workforce?* We do not know whether or to what extent a shortage of skilled people or an underemployment of available skill and competence limits economic growth.

The different sectors in the labour market that are meeting a shortage *might respond in two ways*. The employers can adapt to the shortage of skilled labour by reducing the level of skill demanded in the production process. This would lead to a "vicious circle" with bad jobs and low demand for skill. They could also adapt by educating and training the people already employed. That could in the long run allow for more complex production processes with subsequent higher demand for skilled labour, in short, a "good circle".

There seems to be a widespread awareness that in the future we will be even more dependent than today on the competence and skill in the workforce to support our general welfare system. Industry, health care, public administrations, the private service sector etc. will all have to find out *how* to economize in a better way with human resources. The debate in Europe of today leaves us in no doubt that *the prime factor for future competitiveness is better utilization of human competence.*

One of the *characteristics of the welfare society* is that it creates competence and qualifications in people on a broad basis. One thing seems to be quite clear. No general welfare system can be maintained in the long run if the competence and qualifications of its workforce are misused or underemployed. The guide-lines of the commission express a certain concern about this.

The commission should, given this general background, *analyze the need for competence and skill development in working life*. In doing so, the commission should pay specific attention to *personnel education*, in the light of the general adult education and the labour market education.

3. The Problem of Distribution

The second problem, which could been called the distribution problem, is concerned with another aspect of welfare and working life. Seen from a historic perspective, Swedish people of today are rather equal from economic and social points of view. *In working life* on the other hand, there are *considerable inequalities between people* if one considers freedom at work, health, work content, the opportunities for personal development, education at work and social cooperation etc.

The statistics of personnel education show an expansion during the Eighties but the distribution of that education *shows the same inequalities* between different categories of personnel are maintained. To quote the good book: to those that have shall also been given. A good educational background and a stimulating job is a good start if you want continued education at work. With poor educational background and restricted work content, forget about education at work!

The distribution problem has many dimensions. Even if the work environment in general terms is much better today than in the past, there are indications of an *increase in the number of highly stressful work situations.* In these, women seem to be over-represented. Men and women have different work conditions and opportunities; women are more often stuck in "bad jobs" besides having the full responsibility for the household work at home. *Groups with a weak position* on the labour market such as refugees, immigrants, disabled and other groups with various difficulties often are restricted to the jobs that many other people could chose not to take.

The development of competence and skill in working life is conditioned by many factors, some of which are not manageable either in a work place perspective or in an educational perspective. *Sweden's position in the international economic system*, the investment activities in the society including the companies, the direction of those investments, changes in the relative prices of labour and capital and many other factors affect the composition of the demand for labour and its skill-structure. Thus, in the long run, some of the factors affecting the development of competence and skill have to do with the general economic development and how that could be managed.

The task of the commission is, however, a more narrow one and is directed to *what could be done today*, for instance concerning the ratio between skilled and unskilled jobs that can be affected by measures on a work place level, given that there is always options in work design.

4. The Work Organization as a Bottle-neck

In that view, one important *bottle-neck for the development of competence and skill is the organization of work.* The work organizations determines what you may do and thus what you can do. You can't learn things you are not allowed to do. You can't learn from problems you are never exposed to.

Much is being done about they way in which work is organized. Still, the vast majority of work organizations in Sweden suffer greatly from what could be *the solutions of problems of the past.* The old type of problem might have disappeared. New problems have come to the fore but old solutions still greatly affect the day to day activities. We have to rid ourselves of these deposits from the past. In doing so we have to focus on how problems are formulated and how competence and skills to handle them are allocated in the organization.

The commission should consider *the responsibility of the employers for education and continuous competence and skill formation*. Obviously, that has to do both with the design of work organizations and with education, training and learning activities. There is a general interest that this responsibility should be fulfilled as it is an important part of an overall adult education.

Beside the problem about what we produce in terms of goods and services, the problem concerning how we do it seems to be one of the most urgent questions to be solved during the Nineties. We do have lots of experience of how *to design work for competence development, innovativeness and productivity*. The remarkable thing is that many people seem to know the solutions. Yet they are not implemented. That is a problem - for the society as a whole, for a lot of people at work and thus for the commission for development of competence and skill in working life.

Implementation and Dissemination of the Results of Research into Occupational Safety and Health - Selected Examples from Enterprises

Gisela Kiesau, Dortmund, Germany

The generally recognized fact that the results of government-sponsored research into occupational safety and health are not "naturally assimilated" into everyday industrial life and are not "automatically" implemented via autonomous market processes means that in its work, *the Federal Institute for Occupational Safety and Health (BAU/Bundesanstalt für Arbeitsschutz)* has to process the results of research in such a way that in each case they can be easily implemented by the target groups.

Successful and effective implementation is impossible without involving the relevant offices of the authorities for occupational safety and health, the public health service, further education and advisory services as well as the appropriate professional and trade associations. Since these bodies reach the most important target groups in firms, they frequently have direct access to the firms and, speaking their language, are therefore able to process the results of scientific research and present them in a suitable form. In order to achieve *the goal of preventive occupational safety and health*, the BAU brings together the activities of varied institutions on a wide variety of subjects and coordinates them. For this purpose, cooperative ventures of proven worth have been created between those institutions and persons responsible for ensuring safety at work, safeguarding health and making working conditions more humane, such as for example the firms themselves with their own systems for occupational safety, the Government's Trade Inspectorate, the professional associations or the research institutions.

On a European scale, for instance, there is the following *example from the iron and steel industry:* Many steel-making firms have problems with the destacking of steel bars.

By means of a prototype installation for automatically destacking the bars and feeding them individually to straightening machines it was possible to limit the heavy physical stresses due to the great weights, the high level of noise and the considerable danger of accidents in the form of cuts and hand injuries.

The results of this initial project were discussed with manufacturers and users with the aim of putting into practice the conclusions which had been reached in the discussions and formulated into a rough draft. Once the Working Group of the Iron and Metal--Working Professional Associations had made additional funds available for working out a more detailed design as a basis for the appropriate installation, further funds were agreed within the framework of the 1st Research Programme into Occupational Safety of the European Coal and Steel Community (ECSC), thus enabling a prototype of the installation to be built. The costs and the risk involved would have been too great for one single enterprise or manufacturer to bear alone. The joint completion of the project means that all the companies will jointly profit from the results.

In this way, two birds are killed with one stone: The experience gained by the authors of *"Results of Ergonomic Investigations"* is incorporated into the working result. Furthermore, the *existing transfer channels* of the individual trade associations can be utilised directly for transferring the resulting information into the practical work situations of the member firms. Cooperative ventures of this type, if they are suitably coordinated, can become a network in which the experience gained from working with the individual cooperating partners and the knowledge thus acquired regarding implementation deficiencies will form the starting points for action programmes which will qualify for support as projects within the governmental research and development programme known as "Work and Technology" ("Arbeit und Technik").

Various types of cooperation and new forms of transferring knowledge can be tried out within such action programmes and the factors hindering the transfer of knowledge recognised and limited.

Here is a further example: The Federal Institute for Occupational Safety and Health is working increasing closely with *trade associations* which are themselves in a position to carry out such "implementation projects".

In this way we are able to involve in the project work not only the individual target groups but also the various professional bodies with all their experience and accompanying know-how, by means of workshops and working groups functioning alongside the project. Thus, for example, the German Association for Welding Engineering (DVS), with the appropriate professional associations, industrial medical personnel and experts on industrial safety acting in an advisory capacity, compiled at our request 12 leaflets on the subject of *"Occupational Safety in Welding"*. These were published under a joint venture and a number of them were distributed through the agencies of the DVS. Our main aim in this is to offer assistance to employees in the trades and crafts, i.e., in small and medium–sized firms. At present, we are in the process of compiling an English version to be put at the disposal of the CIS in Geneva.

This form of project planning indicates *an approach* which, in our opinion, is both *feasible and highly efficient*, namely: we organise the cooperation for the implementation project along with the appropriate trade organisation or with a suitable institution which at the same time assumes responsibility for the project.

The decisive aspect is that *use should be made of available structures* which are already accustomed to working in cooperation, that the tasks associated with any planned cooperation should be distributed accordingly and that in this way knowledge be put at the disposal of firms in a better coordinated and more economically efficient manner.

Such *implementation networks* are most clearly discernible among the various enterprises themselves:

Results to date of research within the government-sponsored programme concerning "Making Working Life more Humane" show that an improvement of the immediate *working conditions in foundries* can most effectively be achieved by carrying out innovations on the technical and organisational side of the work. These open up the way towards realising the concepts for a more humane application of technology, especially in small and medium–sized foundries. Unfortunately, too little regarding these results comes to the notice of the foundries themselves.

Over the past 15 years, the Federal Government has supported a total of *77 foundry projects* to the sum of DM 50 million in order to ensure thereby the preconditions necessary for appropriate reorganisation measures. During this period, multi–faceted cooperative relationships have also come into being and the core of their crystallisation process is the Association of German Foundry Experts (VDG).

Seminars held by the Association of German Foundry Experts on its own and in conjunction with the BAU have greatly strengthened cooperation between the firms, the professional association and the Federal Institute for Occupational Safety and Health (BAU) and have furthered the implementation of research results.

In connection with the involvement of the Federal Minister for Education and Science in the programme for "Work and Technology", the Federal Institute is at the moment endeavouring *to extend the network structure to include the* extremely important *field of qualifications*. In the course of this, the existing cooperative contacts are to be expanded to include those organisations and institutions relevant to the gaining of qualifications including, among others, the Association of German Foundry Experts, the Institute of German Company and Industrial Doctors (regd), the professional associations, the Metal Workers' Trade Union (IG Metall) and the Government's Trade Inspectorate, to mention but a few.

My experience is based on many years of working and acquiring knowledge in the application of the results of research. As a final example of the success of implementation activities within appropriate networks, permit me to quote *an exemplary case from the automotive industry* which serves to underline my point:

The authorised workshops of a German motor-vehicle manufacturer were advised by the Federal Institute for Occupational Safety and Health (BAU) to introduce methods incorporating the latest discoveries in the field of work and health protection covering safety at work and ergonomic improvements, organisation of work and qualifications. In an interdisciplinary project covering all these aspects, three research institutes analysed the current state of affairs in the workshops, established the faults and deficiencies and elaborated suggestions as to where improvements could be introduced.

In cooperation with the Central Association of the Automotive Trade and its management consultants, tool manufacturers and producers of workshop equipment as well as a chamber of commerce and industry and an industrial testing station, the results of this analysis were transferred in many and varied ways to practical working systems.

The experience gained in this process was moreover implemented in an ideal manner by the Hamburg Chamber of Trades and Crafts when it introduced *a completely new course of studies* leading to the title "Industrial Assistant in the Automotive Branch". Future participants in this course will find that the knowledge acquired through the processes described above forms an integral part of the material they are required to cover.

In addition, in cooperation with the health insurance societies of the guilds, the research results gained through the advisory measures should be further developed and utilised as *a model to establish a "trade health promotion programme"* for all automotive firms.

Please permit me to make one final point: The implementation and dissemination of the knowledge gained from research into occupational safety and health is increasingly being influenced by the appreciation of factors which go beyond the bounds of the merely technical sectors. *The aspects involving health, organisation and qualifications* are constantly gaining in significance and must be incorporated in formative measures and be allowed to evolve further in cooperative ventures in a more efficiently planned manner.

Due attention must therefore be given to these considerations at this period of national and international activity in the fields of occupational safety, promotion of better health and legislation.

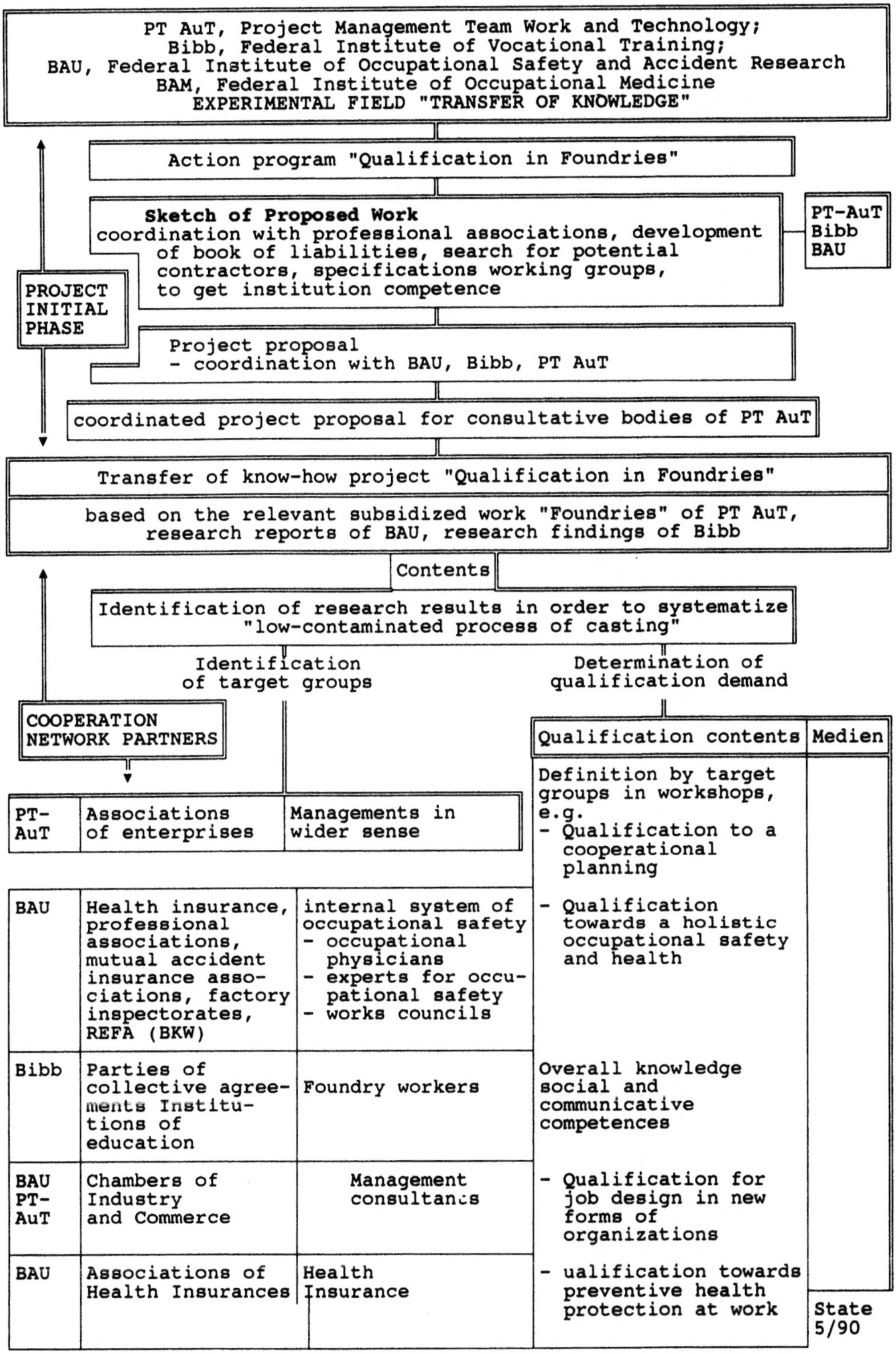
PT AuT, Project Management Team Work and Technology;
Bibb, Federal Institute of Vocational Training;
BAU, Federal Institute of Occupational Safety and Accident Research
BAM, Federal Institute of Occupational Medicine
EXPERIMENTAL FIELD "TRANSFER OF KNOWLEDGE"
Action program "Qualification in Foundries"
Sketch of Proposed Work
coordination with professional associations, development of book of liabilities, search for potential contractors, specifications working groups, to get institution competence
PT-AuT
Bibb
BAU
PROJECT INITIAL PHASE
Project proposal
- coordination with BAU, Bibb, PT AuT
coordinated project proposal for consultative bodies of PT AuT
Transfer of know-how project "Qualification in Foundries"
based on the relevant subsidized work "Foundries" of PT AuT, research reports of BAU, research findings of Bibb
Contents
Identification of research results in order to systematize "low-contaminated process of casting"
Identification of target groups
Determination of qualification demand
COOPERATION NETWORK PARTNERS
Qualification contents
Medien
PT-AuT
Associations of enterprises
Managements in wider sense
Definition by target groups in workshops, e.g.
- Qualification to a cooperational planning
BAU
Health insurance, professional associations, mutual accident insurance associations, factory inspectorates, REFA (BKW)
internal system of occupational safety
- occupational physicians
- experts for occupational safety
- works councils
- Qualification towards a holistic occupational safety and health
Bibb
Parties of collective agreements Institutions of education
Foundry workers
Overall knowledge social and communicative competences
BAU PT-AuT
Chambers of Industry and Commerce
Management consultancs
- Qualification for job design in new forms of organizations
BAU
Associations of Health Insurances
Health Insurance
- ualification towards preventive health protection at work
State 5/90

Feature
On the Way to a New Transfer Network

Dept. Research Application
State 5/90

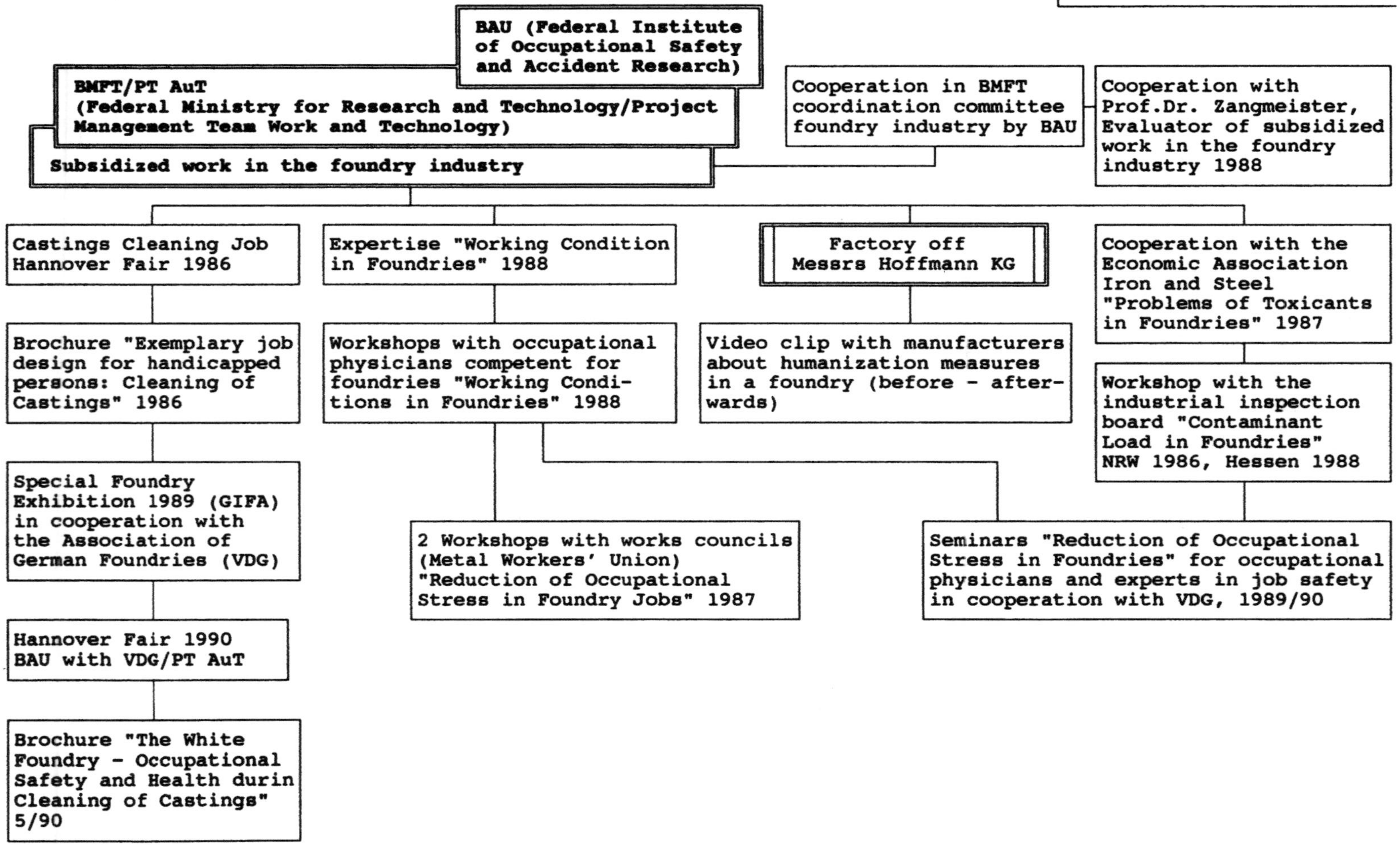

Implementation and Dissemination of Acquired Scientific Knowledge and Practical Experience within Enterprises in the Field of Work and Technology: Progress and Problems (The Viewpoint Held by German Employers)

Fritz-Jürgen Kador, Cologne, Germany

From its inception, *the German research programme "Humanisation of Working Life"* has been accompanied by *critical comments* regarding the failure *to fully implement and disseminate its results* within day-to-day working practice in firms. Moreover, I am certain that this new programme "Work and Technology" will meet with similar criticism. Is this assumption of mine an expression of resignation? Far from it, I assure you that this is by no means the case.

Such criticism originates in the main from *a basic misconception as to the nature of implementation* and the resulting unrealistic expectations as regards the contribution a government-sponsored programme can make towards improving working conditions.Let us consider firstly the misconception as to what *"implementation"* actually signifies and what it requires. People all over the country believe that the acquisition of scientific knowledge is a difficult, long and expensive process. For their part, the "producers" of scientific knowledge do their best to promote this conception. This attitude is neither dishonourable nor is it basically erroneous.

What is erroneous, on the other hand, is the attitude taken towards the implementation of acquired scientific knowledge: *The customary view of the implementation of scientific knowledge* is that it is something of secondary importance, fundamentally trivial or at least simple and "non-priority". Talented journalists are commissioned to present the results of research to the public in a more palatable form. They are printed in a clear, easily comprehensible manner, are made the subject of educational events, are stuffed into consultants' briefcases or they are quite simply prescribed by law or other binding standards - and all is right with the world.

Even those who have made only a cursory study of the subject know that it just does not function in such a way. *Implementation* is just as *difficult, multi-faceted, tortuous and expensive a process* as is research. Sometimes it proves to be even more difficult than research itself.

Implementation and dissemination of knowledge in the field of work and technology take place *in an arena of conflicting interests*, i..e. interests which vary yet remain on the same plane. Those involved in implementation and dissemination have mixed emotions: feelings of both enthusiasm and rejection. In short, this process is anything but a simple one.

Apart from this, the question is not merely one of realising scientific knowledge in the workplace. *Practical in-company experience* is equally important for ensuring that working conditions will be organised in a manner that is both economic and humane. If we wanted to orient our thoughts concerning the planning of working conditions strictly according to "tried and tested ergonomic knowledge", then that would represent a poor lookout for the intention of making the working environment more humane. There are now *two questions* to be answered here:

1. How should examples which function well in firms be disseminated?
2. How do we ensure that their effect will be "infectious"?

What role can be played here *by a government programme* with the title "Work and Technology"? At this point we must appreciate a few basic considerations: Public sponsorship of research into the humanisation of working conditions has proved quite clearly that from *the point of view of business enterprises* it is definitely justified and meaningful.

It is quite true that it has always been *the task of the entrepreneur* to ensure that working conditions are in keeping with human needs. This has always been one of the aims of employers with a social conscience, long before "humanising" became the subject of public discussion and was raised to the level of a socio-political requirement. The further development of humane working conditions within a modern and economically justifiable framework will always remain an essential element of the business policies pursued by independent entrepreneurs.

In recent years, however, the questions posed in the field of "Work and Technology" have assumed *a significance which* is not limited merely to the confines of the company but which rather touches on and *preoccupies society as a whole*. The change in the expectations people place in their work and, furthermore, technological developments are constantly raising new questions of humane ergonomics and require a degree of research which greatly exceeds the possibilities of individual companies. It is the implementation itself of scientific knowledge and in-company experience which poses problems within a firm. These problems affect more than one firm and their solution lies to a great extent in the public interest. In this case, the use of government funds to sponsor research into the humanisation of working conditions can give *valuable support to the entrepreneurial task*. Action from the public sector should, however, play only *a subsidiary role*.

Especially as regards the field of implementation and dissemination of knowledge and in-company experience, the considerations mentioned in the foregoing paragraphs, as seen from the employers' point of view, lead to the *conclusions* listed below.

The fact that scientific knowledge has been acquired in the field of "Work and Technology" does not in itself offer any guarantee that it will be transferred to the shop floor or other areas of the company. Furthermore, *the organisation of work and technology should not be based solely on scientific knowledge* but should also include positive, company-tested examples which do not necessarily make any claim to having a sound scientific basis.

The "classical" approach towards *implementation through laws, regulations and other legislative measures* makes very good sense in cases where ergonomic knowledge contains deterministic statements (for instance, those concerning certain threshhold values) which can be implemented directly in the form of prescribed standards. But also in those cases where they offer a rational basis for clear political action, legislative measures may prove necessary.

On the whole, however, *implementation through legislation is rather limited in scope* because the increasing complexity of ergonomics means that the tasks involved in work organisation have to rely more and more on knowledge and experience which cannot be implemented by means of legislative standards. In fact, legislative measures may even hinder the implementation and dissemination of such knowledge and experience required for solving specific, complex aspects of work organisation. The laying-down of compulsory regulations and duties is rarely the most effective means of motivating those responsible to do more than just the minimum required.

It is eminently more important for all parties concerned to be convinced of the usefulness of the acquired knowledge. The solutions which have the *greatest chances of success* are those which do justice to economic requirements and yet, at the same time, do not have to be achieved at the expense of humaneness.

From the point of view of the entrepreneur, the most important precondition for *organisational solutions* is that they *must be marketable* for them to be implemented or disseminated. In saying that, I do not mean "marketability" in its narrowest sense and so to speak as a "stick" with which the demand for political regulations can to a certain degree be "beaten out" as a preventive measure. On the contrary, I regard those solutions as "marketable" which bring about recognisable benefits for all concerned in everyday working life. This is by no means a matter which can always be settled automatically through market prices.

Before efficient and ergonomic techniques and work-organisation solutions in the field of work and technology become "merchantable" in this sense, substantial innovations often need to be made and other steps taken. I certainly make no secret of the fact that in principle I am *in favour of market forces* being the determining factor as regards the implementation and dissemination of knowledge and in-company experience in the organisation of work and technology.

I shall now outline what I believe are *the most important starting points* for further improvement in methods of implementation, i.e., starting points which lie between standardisation by means of legal regulations, wage agreements or other instruments and exclusive reliance on market mechanisms.

1. necessary requirement continues to be *practically oriented adaptation of knowledge and experience*. This on its own, however, is not adequate for the tasks involved.

There is no need to make special mention of the fact that not only are information leaflets and suchlike available for such adaptation work but also other media which are equally suitable.

2. *Basic and further training* are familiar methods, which are definitely not used to their full potential, for implementing ergonomic knowledge in the field of work and technology. The spectrum of knowledge to be imparted covers the whole range from elementary ergonomic and safety technology to knowledge acquired on work organisation and onwards to methods of professional innovative management.
3. The more complex the tasks involved in work organisation, the greater the importance of drawing upon *external consultants*. This is particularly applicable in the case of small and medium–sized enterprises. Of course, a precondition for this is fully qualified advisers who, with regard to technical and organisational changes, are not only in a position to take all *in–plant aspects of the firm into consideration* from the very beginning but also to accompany the processes of innovation in a competent manner and carry them through to their successful conclusion.
4. A further important precondition for successful implementation is to keep the above consideration in mind right from *the inception of research and development planning* within firms.
5. *The systematic inclusion of existing "vehicles for implementation"* is an essential starting point. Unfortunately, in discussions held up till now this has been reduced, unilaterally, to a question of public sponsorship of so-called "independently supported" institutional projects run by *trade unions and employers' associations*. The latter associations have made explicit statements regarding this, which I shall reiterate here for the sake of concluding my arguments:

   "It is necessary to develop a concept for implementation planning which allows *participation by trade unions and employers' associations with their own implementation projects* but which, at the same time, avoids giving the impression that such organisations are being subsidised unjustifiably. The "complementary regulations" to the general guidelines on subsidies, long considered necessary by the employers' representatives, need to be an integral component of the new implementation planning concept".
6. *A government–sponsored programme on "Work and Technology"* can make an important contribution to the implementation and dissemination of knowledge and in–company experience if it serves to make these more accessible. This, of course, includes knowledge acquired outside the sponsored areas.
7. Such *a programme should*, furthermore, promote the ability of all concerned to take part in discussion and, wherever necessary, should *serve to organise such discussion*. This covers the following "communication levels", for example:

- Everyday working life / Research
- Natural sciences / Social sciences

- Employers / Employees
- Manufacturers / Consumers
- Sponsored sectors / Non–sponsored sectors
- National sectors / International sectors etc.

Progress has already been achieved in all of the areas which I have mentioned, thus providing a valuable basis for the future. I am convinced that, without prejudice to the knowledge already acquired from research, both the programmes of the German government and the governmental programmes of other countries will encourage a basis for open discussion, thereby leading to greater awareness of the problems involved. To some extent, such a basis has indeed already been created. Accordingly, conferences such as this one will continue to be possible and to show promise. Much has already been achieved but a great deal still remains to be done.

Research and Development in the Field of Work and Technology

Jean-Yves Hocquet, Montrouge, France

1. The Context

The reinvigorating of the economic situation in recent times has not necessarily solved the *acute labour management problems*. The downturn in unemployment has been very slow. Lang-term unemployment has even tended to increase.

Meanwhile, in certain sectors in certain regions, employers are unable to find competent employees.

On the corporate level, the upswing of activity has meant *a rise in the number of industrial accidents*. Elsewhere it has meant shift-work and an upturn in overtime.

We have to acknowledge that *the changes often provide employees with only limited prospects*. Despite a favourable economic context, the restoring of corporate image in the public eye, the endeavour to implement non-Taylorist forms of organization, new forms of organization such as the just-in-time and total quality concepts which require a high level of involvement from employees in particular, the impression remains that:

- employees do not fully benefit from modernization,
- current developments concerning employment and working conditions as well as professional life on the whole, are producing mitigated effects.

As a backdrop to this unsatisfaction, there has been *a continued weakening of the trade unions in France*. Employee representational bodies have always been subject to sizable fluctuations. Though this prevents us from referring to an irreversible decline of trade unionism, we have to admit that it is currently at a low. Furthermore, instances of lasting labour unrest have underscored *the role of very demanding employee co-ordination committees* that have been very hostile to the established unions.

We are therefore faced with *a rather negative perception of changes in the organization of work, and especially the introduction of new technologies*. Furthermore, staff representatives are in a relatively weak position when examining the consequences of these developments.

However, *legislation and collective bargaining agreements* constitute a legal context within which to deal with the different aspects concerning modernization and technological change. The first level is the *Auroux law of 1982* which provides the following:

- *negotiations* (at one-year or five-year intervals) at corporate or branch level on issues such as salaries, and particularly the organization of working hours, classifications
- *consultation* of the elected bodies of employee representatives on important changes affecting the company, and the possibility of using instruments such as calling in and expert in the event of new technologies being introduced;
- finally, *employees have the possibility of expressing themselves* directly or collectively in their workshops or offices. This right notably covers working conditions. At national level, the two sides of industry reached an agreement on the introduction of new technologies. This request was echoed in certain sectors (metallurgy, banking, chemicals) and sometimes in certain companies.

There is consequently *a range of procedures* that theoretically enable the relations between technological change and work to be discussed.

Nonetheless, *the outcome* would seem to indicate that the opportunities created *did not produce the results expected.*

The fact that *work related matters are given insufficient consideration* not only entails negative consequences with regard to the situations of employees, but also with regard to the competitiveness of companies. It is increasingly acknowledged that the economy's lack of competitiveness is far from being due only to the factors traditionally put forward (age of fixed capital, cost of labour, etc.), but must also take into account the way these factors are implemented in reality.

This is the context in which *ANACT* has been developing *its role of technical assistant in labour negotiations.*

2. The Aims

ANACT is a public body administered by a tripartite Board (employer organizations, trade unions, qualified personalities). Its activity is organized around the following poles:

- to upgrade the extent to which employees' interests (health, jobs, professional prospects) are taken into account;
- to interactively deal with technical, economic and social problems in anticipation;
- to have work organization given due consideration among corporate preoccupations;
- to reason on the basis of real work with emphasis on employee participation;
- **to fuel labour negotiations.**

The last point is fundamental. It is based on technical instruments. Unlike that of other institutions, *ANACT'S action is mainly political.* It acknowledges from the outset the different interests of employers and employees, but it posits an overlapping of interest zones in many areas. The technical contribution does not deny the conflicts arising from different ideological models, it simply enables better identification of what can be shared.

ANACT'S practice consequently relies on privileged occasions in connection with a technical project. Its aim is to elaborate instruments that can be appropriated by all concerned.

Change is usually an ongoing process in corporate enterprise, and this does not make it any easier to discuss the question of work. It is not even safe to say that what is at stake from the economic and industrial points of view, and all the more so from that of labour management. Whence the intervention at a time when the dimensions that usually preoccupy management will be pretty much identified in order to be more convincing with regard to the need to integrate strategic reflection on work.

Such include new investments, plant construction, computerization, renovation of an installation, etc.

ANACT's point of view is not that of a research organization. Since 1982, ANACT has been *acting in a consultancy capacity* in the field of work. The purpose is to remain in contact with the real needs of both firms and staff, to elaborate methodological elements, and to validate observations pertaining to its watch-dog function.

For instance, ANACT's *operations in the chemicals' sector* have been varied: interventions when investments were being made, a study mission in Japan, assessment of modernization operations in conjunction with a trade union.

Results have been *at several different levels.*

The first level is that of *experience*. ANACT has built up references, pilot operations. Without being exemplary, these cases nonetheless have a legitimating effect within the sector in question. The content of this experience has been enriched over the years. A certain contamination of managers and staff representatives alike has ensued.

The *content of certain rather abstract notions* has become more precise. Precise questions must also be answered when taking work into account in the creation of installations.

When? This means *associating operational staff with projects as early as possible*, while remaining fully aware that it is better to have strong involvement before detailed studies rather than at the preliminary stage.

What themes should be dealt with and at what stage? Employment, team organization and competence, the mental activity of future operators at first, the premises, the layout of the interior, external installations, the system of conduct after the detailed studies' level.

What means should be used? The composition of work groups, use of mock-ups, the headings in the schedule of conditions ... all these topics are dealt with in a concrete manner.

Tools were then generalized for engineering managers and staff representatives.

Discussions with constructors and users led to the improving of numerical controlsys tems.

These efforts facilitated the marking out of *collective negotiations on the introduction of new technologies in the sector* (June 1990 agreement). They also enable employers and employees alike to avail of elements providing a content for a policy agreement of broad remit within companies.

ANACT's concern is to show that, *by clearly formulating work related questions,* both employers and employees increase rather than decrease their room to manoeuvre. It is therefore not so much a matter of the information being deemed neutral or objective, but rather of throwing light on topics that the two sides of industry would not have brought up spontaneously, and of this information being shared by both parties.

ANACT's intervention is not therefore in the form of expertise on "new technologies" requested unilaterally by work councils. *ANACT only provides assistance when so requested* by all the parties involved. The employer has the key to his company and his approval is necessary. But he must also be convinced that the labour aspect of his project is just as much worth studying as the other aspects and that the employees' point is worth a hearing.

In exchange, these conditions imply that employees see the point in the undertaking. *ANACT does not require a formal consensus,* but does seek a favourable opinion from the different levels of employee representation with regard to the aforesaid undertaking.

As a public body, ANACT's calling is *not however to deal with* the whole spectrum of *labour management problems in firms.*

Paradoxically, *it has chosen a market-economy logic* to develop this technical assistance, i.e. to bring out needs, gain acknowledgement for them as a marketable demand and contribute to the structuring of a consultancy offering capable of meeting this demand. That is the ongoing dynamic situation to which we have committed ourselves.

We have *chosen consultancy as our preferred means of action.* Sector by sector action is also a priority, but implies capacities that are harder to mobilize for an organization with political aims but that eliminates itself by means of its technical competence. Reflection with consultants appears to be the best way to gain a broad hearing for our problem. Given their development within the French economic environment, they represent a remarkable dissemination factor.

While *pursuing market-related ends,* they depend on methods being mastered (the consultancy sector is putting the final touches to a quality labelling system) for their survival. This is why *a work sector consultancy network* has been set up alongside ANACT. The condition for membership is the will to deal with labour problems in conjunction with employees.

Admittance simply requires agreement to fuel the methodological debate by giving a certain number of papers. The network members meet at least once every quarter for a day of exchanges attended by ANACT and the relevant research bodies.

By opting for a position other than that of a study group, ANACT is of course gambling. There are many *risks involved:*

- The risk of not being spontaneously seems to be endowed with the objectiveness of research work.

Technical competence mobilized in an operation manner can always be contested by one or other side of industry as being submissive the other party.

- The participants can look upon the role of a public body as a factor of unfair competition.
- ANACT's limited size does not enable it to gather together all the elements required to authentify its contentions.

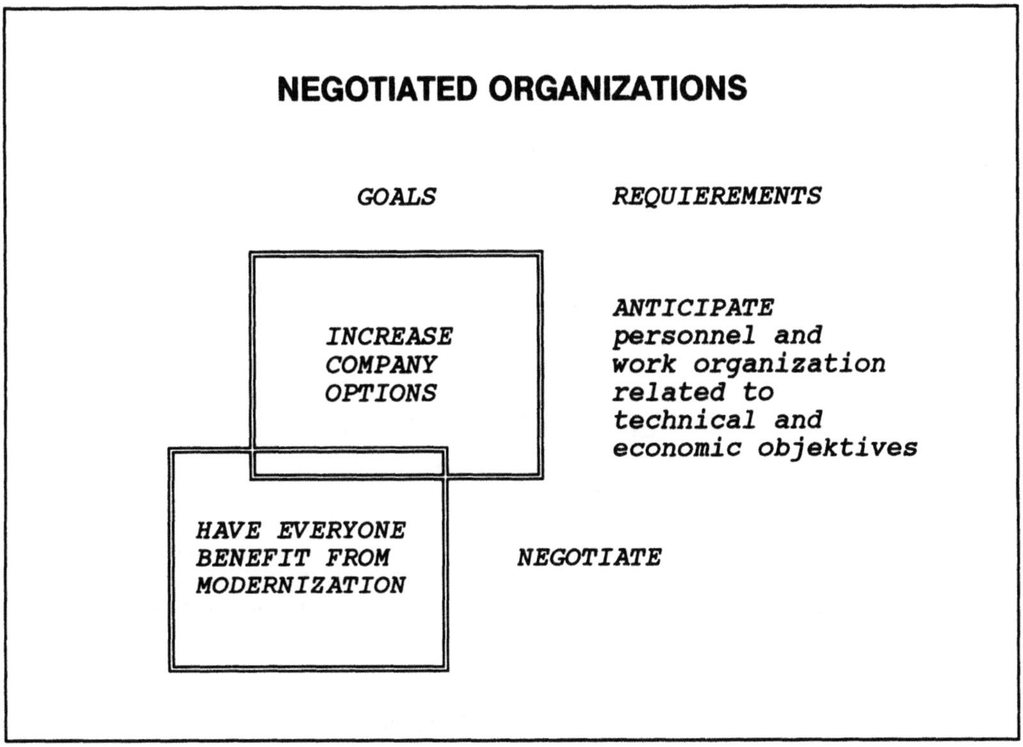

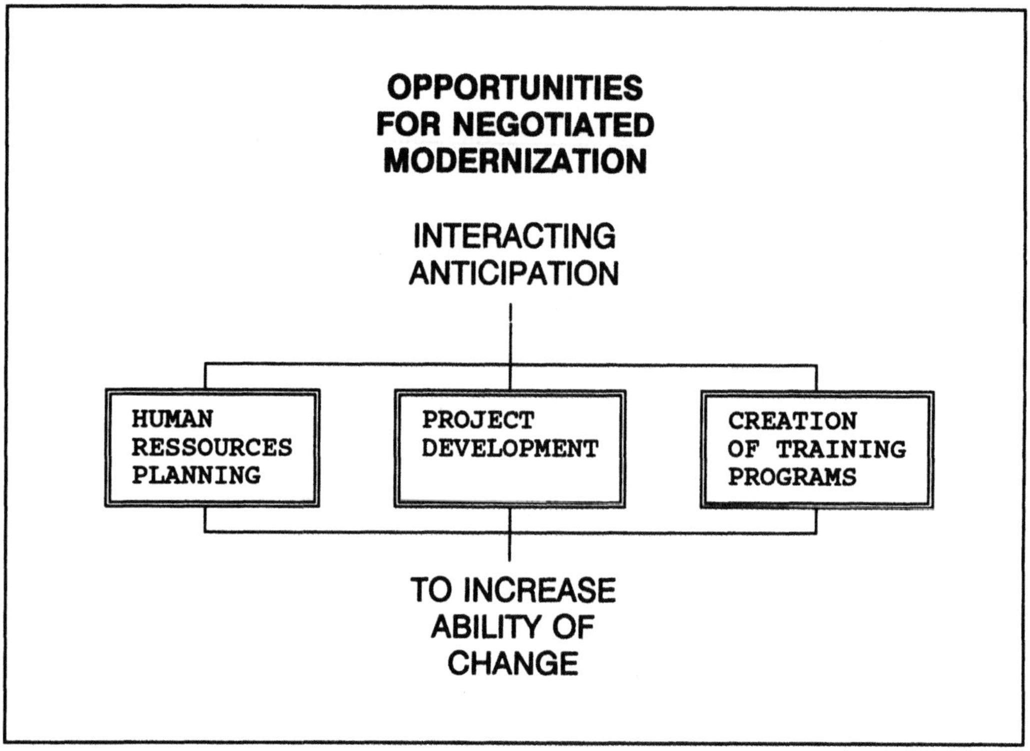
OPPORTUNITIES
FOR NEGOTIATED
MODERNIZATION
INTERACTING
ANTICIPATION
HUMAN
RESSOURCES
PLANNING
PROJECT
DEVELOPMENT
CREATION
OF TRAINING
PROGRAMS
TO INCREASE
ABILITY OF
CHANGE

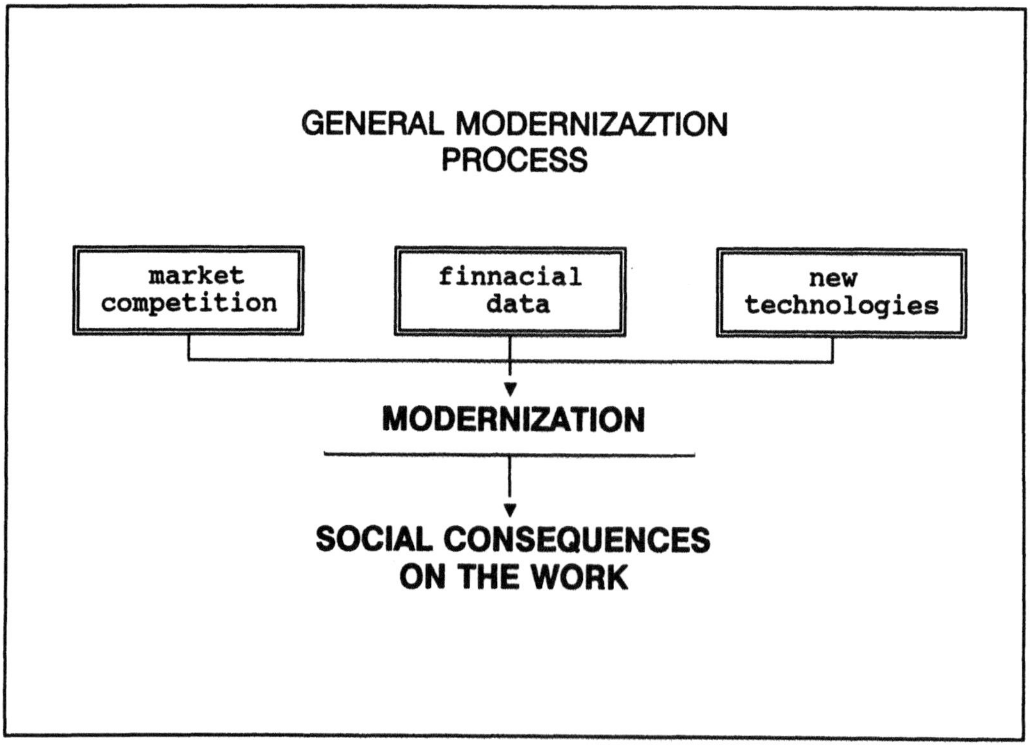
GENERAL MODERNIZAZTION
PROCESS
market
competition
finnacial
data
new
technologies
MODERNIZATION
SOCIAL CONSEQUENCES
ON THE WORK

MODERNIZATION

May have good or bad effects on wage earners

GOOD EFFECTS	BAD EFFECTS
SAFER WORK PLACES RESSOURCES PLANNING	SOME NEW RISKS (related to robots or the diffusion of new products (chemical substances ...) in more and more industries aside from chemical industry
DECREASE IN THE PHYSICAL EFFORT	AN INCREASE IN STRESS
MORE HIGHLY SKILLED JOBS	DECLINE OF WORKER'S SKILLS LEVEL REDUNDANCIES
NEW CAPACITIES REQUIRED (less direct intervention but more understandig anticipation)	INCREASE IN EXCLUSION

MODERNIZAZTION: EXPECTATIONS

COMPANY		WAGE EARNERS
- increased competitivness - New markets - Greater flexibilty	?	- Secured jobs - Career development - Better working conditions - More training - Higher status

Even with heavy investment and technical know-how

IF THE MUMAN FACTOR
IS
UNDERESTIMATED

THE RESULT IS:

Machine Breakdowns
Bad Quality Products
Economic Ineffeviency
Social Tenions

NEGOTIATION

permits

to develop worker's expression in order to understand worker's situation and expactions

to enrich teamwork so that everyone can see options and consequences in modernization

to promote negotiation about job classifications working time, process of change, work classification

not ot underestimate the importance of informal exchanges

The HABUT-Program and its Development Measures[1]

Per H. Engelstad, Oslo, Norway

1. Introduction

The Basic Agreement between the labour market parties in Norway was established in 1935, and it has since then functioned as *"The Constitution of Working Life"* in the country. The Basic Agreement is revised every *fourth* year and it consists of three major parts: Part A regulates freedom of organization, negotiations, shop stewards etc.; Part B regulates cooperative matters like co-determination, works councils, department councils etc.; and to Part B belongs (since 1982) also a number of so-called Supplementary Agreements.

In the 1982 revision of the Basic Agreement between the Norwegian Federation of Trade Unions (LO) and the Norwegian Confederation of Employers (N.A.F. from 1989 the Confederation of Norwegian Business and Industry NHO) the parties included a Supplementary Agreement on "the Development of the Work Organization of the Enterprise" or in short the *"Company Development Agreement"* (CD-agreement). in addition to the CD-agreement there are *seven other agreements* which focus on such issues as technological development, work studies, job evaluation, control measures, equality between the gender, education etc.

The CD-agreement is implemented through a joint company development program, the so-called HABUT program, or the *"Enterprise Development Measures of the Basic Agreement"*. The program is directed by a Board with equal representation from the LO and the NHO sides, and with two professionals (from action research) appointed jointly by the parties. The administrative and practical support-functions of the program are carried out by two small secretariates, which work closely together as an executive committee comprising four persons.

One might say that the HABUT program and the CD-agreement belong to a certain class of national support programs in Scandinavia, within which *the labour market parties and action research collaborate in different ways and degrees*. The purpose of the HABUT program is to encourage and assist emerging initiatives for development in the companies where the management, the unions and the employees engage jointly in company projects that are typically participative in form throughout the project period. The purposes of these projects are, furthermore, *to create "a better company" with "better workplaces" as part of the same development strategy*. In the words of the 1990 version of the CD-agreement

---

1) Revised: May 1991.

this implies that "the objective of developing enterprises is to increase the value creation by means of broad participation".

In the 1986 revision of the Basic Agreement, only minor adjustments had been made in the CD-agreement and no principal changes had been introduced in the program strategy. *In 1990, the agreement and the program strategy were thoroughly revised* on the basis of a careful analysis of the experiences from giving support to some 200 project companies or some 250 company projects, over an effective period of more than six years (1983-89).

The purpose of this paper is *to identify the most important principles* on which the operation of the HABUT program is based. Further, to point out the general requirements that this kind of program context put on all the Project companies that are granted financial and professional support under the program. Some relevant characteristics of the three special development measures that are specified in the CD-agreement, will also be considered. We shall then take a look at the major findings from the two first CD-agreement periods and, finally, make a preliminary evaluation of how each of the support measures has worked in practice. The author has been member of the HABUT Board for eight years.

2. The HABUT Program

2.1 The Principles of the HABUT Program Development Support

In the Supplementary Agreement on company development the parties have committed themselves to give financial support *and* professional assistance to companies that are prepared to put a sufficient amount of local initiative and effort into their own development activities. This implies not only that the external support provided by the program is *supplementary* in relation to the local contribution, but that most of the costs have to be carried by the companies, and they are also themselves primarily *responsible* for their development projects.

A development program is principally a mechanism for concentrating common means into "profitably focused" investments. These means have to be made available from the members of the sponsoring organizations by their obligation to contribute in accordance with some agreed distributive principle. In other words, as a member you have to pay, and in return you obtain a *possibility* to get more than "your share", but far from any guarantee to receive anything. In the short term, you are most likely to get nothing, if you are not yourself prepared to make a special effort, also in terms of living up to a certain set of requirements, which we shall return to below.

The most obvious of these requirements is of course *memberships* in the sponsoring organizations. Hence, the preferred targets for the HABUT support would have to be sought among companies that combine their NHO (or equivalent) membership with the broadest possible employment of LO members, since the money comes from a fund to which both parties contribute. Because both parties had their background in a tradition dominated by "distributive" agreements, it was not always easy for the Board of the HABUT program to keep enough focus on the *developmental requirements* of the CD-agreement. One might call this the principle of disproportional allocations. In addition, they had the problem of the unorganized (companies and employees) on both sides, which also took up much administrative time for the program.

The of the basic ideas of the HABUT program is to combine the financial support with at least some professional advice. However, this combination has two logical consequences with respect to the resources that become available to the program over time. Due to the gradual build up of the volume of projects, the *financial* obligations of the program that could easily be met the first years, will by necessity become a problem of scarcity when there is a fixed supply of money and if the program turn out to be a "success". In this case a total amount of 5 million NOK was set aside each year. (1 DM is approximately 4 NOK or Norwegian Crowns).

With respect to the supply of *professional resources* - which happens to be our main concern - there might also be a situation of scarcity. However, in contrast to the financial circumstances, it was expected that with time, the level of processional competence in the staff was likely to develop for the better. Unfortunately, the professional side of the program proved to be much more problematic than expected not only in the start-up, but also in the long term. A major reason seems to have been the way the whole program was organized. Since the evidence is that this was done differently and with more success in other comparable programs, the HABUT program underwent a complete reconstruction in 1990.

Another basic principle of the HABUT program support strategy in this period was to assist the company Projects primarily on an *individual* basis, and not at the level of company clusters or branch programs The latter approaches would imply a multi-level type of strategy in which mutual support and learning across levels produce supplementary effects, particularly in the longer term. However, it also requires additional, or at least different, forms of support from the program in order to activate and train the intermediate levels of the labour market organizations in relation to company development work.

From what is laid down in the CD-agreement and the program strategy, it is possible to distinguish between a set of *general requirements* that have to be met by *all* project companies, and the special requirements which have to be met by the company Projects separately, and in relation to each of the three development measures that are specified in the CD agreement. The latter measures include the so-called Planning Conferences, development Grants and direct Project support.

Before we turn to these matters it should be noted that "Agreements in Enterprises", which was also specified as an option in the CD-agreement, is not taken up as a special development measure in the following. This is partly because local agreements were not subject to financial support, partly because no assistance from the program was, to our knowledge, ever requested in this matter. This is quite remarkable because, in the 1960ies, the set up of so-called Shielding Agreements was not unusual in this kind of project work.

2.2 The General Requirements to be met by Company Projects

The main purpose of these requirements is to make it *likely* that the project companies will get maximum benefit of an amount of support that in principle have to be relatively small in relation to the company's own contribution. All company projects which are granted support in accordance with the CD-agreement and the HABUT program strategy should comply with certain demands. They may be summarized in terms of a set of four interrelated and compounded quality criteria. These criteria has to do with the "dynamic", the "local", the "balanced" and the "participative" qualities or conditions that characterize the project company as well as the organizing of the development activities that make up the company project.

The *dynamic* criteria implies that the companies are expected to initiate their own projects, that the company will allocate a sufficient measure of its own economic resourses to the project (relative to the external support) and that all parties are prepared to spend time and energy to get results.

The *local* character of the project is expressed by the fact that in particular the content of the development tasks will vary from company to company, and the knowledge about these matters rests with the management, the union(s) and the employees. Thus, external support is expected to focus on assistance with methods for analyses, planning etc. and perhaps some "process consultation".

The *balanced* character of the project has to do with keeping an acceptable balance between the joint and the separate interests of different parties. Here the relevant demands range from having the company project sanctioned by a consorted effort, in order that they also share influence as well as new learning opportunities in the project.

The *participative* dimension of the project is in many ways the most fundamental one, both in terms of linking company effectiveness to workplace quality, and the developmental work to the participative aspects of democratic theory. Thus, broad participation in developmental work may mean that every employee should take part in some kind of development tasks in addition to the daily routine work.

2.3 The Individual Development Measures

The set up of the CD-agreement was of course first of all result of a political process between the labour market parties in Norway in the early 1980ies. New demands for productivity and development and previous experiences with the work environment legislation and from working with action researchers were part of the background for this political process.

The selection of the three individual development measures in the CD-agreement were, however, in a more direct sense based on action research experiences from the Seventies. These experiences included working with redesign of jobs and work organization in production departments and factories, using special process facilitators attached to each project and applying "participative redesign" and "search conference" methods for discussion, planning and implementation purposes. Hence, if we look at the following list of development measures that have been specified in the CD-agreement, it is not too difficult to see how previous efforts are reflected in these measures:

1. Planning *conferences* for dialogues or discussions of company- and workplace development and redesign issues, are here functioning as part of an active adaptive planning process and also as a preparation for participation in implementation activities.
2. Development *grants* (or scholarships) for company project coordinators, being selected from management and the union and working in "pairs", in order to secure progress in the development processes over time. This should also be a basis for building up the competence of the parties in the company.
3. Development *projects* (or direct project support) for participative redesign of jobs and organization are primarily specified in terms of project content and less with respect to methods.

With regard to the obvious qualitative differences that these development measures exhibit, we shall refer to the more detailed descriptions that are given in the 1986 text of the CD-agreement. Most of the specific issues and problems that each of these measures raises when they are applied for company development purposes, will be dealt with in some detail in the final section. At this point we shall, therefore, only take up some of the more principle distinctions that can be made between the different support measures that belong to the HABUT program.

First of all, it may not be accidental that the three development measures, conferences, grants and projects do in fact belong within each of three different *development strategy traditions* that we find in working life. These are concerned with the developments of respectively: communication and language (e.g. "democratic dialogue", "networks"); human resources (e.g. management development, vocational training, education) and organization development (e.g. redesign of socio-technical systems).

Secondly, with the limitations on the external professional support that could be provided by the HABUT program, we would expect this to have an effect on the type of development measure that would be requested by the companies. Our expectation would be that those measures that would in the best way combine the possibility of high level of selfadministration with a well proven systematization of the "knowledge content" in terms of methodological "prestructuring" would be preferred. The reasons being that, since there would be no access to time-consuming process consultation anyway, the companies would go for the second best, which is a well worked out technique requiring relatively little time in direct collaboration with external resource persons.

In order to illustrate our point of view we shall leave out the direct project support and look at the professional assistance offered in connection with the two other measures. These may require different degrees of direct consultation and other less time-consuming, but more pre-structured forms of support. Hence, the grant holders may need a great deal of external training opportunities that require HABUT resources, but only in a concentrated (seminar) form. The much more advanced pre-structuring of the conference method, on the other hand, has made this measure quite resilient and effective as a participative means, even in the hands of the less professional conference managers.

Finally, the three measures can be evaluated not only as separate events in relation to a given company project initiative, but also as something that may function as a part of a more comprehensive support strategy for company development. This gives us at least two possible reference points for our evaluations. In the following we shall nevertheless focus on the former. The main reason for this is that whatever ideas there may initially have been behind the HABUT program in terms of comprehensive strategies, they have only been materializing very slowly.

3. Analyses of the HABUT Development Measures

In 1988 the HABUT program Board decided to carry out a thorough evaluation of the company development program in order to prepare the ground for a possible redesign of the program strategy, well ahead of the 1990 re-negotiation of the Basic Agreement. The point of departure for the following analyses, will be some of the same material that was brought forward by the HABUT secretariate as a basis for evaluating the support strategy and the development measures (HABUT, 1989).

The HABUT strategy was primarily directed to support development projects on the level of *individual* companies. Thus, out of a total expenditure of 23 million NOK (≈ 6 million DM) in the period of 1982-1988, these activities consumed 85 per cent of the total budget. The remaining 3.5 million NOK were allocated with respectively 13 per cent and 2 per cent to development initiatives at the levels of industrial branches and of small clusters of collaborating companies. Calculated in number of cases, there are in total 280 support measures, i.e. conferences, grants and projects taken together. Their distribution on each of the three levels, gives much the same picture as the corresponding cost figures (above): 83 per cent, 13 per cent and 4 per cent.

The HABUT Development Measures					
	Conferences no	Ewants no	Projects no	Total no	Costs Mill Nok
1983-84	25	18	19	62	4,5
1985-86	56	18	14	88	5,8
1987-88	92	23	15	130	9,2
Sum no	173	59	48	280	19,5
Sum %	62	21	17	100	(+3,5)

Table 1

Table 1 shows that out of the total number of measures granted (i.e. 280, across levels) a majority of 62 per cent were conferences, 21 per cent were grants and 17 per cent were projects. Of the total number of 232 measures directed to individual development projects, nearly 180 were allocated to different companies. This implies that at least some 450 companies received support from the program in response to two or more of their initiatives. In these cases it was not unusual to combine the conference method with either grants or project support. We shall return to the support activities on the levels of so-called cluster projects and branch programs below.

The program had an increasing activity level over the period 1983-1988, and this trend was sustained even in 1989. In *Table 1* this development is measured both by total expenditure and by total number of development measures granted in the periods of 1983-1984, 1985-1986 and 1987-1988. The figures in Table 1 also tell us that the total growth in activity level in the period can be ascribed almost entirely to *the increasing number of conferences* that were financially supported. Hence, while the number of conferences almost doubled in every two-year period, there were no comparable developments neither for the grants nor for the projects. It should be noted that the increase in conferences granted reflects a real increase in the requests for conferences, because the average "acceptance rate" for this HABUT measure was close to 85 per cent and relatively stable over the years.

There are, however, notable differences between the three development measures in the program, both in terms of their average level of costs and with regard to the average number of employees of the companies where each type of measure have been applied:

	costs in NOK	number of employees
conferences:	40.000	140
grants:	200.000	480
projects:	85.000	200

The figures suggest that the use of grants was first of all restricted by the cost factor as compared to conferences, but company size is also important. Direct project support is more flexible in these respects, but it has other drawbacks which made the HABUT program gradually reduce its "acceptance rate" for that kind of support.

4. Evaluation of the HABUT Development Measures

The point of departure for our evaluations of the HABUT development measures is the decisions made by the Board of the HABUT program after more than six years of experience with applying each of the three measures more or less separately to support different company development initiatives. The most obvious external criteria for evaluating each measure at the individual level are firstly: to what extent the different measures sustained their position within the text of the 1990 agreement; secondly: to what extent did those measures within the text also survive in an unmodified form.

The conclusions from this simple contest are absolutely clear: the direct project support approach came out as the obvious looser, since it has now been entirely excluded from the specifications in the 1990 agreement. The grants came out second, because they survived the revision only after major adjustments had been made in order to make this grants fit the new branch-based program strategy. The conferences emerged from this contest as the undisputed winner since they were able, almost without modifications, to sustain their position within an agreement text, which has now been made much shorter in total. We shall take a closer look at these outcomes in relation to some of the material that have been presented above, and to some other experiences as well.

4.1 Direct Project Support

Direct project support was first of all granted to companies that made clear that they - in connection with a development project - also intended to involve the employees in the redesign of their own jobs and the work organization on a participative basis. Even if further specifications are given in the CD-agreement, direct support tends to be granted to "promising" projects only on the basis of general declarations of intent, in particular with respect to the participative aspect of the project objectives.

Compared with the conference and grant measures, the direct project support differs in particular with respect to the non-specific way in which the financial and the professional aspects of the support are inter-connected when allocated to project companies. This is remarkable since this part of the CD-agreement originates within the "redesign oriented" development tradition. Admittedly, this tradition has moved very strongly in a participative and almost in a "do-it-yourself" direction, especially in Australia. Yet, it is perfectly clear that to be effective, this approach requires substantial amounts of professional "pre-structuring" in terms of design tools and training opportunities for the design process instructors in order to enable the companies to complete their redesigns (Emery, 1989).

This point is worth noting because in the HABUT program there were almost no professional resources available for process consultation after the start-up of the projects. One reason was that the members of the secretariate were few and that some of them had little experience in the difficult tasks they were facing. Another reason was that no action researchers or other external resources were formally allocated to the *operative* level of the program on a long-term basis.

The direct project support approach was probably motivated primarily by the desire of the parties to support "good" projects in situations where the management and the shop stewards in the companies thought they were able to handle everything by themselves. With the other measures the financing is linked more or less directly to the professional support, through which some of the structur of the company projects take place. Depending on the measures offered, this support may be in terms of discussion methods, work procedures, training arrangements etc.

Under circumstances with very little professional consultation, it is too easy to overlook the fact that the project support measure of the HABUT program may be subject to two entirely different fallacies. On the one side, there are project companies that are so resourceful that they can easily do without any of the professional support that the program is offering, even during the start-up phase. They should definitely not be part of the program and draw upon common financial resources for development that are too scarce in beforehand.

On the other side, there are project companies in which for instance the cash situation is so bad that the management is either unable or unwilling to put the sufficient amount of the company's own resources behind the development activities. In such cases the lack of external professional support in the project tends to effectively prevent the HABUT people to control that their money is used for the pre-assigned purposes. That is why in 1990 this rather nondescript form of development support was removed as a specific section in the CD-agreement.

4.2 Development Grants (for Project Co-ordinators)

The development grants (or scholarships) were offered to company's internal project co-ordinators, and there should normally be appointed one person from the union and one from the management side. There should preferably be two people working in "pairs", usually on a half-time basis and restricted to a couple of years. If the parties (in smaller companies) agreed to have only one grant holder, he/she tended to come, more often than not, from the union side.

Since this measure belongs to the "human resources development" (or educational) tradition, it was motivated by the need to secure that sufficient human resources were in fact made available to project development work. Therefore, the co-ordinator also had to be properly trained in order to acquire the skills and knowledge necessary to cope with their new assignments as "process consultants". The grant holders should in addition be known by the HABUT program, because they were expected to function also outside their own companies, ie to take part in seminars with other grant holders for learning purposes, to contribute in conferences designed to recruite new project companies etc.

The motivation for giving the grants to one from each side were primarily a matter of equal sharing of everything - from the money to be spent to the influence on the development and the opportunities to learn from the project. Perhaps less recognized are the facts that by operating with a "pair", rather that one person, we are confronted with the smallest possible *organizational unit*, not only for work, but also for discussions in a proper sense (ie beyond self-reflection or "talking to oneself"). This adds new dimensions beyond the personal level, which may produce certain "multiplier effects". This again suggests new ways to improve the grant system as a development measure, if one makes it become operationally related to a broader range of organizational contexts.

What had been overlooked, however, was that the pairs of grants were so expensive and so demanding in other ways, too, that they could only be utilized with sufficient "cost-effectiveness" in the larger companies. That is, if they were to "reach out" to a substantial part of the employment, not to speak of a reasonable number of companies. With respect to the latter, the problem for a small nation like Norway is not that our industry is dominated (in number) by small and medium-sized companies, but that we have so few big ones. This first condition we are sharing with most other countries and increasingly the problem of high unemployment, too, and the latter is more likely to be remedied by growth in the population of small companies, than within the big ones. That is why our experiences with the HABUT program might be of some interest also to foreign countries.

Thus, the major problem with the grants was that they had primarily been designed to meet the needs on the level of individual companies disregarding that the small and medium-sized companies were absolutely predominant in most branches. In response to the relatively expensive nature of the grants and the obvious benefits of working only part-time, but over extended periods and using pairs (or small groups) of grant holders, improvements could be sought in two directions: By allocating one type of the grants to several companies in collaboration, and to split up another type in smaller parts, like "one day a week", without reducing the time period. This "design fault" was soon detected, but the agreement-bound HABUT program proved to be surprisingly slow to make the appropriate adjustments.

4.3 Planning Conferences (or Dialogue Conferences)

The planning conference in the CD-agreement is a development measure that grew out of the experiences with "Participative Redesign" (Emery & Emery, 1974; Engelstad & Odegaard, 1979) and with "Search Conferences" (Emery, 1976; Engelstad & Haugen 1979; Herbst, 1980). In the last decade it has been further developed and widely applied under the name of Dialogue Conference (Gustavsen & Engelstad, 1986; Palshaugen, 1989). These conferences are designed to create effective discussions about workplace and company relevant development tasks. In order to achieve this "parallel" discussions of various issues take place in small groups of changing compositions and with a maximum of 9-11 members in each. The conferences may comprise some 80 persons or eight groups in average. They last from one to two days and should include from three to five sessions of 1,5 - 2,5 hours or simultaneous group work, interspaced only by very short plenary reports from the groups. There are no lectures and few comments in the plenary as the group work mostly speaks for itself. Thus, the most powerful variables in the conference design are the selection of the appropriate combinations of "discussion issues" with "group compositions" and the appropriate sequence of the sessions.

In the HABUT program the conference discussions take place among people that come from a setting of one or more organizations/companies. In terms of design a major distinction should be drawn between those conferences where the participants come from "vertical slices" of *different* organizations and those where *all* employees participate in the conference(s). In the first place, the conference design is almost independent of company size except that the number of members in each "slice" should be increased when the organizations are large in size and few in numbers.

When working with design of all-employee conferences a rough distinction has to be made with respect to company size (number of employees) between small (2-20), medium-sized (20-100) and "large" organizations (100 +). The important difference is that all employees can take part in *one* conference, when coming from either several small or from one medium-sized company. With large organizations a "package" of two or more conferences is required in order to achieve the same all-inclusive participation. However, this is not the place to deal with the various ways in which to make up these combinations of conferences.

In summery, the dialogue conference method is in itself undoubtedly a very effective way of initiating and supporting participative projects. In our experience it is flexible, cost-effective, well developed as a method and can be applied under a wide range of circumstances. It should therefore in combination with other measures be suitable as an integral part of any "broad front" participative development strategy.

5. The Use of Action Research Resources

The conferences were the most frequently used support measures even beyond the level of the individual company project, because they could so easily be applied by the HABUT program also for project recruitment purposes across companies. More important though were the action research assisted "experimenting" with grant holders allocated to "company cluster projects" and to "branch level programs" particularly in industries that were dominated by small to medium-sized companies.

After the first steps had been taken in the hotel and restaurant branch, the labour market parties of the car dealers and repair workshops in collaboration with the Work Research Institute launched the first branch-based "network program" related to the Company Development agreement. Gradually the textile,- the fish filleting-, and the catering industries started to move in this new direction. Finally, in connection with the 1990 revision of the CD-agreement, this direction was set out as the core of the new overall program strategy.

At present, the explicit objective of the (HABUT) program is to establish branch-based development strategies in ten different industries before the end of 1993. Needless to say that action research is also back on the scene in the latest program strategy. But this is he story of another paper. (Engelstad, 1922)

References

Emery, F.E., Emery, M. (1974), *Participative Design*. Center for Continuing Education, A.N:U:, Canberra.

Emery, F.E. (1976), *Searching for Common Ground*; in: Emery, F.E. (ed.): Systems Thinking, vol. 2, Penguin Books.

Emery, M. (1989), *Participative Design for Participative Democracy*. Center for Continuing Education, A.N.U., Canberra.

Engelstad, P.H. (1992), *The HABUT-Program in Norway: Experiences and Evolution of Strategy*. Paper presented at the 3 Bremer Symposium.

Engelstad, P.H., Odegaard, L.A. (1979), *Participative Redesign in Norway*; in: Working with the Quality of Working Life. Nijhoff, Leiden.

Engelstad, P.H., Haugen, R. (1979), *Sokekonferansen "Skjervoy, i gar, i dag og i morgen"*. API-publikasjon, Oslo.

Gustavsen, B., Engelstad, P.H. (1986), *The Design of Conferences and the Evolving Role of Democratic Dialogue in Changing Working Life*. Human Relations, vol. 39, No.2

HABUT (1989), Internal report of the HABUT Board.

Herbst, P.G. (1980), *Community Conference Design*. Human Future, Summer.

Human-Oriented Organizational Design and Management in Plants with an Advanced CNC Technology - Research Findings from 29 Swedish Enterprises

Christian Jacobsson, Gothenburg, Sweden

1. The Purpose and Design of the Study

The introduction of *CNC machines* has a rule resulted in a higher productivity, efficiency and quality. We have, however, *a rather poor empirically based knowledge of the new demands* as to the expedient organization, management and the operator's job content, that new technology brings about. We still need more general knowledge about the conditions under which we can master the new technology and obtain positive results both with regard to productivity and to job satisfaction. The empirical studies, the main results of which will be reported as follows, are aimed *at elucidation some of these conditions.* The study was carried out in 29 plants whose production is mainly based upon CNC lathes.

It appeared from a pilot study, that plants of different sizes are often *planning and organizing the work around the CNC lathes in different ways*, have different demands on hard- and software and usually have different kinds of production (length of series etc(. For that reason, *plants with varying number of employees* were chosen. Nine were big plants with on average 3 040 employees. 10 were of medium size (on average 280 employees) and 10 were small plants on an average of 16 employees. The main *objectives of the study* can be summarized in the following way:

- In which ways do *the management style and strategy* affect the organizational structure and the operators' quality of working life and job satisfaction?
- Is there any difference between *the leadership style and role* in plants of different size?
- What impact has *the type of production* on the operators' quality of working life and job satisfaction?
- What *conclusions* can be drawn as to the application of CNC TECHNOLOGY in the future?

In each of the plants, *the production technology* (types of CNC system, machines, ergonomic conditions, length of series, complexity and material) *and organizational structure* were systematically scrutinized. In addition, the *operators were interviewed* about their experiences as to the different job tasks, the autonomy towards the technological and the administrative system, communication and consensus with superiors, stress experiences, training and job satisfaction etc. Furthermore *the managers* were asked about their conception of the job distribution, their belief in the operators' ability and willingness to gain more qualified training and work, their opinion about demands on expert systems and planning terminals in order to make decentralized planning and preparation possible, etc.

Altogether, 41 managers, responsible for the production, and 126 operators were interviewed.

2. The Main Results

2.1 The Impact of the Production Technology

It appeared clear from the empirical data that *production technology and the size of the company* both to a considerable degree *have an effect on the organizational design and management style* and also to various aspects of the operators' job satisfaction. As regards *the production*, the following *four types* were distinguished:

a) *Few recurrent details* (less than 20 per year). The batches are usually large. For instance mass production of shafts for cars.
b) *A larger number recurrent details* and varying sizes of batches. For instance parts of pumps.
c) *New details every month. Setting up new jobs* every week, the batches often large. For instance contract manufacture.
d) *New details every month.* Setting up new jobs daily. Unusually small sizes of batches. For instance a single ring with complex geometry and high accuracy.

The production type D) usually makes the *greatest demands* upon the skill and qualification of the operators, whereas *type A) does not demand too much of operator's skill.*

It was found that *the higher the production type level, the wider became the range of the operator's job tasks:* for instance tasks referring to operations planning, part programming, set up, machinery and machine tool maintenance. As an illustration *the proportion of operators engagement in production planning in* the different production type plants is given in figure 1.

When new machines are installed, the production type D) operators are engaged to a much higher degree than the others. The jobs' design also varies between the production types with regard to shift-work and work at piece rate, as appears from figure 2.

Another interesting finding was that only 37 per cent of the production type D) workers reported that *the delimitations* were rigid between their own jobs and the jobs of the maintenance and repair personnel, compared to 65 per cent, 76 per cent and 68 per cent in production type A), B) and C) respectively.

That *the production technology to a considerable degree affects various aspects of the job experience and job satisfaction* among the operators appear from figure 3. Behind each of the questions are five response alternatives, one very positive (scored 5), one rather positive (4), one neutral (3), one rather negative (2) and one very negative (1). (Note that a mean difference approx 0.3 is significant at the 5 per cent-level.)

It is obvious that the organizational structure and job satisfaction in many respects is experienced in a more positive way among operators in the "higher" production types than in the "lower" ones.

When interpreting these results, the correlation between the production type and plant size must be kept in mind, see figure 4.

Thus to a certain degree the relationship between production type and job experiences may be explained by the plant size factor. Consequently it is of great interest *to analyse the impact of the plant size*, other things being equal.

2.2 The Impact of the Plant Size on the Job Organization and Management Style

The empirical data clearly showed that *the size of the plant to a considerable degree affected* not only the job organization and management style, but also *the job satisfaction* in many respects.

At the interviews, the managers were asked about their opinion about the will and ability of the operators to take responsibility and how meaningful the managers considered it to be to give the operators more advanced training, etc. The results appear from figure 5.

Of course, there are *many contributory causes behind the findings* that managers in small plants usually to a higher degree than in bigger plants have confidence in the operators' capacity. First of all, the production technology in the small plants mostly create opportunities for own initiative and for using aptitudes and talents among the employees. Lacking many of those specialist functions, that are very often found in medium and big plants, the operators usually take charge of all large variety of tasks. In addition the managers in the small companies of course come closer to the operators at the daily work, and give them continuously training.

According to empirically well supported scientific findings, there are *some fundamental job conditions which should be met in order to obtain a positive work environment* with a high degree of job satisfaction and productivity. The most important aspects among these job demand conditions are:

- a satisfactory degree of autonomy and the possibility of self-steering for the individual employee;
- a management style characterized by a basic trust in the subordinates and an open-minded collaboration climate;

- *possibilities of developing competence and qualifications* through the context of challenging and stimulating jobs;
- *an optimal workload* without either over- or understimulation.

As appears from figure 6, *these job demands* are on the average *more satisfied in the smaller than in the bigger plants*. In those medium and big plants, where the organization and management style resembled the average of small plant pattern, the same positive satisfaction could be noted.

2.3 Job Organization and Training Policy

The production technology, plant size and the management style can be considered as important basic structural factors affecting the quality of working life. Together, they leave their mark to a great extent on the job organization. *A prerequisite for a flexible organization framework*, offering the operators tasks, which as regards width and content are experienced as stimulating and motivating, is often that the boundary lines between planning, supervision and production is not too sharp. The correlation between the operators' rating of the rigidity of these boundary lines at their job on one hand and their general job satisfaction on the other hand was - 0.47. *The rigid specialization* proved to be predominant especially at the big plants.

One interesting finding was that in those plants, independent of size, *where the production planning was located at the shop floor* and not at a separate office, *the production went on in a much smoother way* with fewer unexpected stoppages and shorter time for changing batches.

Even if the planning, preparation and programming tasks at most plants are still carried out at the offices, the general trend is *a shift towards the shop floor* in order to enhance the efficiency and job engagement among the operators through job enlargement.

Linked to this organizational question is *the role and function of the first line supervisors*. According to almost all the interviewed managers and operators *this role had changed* to a considerable degree after the introduction of the CNC technology. Most supervisors still have a personnel function, are responsible for the detail planning, job distribution and control. But at many plants it appears that production and planning engineers, programmers and/or operators are taking over many of the supervisor's previous function. As a consequence, *the supervisor is nowadays considered more as coach* with a general responsibility than a production professional. But even for such a new function it is necessary for them to keep up with the new CNC TECHNOLOGY. In some cases this demand is satisfied, but at many plants this is not the case. Especially where operators more or less in autonomous work groups have taken over the responsibility for the planning, programming and production within their own unit, *the supervisors with traditional roles have proved to be redundant*. The general opinion in the majority of the plants is that *the organizational structure will be much more flat in the future*. This change has already started at many enterprises.

A condition of the increased commitment and responsibility from the operators' point of view are possibilities of developing their competence and skill. According to the findings from the study, the possibilities offered at the different plants are to a large extent dependent on the managers' general belief in human capacity. As more positive attitudes were taken up by the managers, as more CNC training was offered to the operators, the more satisfied they got. However, even if many production managers are ambitiously supporting training, *very few plants have a clearly worded training policy*. It is *very seldom*, that *the training aims are explicitly stated*, and hence the training programmes are not always relevant.

3. Some Concluding Remarks

One of the main findings is that the *very design and degree of sophistication of the CNC system in itself affects the efficiency and the employees' job commitment* and satisfaction *to a relatively small degree*. More important is the type of production, the size of the plant or unit, the job organization and the management's attitudes. Among other things it is found that alienation and a low degree of job commitment more often is found in large plants than in smaller ones, and that managers in smaller plants often have a greater degree of basic trust in their subordinates than managers in bigger plants. However, these relationships are only partly related to basic attitudes and size. A considerable role is also played by the production type (few recurrent details and large batches or new details vary often and small sized of batches) and by the job organization.

If work, even in larger plants, is organized in a way that it *allows a wide range of job tasks* and gives possibilities for the operators to use their aptitudes and knowledge, then you will obtain efficient working places with a positive degree of job satisfaction.

In addition, *management plays*, according to our findings, *an important role for the production in plants with an advanced CNC technology both directly and indirectly*. If they are personnel oriented and have genuine basic trust in their collaborators they will directly create opportunities for job engagement and satisfaction. However, they must be able to organize the job so that it provides scope for the operators to use their competence and get challenging tasks covering a wide range of the production system.

Proportion of operators, working with different production types, who as a rule are engaged in the production planning

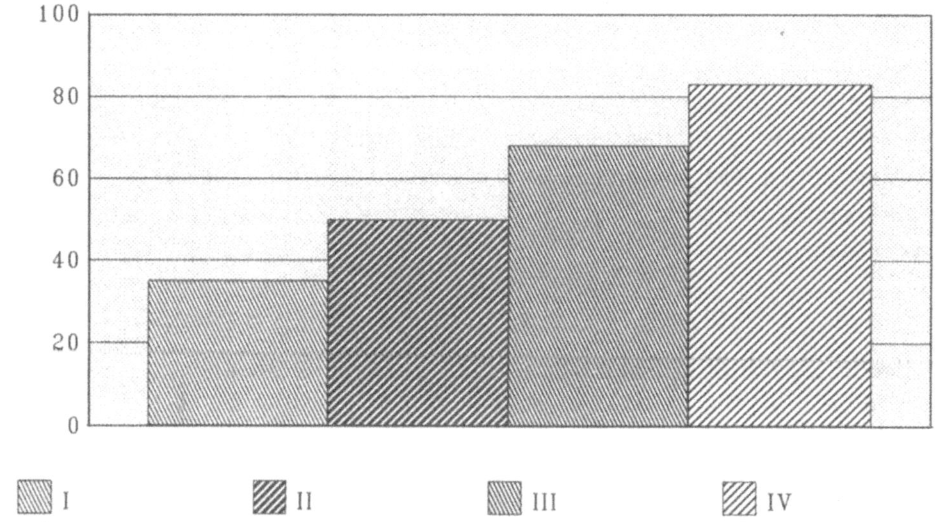

Fig. 1

The proportion of the operators, working with different production types, who have shift-work and piecework

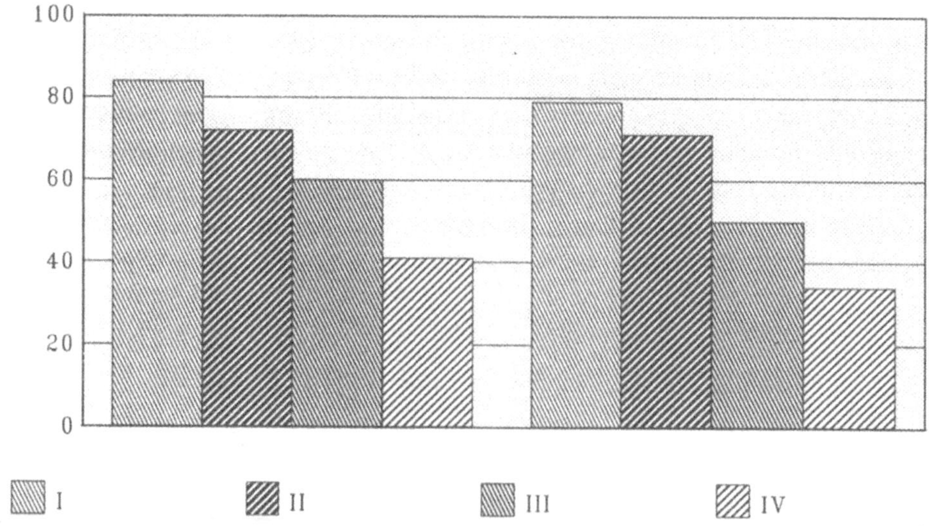

Fig. 2

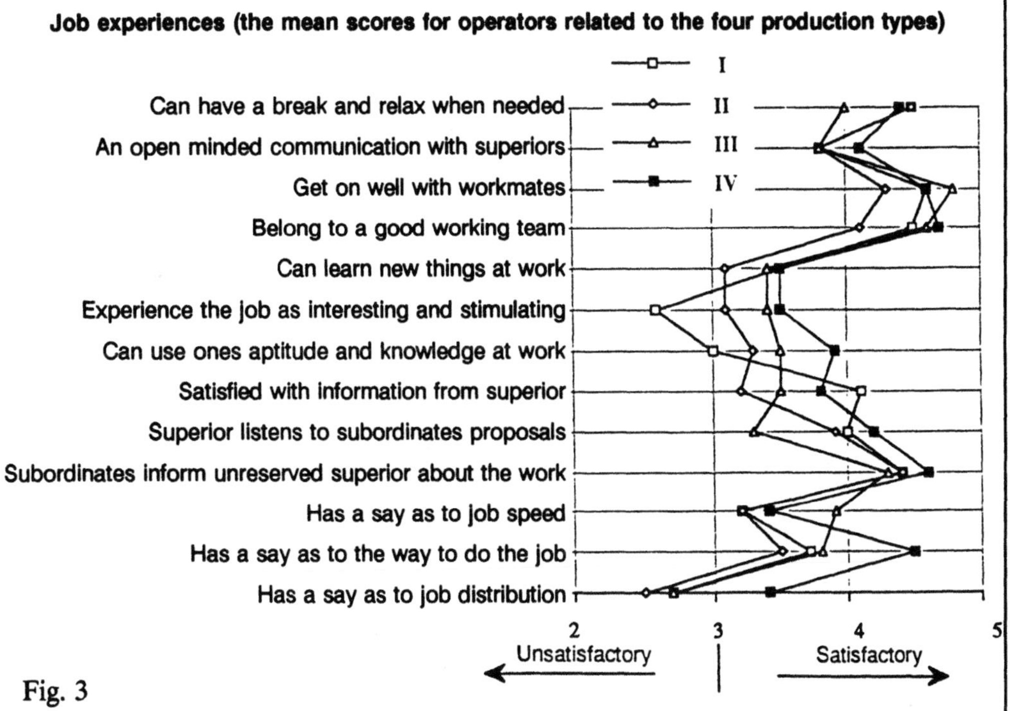

Fig. 3

Plant size and production type

		Production type				
		I	II	III	IV	
Plant Size	Big	6	2	0	1	9
	Medium	0	3	3	4	10
	Small	1	0	5	4	10
		7	5	8	9	29

Fig. 4

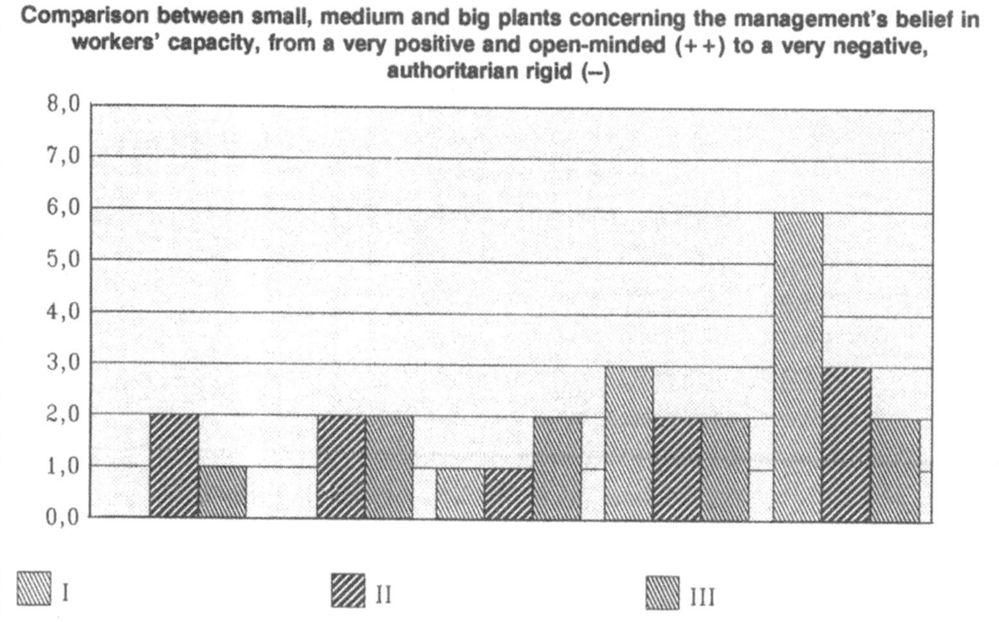
Comparison between small, medium and big plants concerning the management's belief in workers' capacity, from a very positive and open-minded (++) to a very negative, authoritarian rigid (--)
8,0
7,0
6,0
5,0
4,0
3,0
2,0
1,0
0,0
I
II
III

Fig. 5

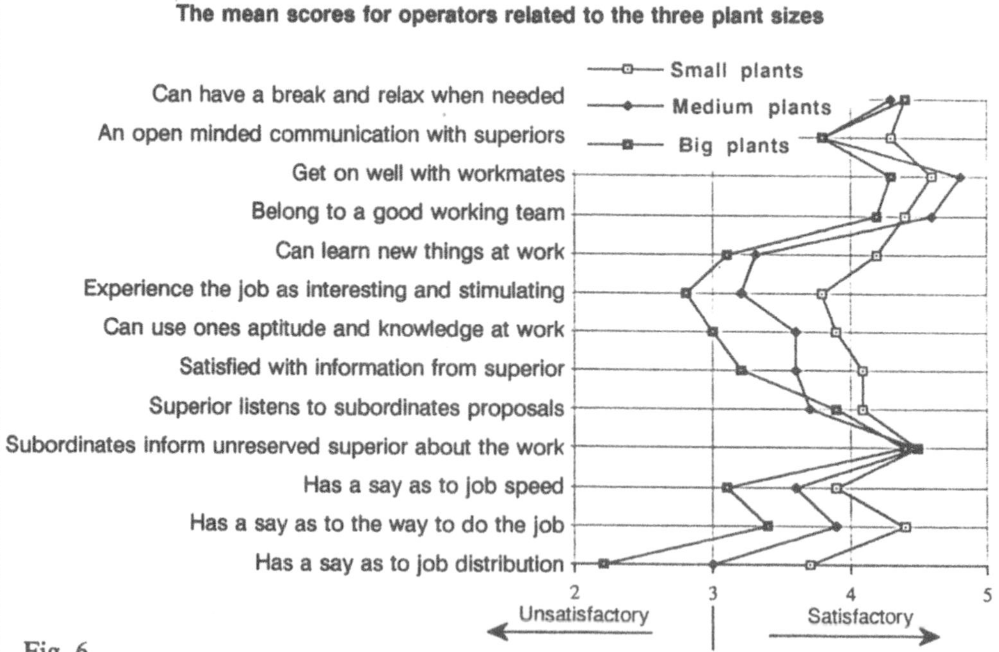
The mean scores for operators related to the three plant sizes
Small plants
Medium plants
Big plants
Can have a break and relax when needed
An open minded communication with superiors
Get on well with workmates
Belong to a good working team
Can learn new things at work
Experience the job as interesting and stimulating
Can use ones aptitude and knowledge at work
Satisfied with information from superior
Superior listens to subordinates proposals
Subordinates inform unreserved superior about the work
Has a say as to job speed
Has a say as to the way to do the job
Has a say as to job distribution
2
3
4
5
Unsatisfactory
Satisfactory

Fig. 6

Reports on the Parallel Sessions

Reports on the Parallel Sessions

Session A

Prof. Dr. Bußmann:

The session was characterized by mutual understanding, a high degree of competence and disciplined treatment of the subject.
I would like to concentrate on certain points. I have summarized a total of eight.

1. The work context is becoming more and more important even if not all determining factors can be quantified, for example, what we call in Germany "internal notice" ("innere Kündigung"). This is a subject that is well-known in Germany and it is very difficult to quantify, sometimes impossible to qualify; at the most, its effects can be measured. Nevertheless, one can identify certain determining factors such as sickness rate, fluctuation and can include these factors in cost load effects ("Kostenbelastungs-effekte"). Thus the work context can be introduced into business rationality. This demands, however, greater involvement of the employees in the shaping of the working context and that the competence to make decisions is more decentralized in many holdings.
2. As the quality of products depends greatly on "intelligent" output and services, the personnel development is increasingly becoming a competitive factor in the field of national and international competition. This means, that the position of the individual in a company, in company goals, must be valued much more highly than has traditionally been the case.
3. Women organize work and solve problems in a different way to men. In order to create the possibility of including this potential creativity and competence in the processes of decision-making, it is also necessary to decentralize decision-making and ensure a greater participating in the work process.
4. We can try to integrate aspects of industrial safety in investment decisions through a broader approach of profitability, i e through a special case of the utility value analysis ("Nutz-wertanalyse"). This was presented us this morning in a somewhat greater context. The problem again is that it is very difficult to qualify the efficiency and utilization of these enlarged aspects, not to mention the problem of measuring, valuing and expressing them quantitatively. One cannot but take recourse to indirect approaches.

   The disadvantages of such approaches were seen in the very high degree of co-operation and time involved. On the other hand, the advantage would be the higher degree of transparency in decisions and decision-making procedures. allowing decisions to be understood at all times and possibly shared by others, i e in an integrative effect. The second advantage is that control in the form of success control is possible. This means that companies obviously have to learn more and more by drawing the right conclusions from mistakes and to gain knowledge of failure rates and sources thereof.

5. There are plenty of chances for a dialogue which deals with the points I have mentioned. Especially in European high-tech companies, both between those who represent workers' participation and the aspects connected with technical rationalisation, the know-how of engineers and the interests of management. The effects of technological change on the personnel sector are not as deterministic as people often think. But technological change, and this is an important point, creates the scope to act and realize both, human and social intention.

   In the session it was pointed out that more humanity, more co- determination in the work process on the one hand, and more technical and economical efficiency on the other, are in no way a "negative-sum-game". On the contrary, if all aspects are combined in a positive way, employers and employees could achieve a "positive-sum-game".

   The indications of research projects were proving that managers often block workers' participation at the reflection stage and of major projects that they do not want to allow co-determination but that these managers very much appreciate the involvement of employees' opinion and experience in matters of information, advice and implementation. This position of the managers also partly meets the interests of employees in certain companies studied, where 60 per cent did not want any participation. The key-word here is the "hostage effect": "If we are involved in a decision we are also its prisoners later and have to support and stand for these decisions."

6. Information and communication technologies are key-technologies at European level and probably at world level, too. The long-term effects are often underestimated, their short-term effects are often overestimated. The problems arising are usually not of a technical nature. This shows once again that there are many options to act on. But it must be pointed out that there is still much to do in this field of research. Research must be deepen and must investigate different aspects such as regional and cultural aspects and aspects of participation.

7. All this shows quite clearly that a new paradigm of modern companies emerged. This was particularly interesting in one example showing that there is a positive difference between the value of a company and the sum of assets. Quality, product quality and services are elements of a company's value which are not covered by the property value. The value of a company can be traced back to personnel factors which became dominant in the formulation of company goals.

8. The slogan : "I no longer know jobs, I only know people" has to be enlarged. This could be done by promoting work and technology, by trying to combine them and by

integrating this approach to smaller and medium-sized companies. These efforts should include public government which is one of the promoters of opinions and decisions. One of the tasks could be to strengthen the role of the parties involved in collective bargaining and thus build up a social network which should be able to support and create social experiments so that they may make a major contribution to public welfare.

Session B

Prof. Dr. Kühne:

The subject of session B was "Integrated Concepts of Health Prevention, Industrial Safety and Ecological Requirements", including the European development. There was a total of seven speakers, only one of whom was a woman. Unfortunately a speaker from the former G.D.R. did not come.

The talks given, reflected different national contexts of Sweden and the former Federal Republic of Germany. Functionally they reflected different contexts of labour, research, state administration, trade unions and finally different approaches and action. Form the seven talks given, I would like to present the following thesis, advice and questions:

1. The structural change in industry or industries since the 70s which has led to certain improvements in work, health and environment would have taken place anyway due to reasons of technical and organizational rationalization. This basic conflict of an ecological reorganization of economies remains. The future will show to what extent the trade unions will take up this conflict and to what results this will lead.
2. In this connection: the reorganization of production towards a greater consideration of industrial safety, the protection of health and the environment demands the mobilization of knowledge and experience of employees. This means that they must have more information on potential dangers and that they are involved in decisions on the use of dangerous substances.
3. Existing legal rules are often too vague to be applied precisely and properly. This was shown in the example of sections 90 and 91 of the German Works Council Act and the terms used there of "established ergonomic knowledge ("gesicherte arbeitswissenschaftliche Erkenntnisse"). The guide-lines of the EC which are existing since 1987, thus not very long, have a Janus face. Elements of legal progress in favour of employees mixed with elements of regression. This gives the process of European standardization even greater importance.
4. Data bases in form of data banks gathering dangerous substances are a basic instrument in the preparation and transfer of urgently required information. In the Federal Republic of Germany alone there are some 254 laws concerned with the

environment and 48,200 substance names and their synonyms. But such dangerous substance data bases should be developed in a decentralized way (possibly at county, "Länder", department etc level) and "wired together" in a network. Central projects of this type have failed at a national level. Thus they should not be demanded at European level.

5. Concepts to develop anthropocentric working methods are confronted with different enterprise cultures in Europe. In the northern part of Europe there is great readiness to integrate human aspects into the development of new methods. In the Anglo-Saxon countries the opposite is true. There is a sort of open class struggle in industrial relations. In the southern part of Europe a patriarchal attitude of entrepreneurs often prevents innovations.
6. Medium-sized companies: In the Federal Republic of Germany 2/3 of all employees are working in such companies. These companies depend on outside advice in questions of more industrial safety and protection of health to a much greater degree than large companies. But, on the other hand, they often do not like to seek such advice and if they do accept it they find it difficult to apply it. Attempts to motivate these employers through printed media do not work. The negative effect of the absence of representation of workers' interest is obvious.
7. Industrial safety, protection of health and the environment already start at the stage of industrial planning. But where architects and entrepreneurs can be found, willing to admit competent experts at the planning stage?
8. Integrative concepts are very important in the removal of the deficits I have mentioned. On the one hand, these concepts must include and co-ordinate the demands of labour, health and ecological policies. On the other hand, they must also consider life outside the company. The general tendency of the discussion was that such integrative concepts must have a preventive character. Only 30 per cent of the employees in the F.R.G. are still in employment when they reach the legal retirement age.
9. The legal regulations on protection of work and health in the former G.D.R. and some aspects of practical labour protection were judged positively.

Prof. Pornschlegel:

I would like to express my sincere thanks to my colleagues for their contributions and I think that also those who were not involved in the session have received a detailed impression of the complex nature of the problems discussed. At the same time it has also become apparent that there are indications which point towards the need for co-ordination of research and the closing of gaps in research. Above all, there seems to be a realistic chance of co-operating beyond traditional borders in an area which is now largely determined on a European level, especially by the interpretation of articles 110,

118, 118a of the Uniform European Act. I think that the most of us are not really aware of this fact. But the train is running fast and the research community runs the risk of missing it.

Session C

Prof. Dr. Warneke:

The subject of analysis in session C was "Holistic and Innovative Concepts of Work and Technology". The main points were the function of qualification, organization and participation in the process of formation of industrial systems and the methodical control of research strategies. I try to concentrate on eight points.

1. Firstly reports, given on various studies and research activities, stressed that there must be a combination between sociological and technological orientated research. A human working world demands comprehensive knowledge of organization theory if problems of organization development are to be solved. Only then can the information process be improved.
2. Knowledge on the development of technology must be improved. On the basis of applications, new methods have to be tested in order to make adoption possible in the future so that it is possible to secure established systems in companies in the long term. This means projects must be defined in the companies themselves and that the various groups in the company co-operate.
3. The practical conditions necessary for influencing the process of change are essential. This means that changes in working conditions must be analysed and that factors which limit the co-operation must be identified. Here we must ask what instruments can be developed and used in order to improve this dialogue.
4. It is essential that competence is increasing and that know-ledge, skills and abilities are developing to set free creativity. It is only under these conditions that approaches and practice will be improved.
5. Competence and training are necessary which allow employees to make abstract use of concrete knowledge and improve the general basis of knowledge. In the learning programmes there must be feedback via goals and measures taken and the realization of these goals.
6. Projects also exist to find better technological methods for improving qualifications and to implement these new methods. But it is difficult to structure organizations accordingly. Not many employers look with favour on extensive employee participation. Thus it is very important that there is an interrelationship between different working groups in a company.
7. The integrative approach to qualification, organization and demands means a

permanent control of methods. The relevance of the logic of the system makes critical consideration of working working processes and methods essential.

8. The function of work organization in the process of technological change also determines the position of the trade unions towards holistic concepts of work and technology. The catchwords are:

   - a new selfunderstanding of a changing company policy;
   - management models for the development and realization of such concepts;
   - the position of the trade unions and their policy towards application of such concepts in the companies.

One field of the discussion was the possible competition that might develop between various government-promoted research programmes and also between government-promoted programmes and private consultant firms. The situation in Sweden and Germany seems to show substantial differences in this respect.

A second point was the establishment of discussions to enable the measuring of cooperation in companies. How can instruments be identified and used? How can they be improved?

Such methods should try to foster changes in thinking; projects should be a contribution to ecological design and corresponding changes in procedures and products.

Session D

Prof. Dr. Böhm:

The subject of session D was "Strategies to Transfer and to Disseminate Knowledge and Know-How in the Field of Work and Technology".

I once had a handbook, a systematic handbook on all the snakes in world, and the most impressive chapter was on Ireland. All it said in this chapter was that there are no snakes in Ireland. I do not know wether that is true but the results of this session could be summarized by saying that in this particular field of work and technology there are obviously no snakes which could be definitely identified. I suppose these snakes still have to be discovered and described. But what has already been found is the general consensus that practical strategies are essential. There were - and this must not be forgotten - various approaches, hints and examples of individual cases.

What impressed me is that Scandinavian is a model for other countries even if it has to struggle with serious problems. For if we are to believe the Swedish experts, and I do, some things which have been stable for years seem to be disintegrating. For example, the welfare system and the production system, the education system and the employment system, the necessity for well-trained people to be increasingly motivated. This can almost be summarized by saying that it is not only a dilemma if highly developed countries have highly developed problems, but that model which was applied for a long time now poses serious questions.

If this view is correct we should not talk about a dilemma but about a trilemma. On the one hand it was pointed out that despite all the bulkiness ("Schwerfälligkeit") of scientific production, the abundance of scientific production in the final result is nevertheless impressive.

- There is much scientific knowledge although it has not always been gained easily and it is very time and cost consuming. But it has been found that there is a real demand for this knowledge, at least objectively. But at the same time there was awareness of the problem that people with practical interests were not always aware of the variety of the problems.
- The last aspect was, and this I found particularly impressive, that the average Swedish employee has a basic education of 11 to 12 years. Basically these employees seem to be predestined by their very culture and education to apply this knowledge in practice.

Yet it seems to be a great problem to coordinate and harmonize these three things. There problems were described very differently. We heard above all employees are in organizations which are involved in the practical transfer and dissemination of knowledge. For example, there were reports on government commissions in Sweden. In France a public corporation has been set up. In Flemish countries they have taken a different path. A corporation with equal representation of employers and employees made offers and all who reported were able to register success on the one hand but, on the other hand, all admitted that the philosopher's stone had not been found.

There is no ideal solution. It may be that Professor Edquist from Lindköping University, who does not work in such an organization but who observes the process from his safe professor's chair, gave us key advice. He investigated the diffusion of process technologies, the technology of products and new socio-organizational forms and he was able to present us interesting tableaux. They showed that there is a broad diffusion of new technologies in the leading industrial nations. But Professor Edquist was not able to give us an overview of socio-organization. He referred to the fact that a great deal is said about "just-in-time" organization; even whole congresses are concerned with this. But if we try to gain an overview of the actual diffusion of the organizational concept, we just find a white patch. No reliable information is available. And Professor Edquist concluded that it is very difficult to give practical advice and offer a remedy if there is not even adequate information on reality.

But there were also contributions of a very different kind. It was not asked what commission, what organization, corporation etc. had been constituted to solve these problems; these experts had simply reflected pragmatically on the problem of how individual cases could be solved. I would like to give one example from the "Bundesanstalt für Arbeitsschutz". The problems of wood working machines were listed, dealt with and published in a leaflet. This publication met with so much interest that it has even been translated into English. But before anyone adopts the opinion that this strategy of taking small steps may be the only one reliable in a situation lacking major concepts and strategies, they should consider what Mr.Asklöf said. He said the problem is not to supply big companies with information and to make them aware of problems; the problem is to sensitize small and medium-sized companies where the great majority of people are employed. If you hand a piece pf paper to the owner of a small firm , he'll just throw it into the waste paper basket. You have to talk to him personally because he only believes what he hears and not what he reads, and he only believes information he gets from somebody he really knows. Only then does he have the guarantee that the information is true.

In view of the dilemma that small steps are probably the only practical solution at present and that a different form of communication has to be developed, the catchword of dialogue capability emerged. There was a general consensus of opinion that this competence has to be improved. But how? When I asked this question a representative of the German employers' association replied: "There is not one right solution. For every target - and there are different targets - there are alternative correct and appropriate ways. And then he said something which I would just like to quote here and which was applauded by everyone: "Every problem has to be solved in the best and most sensible way!"

Discussion

von Keiz, BAU:

I would like to draw your attention to a certain aspect that we will have to consider much more in the future: the development of qualification and competence. The work and experiences of the "Bundesanstalt für Arbeitsschutz" show that as responsibility increases, the demands made on the individual also increase. The responsibilities will increase because the product liability will be introduced by the development of the EC guide-lines. By doing this the quality assurance is becoming more and more important. Enterprises, especially small and medium-sized ones, will be confronted with this demands. The same applies to the employees. Traditional forms of responsibility will not meet these demands. It will be not enough that a product is simply used normally, it must also meet all ecological requirements. This points to a new dimension the demands on individuals concerned must not be underestimated. If they want to maintain their personal capacities the employees will have to be responsible for their own health.
The connection between the quality assurance, responsibility and individual demands made, will have to be taken into consideration with regard to preventive formation of the employees' abilities as well as to social relations.

Summary and translation by K.W. Weir

Closing Session

Work and Technology Programmes
- The Need for Adjustment and Innovation

Today's Work and Tomorrow's Work - Lessons from an ILO Project

Joseph E. Thurman, Geneva, Switzerland

1. Introductory Remarks

One of the reasons for this workshop is the growing recognition that we have arrived at a *critical time for policy towards working life*. The International Labour Office is involved in the search for new ideas and new research strategies in this field. There are many relevant ILO activities in fields such as technology, training and labour relations. Rather than attempt the impossible task of saying something about each of these, I would like to describe one specific project that integrates many separate concerns.

For two years now, a small international group of experts has been preparing *a report on "Working Life in Transition"*. We have received support from the Swedish Work Environment Fund that has permitted several meetings and the development of a genuinely interdisciplinary and collaborative approach. It is important to add that the group includes members with employer and trade union perspectives, which keeps our meetings lively and our work focused on practical as well as theoretical issues.

The report is in an advanced stage of preparation, and will be considered at an international meeting to be held in one year's time. The purpose is *to stimulate discussion and action* among policy makers in governments, employers' organisations and trade unions. We hope to revise the report on the basis of the results of the meeting and to circulate it widely.

We have not yet decided on the final structure of the report, but its current form provides a view of the themes and developments that we will cover. I therefore would like *to summarize the report as it now stands*. It is divided into four main parts, on context, experience, policy and design. I must warn, however, that this is necessarily a superficial summary of a very complex report, and that our goal is to challenge policy-makers, not to prescribe policy.

2. Context

We begin with the *context of change*, that is the major trends and developments that shape the options available at the workplace. Here there are *four main categories of developments.*

First the enormous structural changes in Eastern and Central Europe illustrate the very important developments that are occurring in *the role of the state*. Working life will be very different for those workers faced with changes from centrally planned to market economies. However, it is also important to remember the very important changes that are occurring in the market economies themselves, through such developments as privatisation; fresh conceptions of the state as employer; challenges to regulatory functions; new promotional activities on many fronts; and the separate debates that characterise state policy in such matters as infrastructure, labour mobility, training, worker participation and technology.

A second key category of contextual change is business. Modern enterprises operate in a highly competitive, global environment. The life cycles of products and business ideas are becoming shorter, bringing pressure for rapid renewal of business structures and business systems. One result is the development of forms of organisation that are flexible, decentralised and close to the market. From the point of view of working life, these new organisational forms tend to be competence-based, and thy require constant renewal and adaptation of resources, organisational systems and competences.

A third contextual category is developments on the labour market. Here several long-term trends are well known, such as the shift in employment from agriculture and manufacturing towards services; the growing proportion of women workers and, in many countries, of older workers; and rising levels of education and qualifications (mixed with some skill obsolescence). The unemployment picture is less uniform, with persistent problems in most of Western Europe, the potential for massive increases in Central and Eastern Europe, and some relative success in parts of Europe, Japan and North America. There is evidence in several countries of the growth of various types of jobs that are precarious or poorly protected, in particular in temporary, clandestine and part-time work. There are also important changes in labour market institutions, especially those related to collective bargaining.

Finally, technology is important to the context of work. New technologies are pervasive and rapidly evolving; they are closely linked with changes in organisational forms; they result in a reallocation of knowledge between people and systems; and they can support new forms of social relationships and enriched work. There are specific implications for task interdependence, the size and degree of integration of organisational units, decision making and control mechanism. The basic ways the organisation is run (through a market, as a hierarchy or as a team) can be affected, which in turn has implications for the tasks, roles and quality of life of people at work.

In analysing the impacts of all these contextual developments, especially new technology, it is important to keep in mind *the constant interactions* the mutually influence the course of events. *The future is about choices, not determinism*.

3. Experience

The second part of our report on "Working Life in Transition" is concerned with *experience*. In workplaces of many kinds we can see that the importance of capital and technology for innovation and competitiveness is diminishing relative to social factors related to the roles of workers, work organisation and labour relations. This can be recognised through *a series of developments*, including:

- *organisational innovations* introduced to promote or permit technological change;
- *factory planning* which gives human resources development an importance equal to capital from the earliest stages;
- *dominance of multiskilled group work* as a form of organisation;
- *use of networks and cooperation* instead of hierarchy.

Experience also shows, however that there are *a number of bottle-necks and obstacles* to development. In particular, organisational and social conservatism has resulted in many cases in the continuation of traditional hierarchies, limits on technological innovation and forms of participation or decentralisation that retain central control and do not realise their potential. In addition, flexibility and just-in-time systems often export uncertainties and problems to subcontractors or others outside the firm.

New production systems mean considerable increases in productivity and social improvements as well. However, they are associated with considerable physical, cognitive and emotional strain; with stronger differentiation between core and peripheral workers; and with continuous restructuring and reductions in manpower requirements. New demands on workers, management and trade unions have created conflicts that are not yet resolved in industrial relations systems.

4. Policy

The third part of our report is about *policy*. *Governments, employers and trade unions* now need to develop effective responses to the many developments I have described. Our report attempts to stimulate policy development by describing the challenges that face each of these three groups, both in terms of the substance of the problems and in terms of possible action.

Governments have sponsored enterprise development in several countries, often by supporting research with a view to proposing "models" for good practice. Today we know from experience that it is not so simple to diffuse improvements in working life. Each enterprise has its own context and must seek its own solutions. On the other hand, governments can do much to help create the conditions in which people in the enterprise can create their own changes. Such programmes could operate at three levels: improved relations and conditions within the enterprise, improved organisational performance and improved economic performance at the level of the society. These are interrelated long-term goals that require a planned, coherent approach. However, there are many options or pathways for the achievement of these goals.

The policies of *employers* are of obvious and immediate importance to the development of working life. In periods of rapid change, demands on management increase dramatically, firstly to create new businesses fast enough and secondly to handle the innovations required in existing business systems. The changes in business described under "Context" require strategies for fast, effective, flexible and highly efficient production processes. Managers are the leaders in introducing such strategies and their work is rapidly becoming more difficult and intensive. Employers insist on their need to be left alone to face these challenges without excessive intervention from outside.

Trade unions have a rather different perspective. Since their objective is to improve the lot of workers, they evaluate changes in work organisation in terms of economic, social and control impacts, more specifically in terms of pay, employment security and performance standards. They do not wish to see a separation of so-called working life improvements from traditional union goals such as reduced hours of work, nor can they accept forms of "numerical" flexibility that push problems to workers outside the enterprise. New technology, new management techniques and new forms of organisation of work are not opposed by unions as such, but they can only be accepted after a careful evaluation of their impacts on workers. Numerous challenges exist for trade unions, especially where their role needs reconsideration, as in Eastern and Central Europe, or where their approach to organisation of work is a part of their struggle to increase their strength and influence.

5. Design

The fourth and final part of our report concerns *design*. In both the manufacturing and service sectors, *traditional Taylorist-Fordist assumptions* about design have been eroding for at least 20 years. This has resulted from numerous factors, which include a consumer rebellion against the mediocre quality and lack of product variety associated with mass production; experience with other organisational forms that have proved highly effective, such as just-in-time, flexible specialisation of small enterprises within regional integration, or multiskilled group work; technological developments that have facilitated and in some cases required organisational adaptation; and the opportunities and pressures due to a more educated and demanding labour force.

Optimism about what we have learned about design should be tempered by recognition that many current examples of design are characterised by *very serious problems* which have been mentioned under "Experience". In particular, approaches to design that address societal problems are difficult to find.

Any policy or intervention that shapes *the development of technology at the workplace necessarily involves an element of design.* The process of design is particularly important: it determines both the effectiveness of the design and the human tasks that are created. Design itself is being partly automated, and the role of specialised designers is being challenged. New user-friendly systems allow some design tasks to be shifted directly to users, which opens important possibilities to enlarge the social basis of design. The exploitation of these opportunities depends on the conception the parties concerned have of design as an activity.

As you can see, the scope of our report is daunting. We are conscious that it necessarily has many limitations. There are, however, *several themes* that run through the report in which *we have considerable confidence*. Among these, I would mention three. First is the growing *importance of workers to productivity, quality and effectiveness*. Second is the *key role* that the organisation of work plays in both *economic efficiency* and *social considerations*, whether for individuals, organisations or society. And third is the rejection of pre-packaged *solutions to the search* for appropriate forms *of work organisation.*

I should immediately add that the absence of easy solutions does not mean that there should not be policies towards working life. The conclusion should be exactly the opposite: we should invest the *necessary effort* to find policies that *support the kinds of changes* we favour. It is to this effort, which ultimately is the responsibility of various governments, employers and trade unionists, that we are preparing our report as one contribution. I hope that we are also able to stimulate groups like this to expand their research on working life, because such *research is urgently necessary as the foundation of future* policy and action.

Work and Technology Programmes
- The Need for Adjustment and Innovation

Eckhard Lübbert and Jost-H. von dem Knesebeck, Bonn, Germany

Ladies and Gentlemen, You are most probably expecting just *one* paper from us on the subject stated in the programme, and the organisers of this event too have allocated only a very short space of time for our presentation. However, I do feel that we are faced with *two separate topics*: "Work and Technology Programmes - The Need for Adjustment and Innovation". The fact that the two concepts are separated by a dash means that we can interpret these in two different ways:

1. "Work and technology programmes *in view of society's need to adjust and innovate".*
2. "Work and technology *programmes; the need for these to adjust and innovate".*

The first of these topics involves a wide range of theoretical and pragmatic aspects of society in that it relates work and technology programmes more closely to developments in our industrial and working society. The second topic, on the other hand, means that the programmes are restricted in their practical dimensions in that discussion is required of the need for adjustment and innovation when implementing work and technology programmes. The possibly intentional *ambiguity of the title* offers the speaker the chance to be selective. This choice, however, is not an easy one to make when one is addressing a European audience at a European workshop because the first topic would make it possible to draw attention to common elements within European developments and to call upon these common elements to point out possible solutions for the organisation of work and technology. The second topic, on the other hand, is advantageous in that it would serve to provide our guests from other European countries with information on *the development and administration of our research programme on "Work and Technology".*

I shall attempt to cover both topics by identifying parallels between them. To discuss the pros and contras of both topics would be going too far. Lack of time forces me to narrow down my description of the *similarities between innovations within society* as a whole *and* those *within the programmes.*

The relationship appears to me to be as follows: The tasks involved in *developing and managing programmes of this type* become particularly apparent when it is recognised and acknowledged that current programmes aimed at providing an ergonomic framework for the organisation of work and technology cannot be implemented in a haphazard or piecemeal manner, e.g., by dealing exclusively with areas of extreme pressure within our working environment and applying the results of research and development simply to reducing such pressures. Instead, it is important to recognise that such programmes are integral to the processes of development and reorganisation within our industrial and working society and that they have historical relevance within these processes. In other words, work and technology programmes not only represent research-based solutions to isolated problems but are *part and parcel of long-term processes of development* within industrial society. These processes are decisive in providing a future framework for the organisation of work. In view of this, it is essential to organise research and development programmes in such a manner that possibilities and risks inherent in these development processes can be recognised and viable solutions found which in turn will allow these possibilities to be utilised and risks avoided.

Development processes of this kind, which are instrumental in the future organisation of work, *will* sooner or later *affect most European countries* to a greater or lesser degree. Here are some examples:

- *Fundamental changes in the age structures of populations* as a result of demographic change. The growing numbers of older employees within the working population, and the likelihood of it becoming necessary to attract new sectors of the population into the working environment, place specific demands on the organisation of work and technology.
- *Radical changes* are taking place *in modes of behaviour* in highly industrialised countries *and in the concepts of values* by which people are guided. As a result, new expectations arise with regard to conditions and types of work; at the same time there are far-reaching consequences for education and training, for decisions as to choice of career and for participation in working life.
- Technological changes are affecting all areas of society to an ever increasing degree. The *increasing speed of technological change* is becoming particularly evident in the world of work, manifesting itself in the form of new production technologies, new information and communications technologies, development of new materials and in the application of the results of pure scientific research conducted in areas such as biotechnology and genetic engineering. It is in these fields that the possibilities and risks for the organisation of work are particularly manifold and extensive. Here are *some key examples*: The occurrence of new forms of stress and health hazards; changes in the hierarchical and lateral distribution of responsibilities within companies (in some areas it is also becoming possible to move away from traditional principles of the Taylor System, i.e., from those which involve sharply demarcated division of labour and responsibilities among employees); modification of existing professions and appearance of new ones; special demands on the fields of basic and further education and training, with "lifelong learning" becoming a necessity in many areas; integration of more and more companies into networks which serve to link ancillary suppliers with customers (this may result in narrowed scope for company decision-making, also with regard to the organisation of work and technology).

- *Changes in the international division of labour*, particularly between high-wage and low-wage countries, with altered standards of competition in highly industrialised countries: Shortened product cycles, higher flexibility, ever higher demands on quality. Possibilities and risks for the organisation of work are also influenced by these factors.

If research and development is to play a valuable part in the more long-term processes of social development and structural change as described above, then *several preconditions* will definitely have to be satisfied. This also applies to attempts to actively shape these processes by means of certain types of comprehensive research and development programmes, which may be run by the state or the private sector. These preconditions are as follows:

a) Without doubt, the basic precondition is that *society itself becomes aware of these processes* of development and structural change. In fact, this awareness must also have evolved sufficiently to allow recognition of problems which arise whenever these processes can be actively influenced, above all in the context of work and technology. It is my impression that awareness of such problems is more highly developed in European countries than in other countries, including highly industrialised ones. This may be connected with the fact that, speaking from the viewpoint of economic systems, the work ethic has always been of particular significance in European culture, with working people already having become aware of the value of their work and aware of their influence on working conditions at an early stage of the industrial development of Europe. They also created appropriate organisations for consolidation of this self-awareness.

   This shared awareness of problems is a further decisive requirement for harmonising societal structures throughout Europe. Research and development programmes for the organising of work and technology can also contribute to harmonising European society by elaborating, or identifying, solutions which serve both to protect working people from dangers and to benefit them by using their creative abilities and professional experience.

b) A further precondition for participation of research and development programmes in the task of long-term organisation of work and technology is that *political and governmental institutions should play an extensive role* in this task. Enormous research and development efforts are definitely required, along with co-ordination of the research, if the processes of development and structural change are to be influenced successfully at an early stage in order to utilise possibilities and avoid risks. Combined efforts at national level would appear to be called for in view of the wide-ranging significance of the tasks and problems which need to be solved. In addition to this, however, further efforts towards standardisation are needed throughout Europe. The Single *European Market* will become reality as from 1993.

The creation of a harmonised European society is on the agenda for the more distant future. Moreover, *European Law* is increasingly being substituted for national legislation and is exercising a strong influence, for example, on occupational safety and public health regulations.

European research initiatives will first be concentrated in those fields where it is possible to standardise the results obtained. At the same time, however, central importance must also be attached to those national programmes in which the results and experience gained with regard to the structuring of technology, organisation and qualifications do not always lend themselves to standardisation. They may well prove valuable for highlighting national and cultural peculiarities as well as specific areas of institutional competence in cases where it is necessary to take account of these when implementing the results of research. Heightened European cooperation is, however, desirable in this instance also. Such cooperation would above all allow research funds to be utilised more efficiently, should it be possible to reach agreements on the division of work and responsibilities and on the allocation of priorities within different areas, aided by the mutual exchange of knowledge and experience gained.

In the Federal Republic of Germany, the "Work and Technology" research programme has been successful in ensuring that governmental responsibilities, and the co-ordination of these, encompasses three essential political spheres: research and technology policy, social policy and educational policy. The research programme is being supported accordingly by, among others, the Federal Ministers for Research and Technology, for Labour and Social Affairs and for Education and Science. We regard this wide basis of support for the research programme in the Federal Republic of Germany as important also in view of the tasks and responsibilities involved by the future-oriented organisation of work and technology. In other countries, however, such tasks and responsibilities may assume completely different forms and may be quite usefully combined in a completely different manner.

Irrespective of which authority is responsible for such programmes in any one country, the most important aspect is that research and development efforts concerning work and technology should be understood as representing a *comprehensive and generally applicable task* and that a wide forum should be created for discussion of this task, with equal involvement of economic and scientific spheres.

c) *The manner in which a research programme on work and technology is developed and administered* is, in our opinion, the third precondition for ensuring that such a programme shall be relevant to and exercise influence on the more long-term processes of development and structural change in our industrial society. In making this point, I have now arrived at the second of the topics referred to at the beginning of my paper.

With regard to *the development and administration of a work and technology programme*, there is a difference between participation in processes of structural change within society, where the progression and outcome of these processes remain open, and evaluation of and decision-making on isolated questions which need to be dealt with by means of research and development:

- It is necessary *to engage in strategic and dynamic research planning*, as opposed to simply making isolated decisions for which static evaluation criteria normally suffice.
- It is necessary *to appraise the consequences of technological and social processes* and *to develop preventive measures* aimed at averting undesired consequences, as opposed to merely attempting to subsequently rectify misconceived developments after these have already proved to be detrimental to working people.
- It is necessary *to recognise at an early stage* and to utilise positively those *possibilities* which arise in connection with the processes of structural change, whereby completely new approaches often have to be followed in order to amalgamate production factors.
- It is also necessary to increase the efficiency of programmes by *attending to problems of present and future relevance* concerning processes of structural change, particularly as such problems will sooner or later need to be solved in everyday commercial and industrial life.

Together, these requirements represent *a complex task of development and administration of research programmes* which make specific demands in that they cannot orient themselves exclusively towards manifest needs but have to proceed on a trial and error basis which often requires the appraisal and testing of possibilities which involve a high degree of uncertainty and risk.

The structuring and organisation, progressive development, and administration of programmes can be appreciated and implemented in a highly differentiated manner. The ideal programme can be administered by concentrating either on *the aspect of supply or on the aspect of demand.* In the case of a programme which is administered according to demand, its organisation and development proceed according to a basic research programme covering a wide range of subjects based on either scientific or industrial requirements. At least with regard to industrial demand, this does offer the basic certainty that the programme will really be of use for an existing need. It does not, however, offer an adequate guarantee of success for future-oriented research and development projects in the sense that the opportunities and risks connected with radical processes of structural change will be studied as deeply and comprehensively as possible and organised in such a manner as to have a preventive effect. This is particularly the case when the research is to be promoted primarily for the benefit of small and medium-sized enterprises.

In the case of a programme administered according to supply, the programme itself provides *guidelines for the making of applications for research and development*, in addition to the basic research offered. Such guidelines serve to determine content and aims of such programmes, e.g. when advertising for research work or when further developing those work areas which should be concentrated on. On the one hand, they may give applicants greater confidence when making their applications and, on the other, they draw attention to important questions relating to processes of structural change and thus encourage the making of research and development applications which are suitable for dealing with such questions. The question which first springs to mind *in the case of predominantly supply-oriented programme* administration is this: Where do promoters of research obtain the wisdom to stipulate guidelines which are correct and relevant in the context of continuous processes of change within society? Above all, it is necessary to proceed with the greatest caution if it is assumed that the State possesses such wisdom or, even more so, if the State itself assumes that it possesses such wisdom. There is therefore no justification for a situation in which governmental promotion institutions stipulate such guidelines on the basis of their own (absolute) authority.

Guidelines intended for directing the content of the offered research towards future-oriented organisation of work and technology can only be obtained through *dialogue between science and practical everyday life*. Here, *the scientific side* should be constituted by the various disciplines relevant to the organisation of work and technology and the practical side should be represented not only by *experts from companies* but also by *labour market parties* and by other relevant institutions.

Provision for *such dialogue* has been made *at several levels*, and an institutional framework created for it, within the Research and Development Programme "Work and Technology" of the Federal Republic of Germany.

I should like to emphasise *three of these levels* at which the above dialogue between science and practical everyday life is being conducted unceasingly and at which fruitful discussion is also taking place on possible further developments and administration criteria for programme sections (areas where work should be concentrated), for subject areas or for the programme as a whole:

1. With regard to existing areas where work should be concentrated, *"expert circles"* and *"coordination committees"* are assessing not only research applications but also the structuring and organisation and the further development of such areas of concentration.
2. *"Status seminars"* are examining not only the presentation of research results from selected subject areas but also the future requirements for research and development and the urgency with which individual subjects should be handled.
3. The *"Discussion Circle on Work and Technology"* is assessing the structuring and organisation and the further development of the research programme as a whole.

The *numerous channels for intensive feedback* between programme administration on the one hand and science and practical everyday life on the other, as well as the breadth of the above forum for discussion, are absolutely essential for structuring, organising and further developing the "Work and Technology" research programme in keeping with the processes of social development. At the same time, the opportunities for discussion and the feedback mechanisms allow the analytical differentiation made above between demand-oriented and supply-oriented programme administration to be seen in proper perspective.

The need for adjustment and innovation in a programme for the organisation of work and technology becomes obvious on examining the relationship between such a programme and the processes of structural change within society.

I would like to refer just briefly to *a few examples* from our German programme which illustrate how knowledge concerning structural change has led to adjustments and innovations in the programme:

- *The great importance of qualifications* as a result of both technological and structural changes on the competitive markets, and also their importance as a factor in the development of such changes, has only been recognised during the course of the programme and has led to corresponding adjustments and innovations in those areas where work is being concentrated. The most visible expression of such innovations is the fact that the Federal Minister of Education and Science is one of the sponsors of the "Work and Technology" programme.
- *Research into demographic trends* and *the effects of these on working life* have been included in the programme in the form of new subject matter relevant to a cross-section of the topics dealt with and particularly with regard to preventive measures in the field of occupational safety and health.
- It is only now that knowledge has been gained about the potential benefits of technological change for the structuring and organisation of work that *the utilisation of opportunities offered by such change* has become an objective of equal importance, in the programme, to the hitherto dominant objective of avoiding risks.
- The increased orientation of the programme towards *preventive approaches to structuring and organisation* essentially means that adjustments are being made within the programme to take account of accelerated technological change.
- As a result of the insight gained into the intensified processes of structural change taking place in *the trades and crafts sector*, the research programme intends to devote more attention to this sector of the economy by applying suitable means of promotion.
- The programmme will now need to include special forms of adjustment, and possibly innovations, in view of the developments and needs now becoming apparent in *East Germany*.

We could continue with this list but we feel that the need for adjustment and innovation in a "Work and Technology" programme is now obvious.

In our opinion, the subject of further development and administration of a programme is not exhausted by discussing only the need for innovations. In addition to this question of *the need for innovations*, we must also consider the extent to which they are feasible and, furthermore, permissible. I can only mention these two points briefly here.

With regard to the question of *feasibility of innovations* or, as one could say, of their ability to be implemented, one of the important aspects is to make provision for innovations in the programme in such a manner that the target groups will not become divorced from the purpose of the programme or that the users of the research will not be expected to achieve more than they are capable of achieving.

This problem usually arises at *the interface between research and implementation*. It is necessary to clearly establish the limitations or preconditions which affect the implementation of the results of research and development when, for example, new areas, new topics, expanded goals or higher expectations are being formulated as innovations in the programme. It is precisely in view of accelerated processes of structural change that there is a danger that a programme may lose its relevance for at least certain sections of the target groups and users. Such processes may also generate an accelerated need to make adjustments and adaptations within the programme, even though sufficient demand for these may not yet have arisen.

When questioning *the permissibility of innovations*, it is necessary to check whether essential components of programmes, such as targets, guidelines or principles, may cause requirements to be made on projects, especially development and organisational projects, which may lead to unintended consequences and to risks which cannot be assessed, particularly when such requirements are fulfilled because the institute concerned values its reputation and is encouraged by the prospect of financial incentives. The more a programme is involved in an continuous process of structural change, thereby making it necessary to attempt to organise work and technology on a risky trial and error basis, the greater becomes *the danger of faulty or excessive adjustments* or of high-risk innovations in company-based research and development projects. This places a high onus of responsibility on those who are charged with implementing the programme, particularly as the main recipients of promotional incentives are small and medium-sized enterprises.

This question as to the permissibility of innovations in a programme also arises because some of the aims and requirements developed in the programme will serve to provide orientation as to the direction which should be taken by company-based innovations or company-based research and development projects.

This applies, for example, to *the comprehensive approach to innovation* developed or represented in the programme. Under this comprehensive approach, consideration should be given not only to technological questions but also, and on an equal basis, to organisational structuring, to development of personnel and qualifications and to extensive measures for occupational safety and health. Such highly demanding recommendations for the making of innovations within companies cannot be formulated in a programme in the form of clearly established tenets to be obeyed from the outset but can only be elaborated stage by stage during the programme. Such recommendations need to be founded on intensive company-based experience gained from sponsored and non-sponsored projects and involving the above-mentioned comprehensive approach to innovation. Moreover, this approach to innovation will first need to have been formulated through broadly based discussions with representatives from industry, industrial associations and science. One could therefore say that a *process of evaluation and assessment* will have taken place under several auspices before a set of recommendations can emerge for the making of innovations within a programme.

The incorporation of a programme for the organisation of work and technology into the processes of structural change within society causes specific problems and requirements to emerge with regard to development, structuring and organisation, and administration. We have attempted to give you a clear idea of these by using the key words "need", "feasibility" and "permissibility" when referring to innovations. I would like *to summarise* this quite simply as follows:

- The question as to the *"need"* for innovations is directed primarily towards programme development;
- The question as to the *"feasibility"* of innovations refers also to the dynamics of programme development;
- The question as to the *"permissibility"* of innovations refers essentially to the distribution of responsibilities involved in programme development.

Obviously, the above summary *cannot provide overall remedies* for dealing with these requirements and problems, and definitely not remedies which could claim to be applicable for other European countries with their different traditions and requirements.

However, with regard to the German research programme on "Work and Technology", it can be concluded on the basis of experience that the questions, constantly recurring in new forms, as to need, feasibility and permissibility with regard to innovations can only be answered within the framework of wide-ranging and intensive dialogue between science and practical everyday life, conducted against the background of the developments which can be seen to be taking place in the processes of structural change. In its capacity as the main sponsor of the research and development programme, *the State acts essentially as the "coordinator" of this dialogue.*

This brings me to the end of my short discourse on the two topics between which I was unwilling to choose. If you have been expecting me to deal with a third topic, namely: "Work and Technology Programmes - Future Needs with regard to Adjustment and Innovation", then my contribution will at least have provided you with an answer as to why I have been unable to deal with this particular topic in my capacity as government representative for the research programme.

Some Concluding Remarks

Carl A. Asklöf, Stockholm, Sweden

This workshop has dealt with how to achieve important goals at the workplace, such as improved efficiency and productivity, good use of technology and a work organisation in which people can use and enhance their knowledge and experience.

Perhaps we should have discussed more about *how to support women at the workplace,* so that they may play their legitimate and necessary role in working life.

We should maybe also have focussed more on the *evaluation of changes at the workplace.* To what extent do we reach the goals we set up for ourselves?

There has been full agreement among the participants in the workshop that *a prerequisite for change processes* at the workplace is *active participation.* We have in fact here a win-win situation for management and labour.

During the workshop important trends in working life have come to the fore. The *business idea is now the governing concept,* not the production, and *the customer is in the focus.* High flexibility is needed at the workplace and how to best use human resources is a major challenge. The *individuals doing the job* are more important than the job structure at the workplace.

The *workplace must be looked at in a wider context* than before. It is *linked to its suppliers and customers* in a complex often international system. Many conditions outside the workplace as e.g. the external environment must be considered. The *workplace is dependent on society in a variety of ways,* for instance in the supply of well educated people. The *division between working and leisure time* gets more complex through people's requirement of more flexibility. *Information technology* gives the means to do work outside the workplace.

So much about trends that must be taken into consideration when creating the workplaces of tomorrow. Which is then *the role of researchers and research* in this connection? This question was an important theme in this workshop. *Research can contribute* in principally very different ways:

- First, knowledge from research should of course be *used to change working life*. One way to accomplish this is by the *participation of researchers as consultants* in the *development at the workplace*.
- Second, researchers can participate in the *development at the workplace using established research methods* e.g. for *generalisation or evaluation purposes.*
- Third, research may b done *at workplace with the purpose of moving the frontiers of research* i.e. producing qualitatively new knowledge.

It goes without saying that it is difficult in practice to separate these different researcher roles from each other, especially in so called *action research*, used in many of the cases presented at this workshop. But still it is important to make the necessary distinctions.

Another focus of the workshop concerned *dissemination of results.* Research gives us knowledge but does this knowledge induce any changes in working life? Here we must *admit that we know too little*. We need more research on dissemination processes.

We have strong indications that *established channels should be used* in order to inform the target groups. It is then natural to cooperate with the *social partners* and other *sectoral organisations.* Of course it is advantageous if one can use those channels in which the most important information flows, e.g. regarding business, technology etc. Especially *for small companies it is necessary not to separate* information on working *life questions from other topics.*

A Short Summary and Tentative Conclusions of a European Workshop

Hans Pornschlegel, Dortmund, Germany

1. General Perspectives of the Workshop

In a double sense, the *workshop* functioned as a catalyst in the field: it served *its purpose as a platform, as a "forum"* for research and development agencies, agents and policy-makers, for representatives from the labour market parties as well as from the scientific community in the field of national and European activities concerning work and technology. This appears to be true for a number of objectives: The *institutions and persons involved* represented in fact most of the European activities:

- It included the German "teams" attached to the Federal Ministers of Research and Technology and of Labour and Social Affairs, from the Federal Institute for Industrial Health and Safety (Bundesanstalt für Arbeitssschutz), the Project Administration for Work and Technology (Projektträger Arbeit und Technik), and also from RKW (German Productivity Centre), the REFA Work Study Association, the labour market parties and related organisations, as well as from the research community. The Federal Minister for Education and Science, together with the Federal Institute for Vocational Training, will soon be adding their participation.
- The Swedish community was very well represented by various agencies and programmes, especially through the co-organiser, the Swedish Work Environment Fund (Arbetsmiljöfonden), Stockholm, with its MDA Programme (People-Computers-Work), LOM Programme (Programme on Leadership, Organization and Participation), Programme against Musculoskeletal Injuries, and high-ranking representatives from the labour market parties, from the researchers' side and from business life.
- Also present were representatives of the National Institutes for the Improvement of Working Conditions, Brussels.
- The Commission of the European Communities, Brussels, with the MONITOR/FAST Programme (Forecasting and Assessment in Science and Technology).
- The International Labour Office, Geneva, "Working Life in Transition".
- The Nederlandinstituut voor Arbeitsomstandigheden, Amsterdam, and MERIT, Maastricht, with the TAO Programme (Technology, Work and Organization).

- The USA were represented by the Office of Technology Assessment of the Congress of the United States of America, Washington.
- The European Foundation for the Improvement of Living and Working Conditions, Dublin.
- The Work Research Institute, Oslo, with the HABUT Programme (Bipartite Body for Development of Work).
- The European Association of National Productivity Centres, Brussels.
- The Flemish Foundation for Technology Assessment (Stichting Technologie Vlaanderen, STV), Brussels.
- The Ministry of Research and Technology, ANACT (Agence Nationale pour l'Amélioration des Conditions de Travail), Montrouge.
- The Social Studies University of Oxford Faculty Centre with the Programme on Information and Communication Technologies: PICT.
- The Work Research Unit, London
- The ESRC-SERC Joint Committee, London.
- Representatives of the EUROJOBS Working Group on Job Design and Work Organisation, Stockholm/Brussels.

The workshop is also the result of a process which started in 1986 in Munich within the framework of the International Productivity Symposium III, when *national programme directors* in the field of work and technology from Europe met for the first time. This network has been growing in the meantime. It met again on the occasion of the workshop.

There was also a meeting of the *Swedish-German steering group* of the Swedish Work Environment Fund and the Project Administration for Work and Technology, DLR, preceding this event. Under the leadership of Prof. den Hertog of Maastricht, the *MODEM group*, a loose European grouping of research and development institutions in our field of activities, also met to discuss their past and future activities, again within the context of the workshop.

These are strong indicators for the *building up of a European community of agencies and actors* with increasing mutual information, coordination and cooperation. In various contexts, there was therefore a strong demand to strengthen such networks, to foster bilateral and multilateral "joint ventures" without giving up the respective autonomy of institutions and programmes. *The network idea seems to be flourishing.*

The *synopsis of national and European programmes* in the field of work and technology by A. C. Hubert drew widely on den Hertog's earlier studies and its interpretation made it clear that Southern and Eastern Europe are not yet part of this community. In spite of many common areas of interest, the relative importance and direction of the programmes still varies considerably.

Some of the highlights of the workshop were the contributions of R. Lindholm and S. Bleicher: They both sketched their *visions for emerging future roles for management and business, for the labour movement and the workers,* instead of drawing rather static and conventional status quo pictures. For the formulation of future strategies for research and development such visions, though in some respects with utopian features, are fitting guidelines for adjusting programmes, programme areas and research to such changes in basic attitudes of the labour market parties. An astonishingly high degree of *common opinion and even similar approaches* was apparent in:

- the definition of basic problems;
- the approaches to problem areas discussed here;
- concepts for solutions (e.g. holistic approaches, economic assessments, health, industrial safety, ecological aspects);
- the rather pragmatic approaches, sometimes by trial and error, in the field of transfer and dissemination of knowledge.

On casually observing the scene around the workshop, many *personal and institutional contacts* for the further development of mutual information, coordination and cooperation in different forms took place - they seem to be at least as important as the professional contents presented.

With the institutions and personalities present at the workshop, it came as no surprise that the *creation, strengthening and functioning of networks* appears nowadays to represent a general approach to many problem areas discussed. *Networking has become a leitmotiv* for future work in the field.

Research and development in the field is still very much determined by the *national characteristics* in the social framework, traditions and cultures. But the necessity of integrating *the European and global dimension* more and more into the work of the programmes and agencies has become evident in nearly every context. This reflects the impacts of European harmonisation and standardisation, and also of globalized trade and business.

Without giving up their specific national contexts, a strong need for *more efficient interaction between national and European programmes* was stressed during the workshop. The innovation and productivity components in these programmes make such an effort desirable in the interest of:

- strengthened competitiveness;
- systematic orientation towards innovation;
- human and social goals pursued by the democratic and pluralistic societies concerned in the respective national interests.

2. Some Further-Reaching Implications

The final session demonstrated clearly the urgent need to find ways and means for developing in adapted form concepts for the improvement of work and technology in order to *transfer experiences in R&D and in know-how* to:

- some *European Mediterranean countries* which so far lack efforts to systematically improve conditions of working life;
- *Eastern Europe*, where dramatic changes towards democratic and market-oriented societies are taking place without any policy for restructuring working life accordingly being hitherto in sight;
- *developing countries*, especially poorer ones, where the International Labour Organisation is urgently asking for assistance in this field, with emphasis on industrial health and safety concepts.

Since existing patterns and concepts of the countries already active in this field cannot simply be transplanted, a *process of adaptation* to the respective needs and circumstances will be indispensable.

New forms of cooperation are needed to tackle these tasks in the future. Again, the idea of networking and joint ventures between national and European programmes and institutions, as well as with counterpart organisations, appears to offer optimum chances to make these efforts more effective. Since capacities in this field are limited and scarce in terms of personnel, finances and materials, and since they are also competing with other, often overriding priorities, such a process can only move carefully, step by step, using past experience.

3. Some Tentative Conclusions for the Further Development and Implementation of the German Programme "Work and Technology"

After some short, turbulent periods during the last fifteen years, which were reflected in parliamentary controversies, the German programme "Humanisation of Working Life", now reorganized and known as "Work and Technology", is on the whole running smoothly and successfully.

For the moment, and for some time to come, there are some major *centrifugal forces on the German programme*, which present both *opportunities and risks*:

- There is the *German unification process* which will absorb relatively high capacities of the programme and will require adapted and sometimes new approaches and concepts for the five new Federal States (Länder).

It could endanger future-oriented, innovative approaches unless the chances for an entry of industries and companies into high- tech and highly developed organisational concepts are used.

- There is a strong factual demand and justified parliamentary pressure to let *medium and smaller-sized companies* participate more extensively in the programme. The concept of *"indirekt- spezifische Förderung"* ("indirect specific promotion") is being fostered with a more general, less systematic approach to R&D in favour of more professional consultancy on work and technology.

Both trends tend to endanger the basic character of a research programme unless counteracted by a fruitful balance between proper research and other activities in development and consultancy which are, of course, justifiable. The safeguarding of a broad research base therefore appears to be urgent.

Another important problem is that the German programme seems to receive *insufficient feedback as regards European harmonisation, standardisation and R&D programmes.* There are also *some deficit areas* within the German programme:

- The German *input into international activities* appears to be too weak, except at the ad-hoc level of a very hasty process of European and international standardisation.
- *Findings, results and experiences from other countries and Europe as a whole* in the fields discussed have so far not been systematically analyzed, evaluated and used in German research programmes.
- Compared especially with Scandinavian practices, the *role of labour market parties* and of the *research community* appears to be less strong and is possibly politically underrated (perhaps as a result of the general political patterns).
- The German programme "Work and Technology" seems to need *clearer distinctions between its activities* in research, development, consultancy, transfer and combinations of different methods of promotion. A misleading and sometimes erroneous labelling of activities within the programme increases the danger of losing identity and legitimacy. A correction is a prerequisite for a sound balance between research and related activities.
- In spite of an internationally respected German programme, the *danger of national provincialism* exists where scientific standards are not sufficiently maintained in the core of the programme. Den Hertog pointed rightly to another aspect of such a danger: the *language barrier* of German in a European and international professional world.

The question remains: What *possible steps within the activities of the German programme* might be undertaken in the light of the results of the workshop ?

- A careful *analysis and evaluation of the workshop results* should be undertaken as a possible input into the further development and finer structuring of the German programme.
- *The practical realisation of European and international perspectives* within the framework laid down in the German programme of 1989 (Ch. 6) appears to be urgent:

  - for its further *implementation and development*;
  - for its *sub-programmes* (*Arbeitsschwerpunkte*);
  - for current and future *projects*, and
  - for new *networks*.

- Some of the results might be used as a *possible input* into the development of European and international programmes with improved forms of cooperation between different institutions, ministries and agencies.

It is obvious that the German programme cannot be overextended in its scope and potential demands: It is a government programme with basically national objectives embedded in a European and international context and already taxed by the repercussions of German unification. Within these limits and with the capacities available, the programme could enter networks where activities of third parties would be funded, for example, by the Federal Minister of Development (BMZ), by European or international institutions or by powerful private foundations.

4. A Final Note: The Wider Frame of Reference

The German programme aims expressly to combine the fields of research policy with those of social and societal policies, as well as with training and educational policies of the Federal Government, as represented by the ministries participating. It aims at fostering economic strength and competitiveness. It aims also at the optimum pursuit of human, social, economic and technological goals alongside each other.

As Europeans, however, we must not forget the final basis and reason for all of the programmes present and presented here, including the labour market parties as independent agents in the common concept of Western democracies united within Europe: *Human dignity and human rights* in the field of *work*.

In the countries participating, the fundamental urge for human rights in the relationship between state and citizen has slowly found its way in a long, historical process into the world of work and into the contractual rights of the workers. This process is still going on. May the programmes discussed here contribute to promote it effectively and lend it a longlasting perspective!

Plenary Discussion

Plenary Discussion of Work and Technology Programmes

Mr. Carl A. Asklöf:

I would like to begin by asking a question: "Is it sufficient to focus our attention on the individual work place?" We have heard a lot during this workshop of the changes that are occurring in society. At least two speakers have described the radical changes involved in becoming customer-oriented. They spoke, not of jobs, but of individuals with their own careers. Human resources have to be used. The world of business is changing. This leads to another question: Do we have to change the whole concept of the work place?"

New Accounting Methods

The same applies to accounting methods. Some people in this workshop asked whether the accounting methods that we use today are perhaps not the best ones. We must examine them from a completely different point of view.

Why is this important? - Because these are the principles through which people invest both in human and in physical resources. How otherwise could they make their financial calculations?

Integration into the Broader System

There is also the problem that the work place is becoming more and more integrated into the outside system in different ways. The surrounding system, the supplier-customer system cannot be changed very much at the individual work place, because employers and employees are part of a larger system. The broader system can, of course, be changed but that would be very difficult. And there is also a need for flexibility for the future, which has to be examined in the context of the present.

Working Time and Leisure Time

The next point is the question of working time and leisure time. Information technology and all the possibilities that technology gives us will blur the barrier between working time and leisure time. Currently only 5 per cent of workers are involved in real production using machines, producing something. Even those are mobile. They may use information technology to see that the machines, producing something. Even those are mobile. They may use information technology to see that the machines run properly. Then you have the service sector where people move around and you wonder what they do in their leisure time. Another indicator is that all of us seem to have mobile telephones. We do part of our jobs at home.

This all goes to show that we have to look at the work place from another point of view and in a wider context. It is important to think about this question and what it implies for future programmes. I also think that it is significant that the German programme is supported by many ministries and not just one ministry.

Research and Development

We then have the question of research and development, of development without research, because development takes place every day with no researcher involved, while in some places where researchers are involved there is no development. That is quite legitimate. I think it is very important to discuss all the respective roles. I have noticed on at least three or four occasions that the current role of researchers is research, but that they take on the role of consultants at the same time. Some researchers are normal consultants, other researchers use research methods at the work place and others are involved with basic questions. This brings us to the difficult question of action research which also features in Swedish programmes. Are they intended to be research programmes?

We have heard of many programmes. For example, the two English programmes that were presented are research programmes. Many of our programmes are intended as research programmes. They are in fact development programmes used mainly by the social partners on the labour market. We also have other programmes where we have tried to incorporate both research and development at the work place.

The LOM-programme, which has been discussed here is one of them. But I think, when we look at it, we see that the share of research has become smaller than we at first thought, because it has been rather difficult to do the research. Everyone was so deeply involved in the development question at the work place that they did not have the time for research, perhaps they will have to do it later. We now have a two-year period during which we will bring some of the researchers in to continue in some ways and to stabilize what we have learnt and to do a little more research on it.

We in Sweden intend to examine our programmes from many points of view, but also from this point of view. The first step is for the LOM-programme to be evaluated with the help of Wissenschaftszentrum in Berlin. We will have to look at this programme in comparison to other programmes, because we have now started our sixth or seventh programme.

Dissemination

One point that I would like to mention at this stage is dissemination. Professor Allegro said: through research you gain knowledge, but whether that knowledge induce change is not quite clear. We know very little about this. We should carry out more research into the process of dissemination. We do not even know the state of the art in some areas. Charles Edquist mentioned, for example, that even rather simple statistics are missing regarding FMS systems and CAD systems, the extent to which they are diffused and the channels through which they are diffused?

Dissemination has to be regarded in a broader context. We have to create new networks, networks between researchers and developers. Programmes on this are very important. Indeed many programmes that have been discussed here are programmes on networks.

We also need broader dissemination which will probably not disseminate such high quality information. Moreover we must use established channels. Because only through established channels will people believe what we say. If a researcher goes out for example, from Stockholm to a small company in the North of Sweden it is completely impossible to convince the people what they should do there. They need persons that they trust.

Programmes like this must work together with those of other organizations, the social partners in the labour market and of course other organizations.

One might also wonder where the main flows of knowledge occur today in working life? We should find out a little bit more about the main streams of knowledge transfer generally. Business changes, structural changes: how do they come about, where does the information come from and where does it go to? Our information has to go through these channels, then we have a chance of influencing the process where it starts.

Importance of Small Companies

The final point concerns small companies. I think it is extremely important for programmes such as those we mentioned here to focus also on the problems of small companies; much government support regarding research and development is for large companies. We must do much more for small companies.

Now I am back where I started: change at the work place. You cannot only discuss work and technology with a small company. You must discuss the whole of its business situation, its surroundings, its work environment and its financial situation. That is the challenge: How should support and advice be arranged? How do you give advice to small companies so that they can accomplish what they really need, without trying to impose on them what we think they need.

Annex

National and European Programmes of Research and Development in Work and Technology

Tony Hubert, Brussels, Belgium

1. Introduction

The purpose of this report is to present the current state of development of national research programmes on "Work and Technology". In the course of the quest it became clear that each nation state examined has its own concepts and structures. Thus, apart from Germany, the five key words cover the general field in differing ways and to different extents:

- "national" programmes sometimes only cover a part of the nation - one segment of the population which can be, for instance, a geographical area (eg. Flanders) or a profession (eg. researchers of the French CNRS);
- "research" programmes can be designated for a clearly defined group of researchers (eg. the TAO programme in the Netherlands) or very broad sets of projects financed on the basis of their scientific or (more generally) practical merits (eg. AuT and AMFO in, respectively, Germany and Sweden);
- "programmes" can be reactive (with funders waiting for submissions) or proactive (with funders looking for topics in clearly specified fields decided on on the basis of needs), sometimes with contributors sought out or themes put out for competitive bids through tenders (eg. the German Arbeitsschwerpunkte and the AMFO programmes);
- "work" is essentially the soft domain of one ministry (labour or employment), which hardly has a high profile in society. "Technology", on the other hand, is the "stuff of future national competitiveness", promoted through vast dotations by sexy (though still bureaucratic) ministries of technology, research or industry. And never the twain shall meet ... well, just in a few countries and programmes.

Only the Germans and the Swedes really do have continuing, government financed joint work-and-technology research programmes. Thus, overall this report should consist of descriptions of two countries. Since it is destined to be a background document for the Dortmund workshop where most western European countries will be present, a relatively broad interpretation has been given to the area.

The information provided has drawn on a variety of sources (over some delightful refreshments); but none more valuable than the publication of MERIT, Maastricht Social research for technological change by J Friso den Hertog & Peter Schröder (1989).

2. Germany

2.1 Arbeit und Technik (AuT)

AuT ("Work and Technology"), launched in 1989, is the follow-on programme to HdA, the "humanisation of work" research programme. It runs for the quadrennium 1990-93, having a total budget of MECU 195 (DM 400m).

The AuT programme strives to prevent, rather than cure. Thus, its stated basic aim is to protect health by reducing, avoiding and preventing deleterious stress factors through the innovative design of work and technology in accordance with human needs, taking at the same time environment protection into account. This aim is to be achieved at the same time as enhancing innovative organisational potential by raising efficiency, productivity and flexibility: worker demands and corporate goals need to be in harmony. The key words are: innovation, prevention, joint and holistic, designers-manufacturers-users cooperation, alternative solutions, the market, active transfer and implementation and linkages between technical, social, and organisational change.

The programme has a multi-partite advisory committee with representatives from research, trade unions, employer organisations and government, under a neutral chairman. This committee has drafted a policy document to guide in the choice of AuT projects based on the principle of "productivity increase in accordance with human needs".

The HdA had two sponsoring federal ministries: BMFT - research and technology, and BMA - labour and social affairs. To these, AuT has added a third: BMBW - education and science. The aim in this respect is to enhance the impact of the programme results on the educational and training systems in a broad sense. This means not only recurrent vocational training, higher education and other "knowledge-transfer bodies", but also developing individuals' self-learning and increasing the numbers of jobs that promote continuous learning.

The main role as regards both funding and management is performed around BMFT. Thus management and administration are devolved to DLR, the German Aerospace Agency, AuT being one of that agency's three divisions.

The programme is meshed in with the other federal research and technology programmes which should be able to suggest research topics and in turn be influenced by AuT's project results. Through the focus on accident prevention, it should help reduce the cost of occupational illness and accidents and contribute to improving the industrial relations system.

HdA was particularly successful in the areas of reducing physical and mental workloads, the development and use of industrial robots, joint projects for branches of industry and developing measurement instruments. Since the mid-1980s priority programme areas have emerged:

- the identification and prevention of occupational diseases and work-related health hazards;
- designing user-friendly and locally-adaptable software
- devoloping integrated planning processes; and
- operator involvement in problem-solving.

To avoid any discordance between the stated preventive goals and corporate reality, the AuT has been assigned the following goals for transferring available knowledge:

- the regular compilation of project results for the benefits of a variety of user target groups;
- the greater mutual transfer of the available knowledge and experiences to other policy areas;
- more demonstration projects need to be carried out which are more convincing;
- emerging societal, economic and technical trends need to be assessed to determine their possible effects on work and technology; and
- new networks have to be tapped for the transfer process.

"Transfer" of results and know-how has to be designed into projects from the very beginning. The participation of key individuals and groups in all stages is essential in this respect. Standard-setting (DIN norms) can be an appropriate way disseminating some results, ... but not others. This is particularly the case of such softer issues as organisational design, skills, ... Experience indicates that for results to be used, programmes and projects have to be consistently focused on the needs of the organisation and the workplace; the positive financial returns on projects need further emphasis. And experience also indicates that branch projects are particularly effective.

Results also have to be built into specialised and general training programmes and exhibitions.

2.2 Priority Areas

Given fund-limitation, priority areas are singled out for particular focus of projects over a 5-12 year period. The table at the end of this section provides an overview of priority areas.

2.3 Funding Conditions

To receive funding, projects must:

- be economically or scientifically risky, yet be likely to come up with results: the nearer to the market, the less the funds;
- be likely to come up with resultsof interest to a large target group;
- relate to AuT aims; and
- demonstrate (for company projects) the novelty and demonstration value.

Furthermore, all corporate projects must demonstrate that:

- attention is being paid to the secondary technological and employment impacts;
- the results can be used in other firms and branches;
- the project results in no dismissals or negative displacements for those involved.

Subsidies (grants), whether based on costs or expenditures, amount at most to 50%. Larger firms (>DM 200m - MECU 97 - turnover or >1,000 employees) can receive no working capital. Recognised scientific institutions can have 100% of their expenditures financed.

Grant-receivers have to produce interim and final reports in a language understandable to normal human beings! In addition, articles in professional journals should be produced.

2.4 Finance

The following financial resources have been earmarked through to 1993:

1. Health protection through reduction of and protection from stress and strain	9.9	9.9	9.9	9.9	9.9
2. The human use of new technologies	9.9	10.2	10.4	10.7	
3. Transformation of scientific knowledge and corporate experience into the human design of working conditions	4.4	4.4	4.4	4.4	4.4
4. Humanly designed working conditions in specific branches	16.3	16.5	16.5	16.7	16.9
5. Basic and cross-sectional research, project management	5.8	5.8	5.8	5.8	5.8
Total	46.1	47.3	47.6	47.8	48.3

2.5 Promotion Instruments

Three sets of instruments are used

- *Individual projects*. The range of projects financed extends from basic research relating to work, through the design of new technological solutions, innovative in-company pilot projects, the transformation of research results into organisational practice (especially for smaller firms), in-company training and development, evaluations and appraisals (especially drawing on research from abroad) to analyses synthesizing a series of projects for the benefit of information mediators.
- *"Joint projects"* (Vorhabensverbünde). Third party establishments can undertake projects for groups of companies having similar needs or belonging to a single industry (for which a number of priority areas have been singled out). "Transfer projects" focus on overcoming the problems of transferring approaches and results from one area to another.
- *Meetings*. A variety of means are available for enhancing learning by bringing together similar and dissimilar specialists both for exchanging experiences and enhancing solutions to common problems.

2.6 Projects Funded

Since the establishment of the HdA, some 1,600 projects have been funded, including the on-going finance of separate trade union and employer groups providing information and advice on implementations relating to work and technology to their members.

2.7 Focal Areas

Special research areas are designated according to needs assessed by the consultative committee. Research proposals made in these areas are publicly announced for tendering usually for a period of 5 years. The current areas are:

- applying new technologies in manufacturing in accordance with human needs
- designing working conditions in accordance with human needs in specific industries: the foundry industry ...

  - ... the forging industry ...
  - ... road transport ...
  - ... coal-mining...
  - ... public commuter transport;

- noise abatement in sheet-metal working ...

  - ... the textile industry ...
  - ... the printing industry ...
  - ... in software design and planning;

- working conditions and the health of the working population;
- health protection at workplaces involving new information and communications' technologies, ...

  - ... for applying new technologies in production, ...
  - ... at workplaces in air-conditioned rooms;

- cancer risks at the workplace;
- measurement techniques for monitoring noxious substances at the workplace;
- basic principles and issues of the programme;
- allergic reactions and diseases;
- the design and application of new technologies in the office and public administration;
- improving working conditions in the building industry; and
- work and technology in municipal administration.

2.8 Fertigungstechnik - The Production Technology Programme

This programme was launched in 1980 for a 4 year period. It was subsequently extended for further 4 year periods in 1984 and 1988. The programme focuses on the capital goods industry. The two central technological themes of the first 4-year programme were:

- CAD/CAM, which moved on to CIM in the second 4-year programme;
- Industrial robots, which moved on to new production and assembly technologies in the second 4-year programme.

Projects have two stages: the first stage consists of the analysis of the production system, the technological requirements and the organisation. For this stage, grants cover staff costs for systems analysis, retraining and external consultants. Stage 2 covers development as such, the acquisition of software modules, which couple existing CIM components and includes retraining. Companies can have up to 40% of the costs they incur covered, but this cannot exceed MECU 0.14 (DM 300,000). Total finance for each 4-year period, including the current, is MECU 180 (DM 400m).

There are four types of projects:

- *Individual company projects* (Indirekte spezifische Förderung), the bulk of projects;
- *Joint projects* (Verbundprojekte) covering specific topics of interest to a number of companies such as "quality maintenance in flexible manufacturing", "reducing waiting time" and "new adhesives". Such projects can also link small and medium-sized companies with large companies and research institutes.
- *Technology assessment*. Projects cover such matters as organisation, training needs, qualifications and work contents.
- *Knowledge and experiences transfer* - through publications, workshops, training courses and demomonstration labs.

2.9 Rationalisierungs-Kuratorium der deutchen Wirtschaft (RKW)

The RKW is a tri-partite body originally established in 1921 and reconstituted as the national productivity and management centre in 1951. It has three main domains for helping industry, particularly medium and smaller firms:

- technology
- labour and social economics (ASW), and
- business economics.

All three departments carry out and commission research and studies. The ASW focuses notably on the interactions between technology and work, a subject of major concern to its scientific advisory council (Beirat). It acts as an agent for transforming and transferring the results of research carried out in the AuT programme to work *organisations.*

Addresses

AuT, Südstrasse 125, D-5300, Bonn 2

Produktionstechnik

RKW, Postfach 5867, D-6236 Eschborn

Priority areas of the "Work and Technology" research Programme (as per July)

Status	Basic principles/cross-sectional issues	Production	Services
expering		Reduction/Elemination of stress factors and development of new working structures in the clothing industry	Reduction/Elemination of stress factors in the hotel and catering business
		Reduction/Elemination of stress factors and development of new working structures in the packing industry	
		Improvement in working conditions in the joiner's craft	
taken over from the HdA programme and updated	Research on stress and strain factors (e.g. noise, vibrations, combined and mental stress factors)	Noise abatatement using the textile industry and the sheet metal processing industry as examples	Planning of working conditions in road transport Research and development
	Working conditions and health	Application of new technologies in production in keeping with human needs	serving the design and application of new technologies in offices and administration in accordance with human needs
	Cancerogenic risks et the work place		
	Allergic reactions and diseases	Improvement in working conditions in the forging industry	
	Software design in accordance with human needs		
	Holistic planning processes including extended profitability calculations	Improvement in working conditions in the foundry industry	
	Dissemination and implementation of new scientific findings and operating know-how including accounting	Improvement in working conditions in the mining industry	
	Programme-related and con-comitant research	Application of new assembly line technologies in production in accordance with human needs	
new/planned	Simple measuring processes for noxious working substances	Application of new technologies in the printing industry in accordance with human needs	Design of new technologies in commuter traffic in accordance with human needs
	health protection at the workplace in air conditionend rooms	Improvement in working conditions in the building industry	Planning of working conditions of office workers in accordance with human needs
	Work-orientied design, health protection and acquisition of skills when new technologies, processes and materials are used	Improvement in working conditions in the food industry, agriculture and forestry	Planning of working conditions in the trade in accordance with human needs
	Work-related design of networked systems (e.g. CIM, Logistics, JIT, etc.	Transfer of design know-how in certain industries - taking the example of the shoe industry	Improvement of working conditions in nursing and providing care for old, sick and handicapped people
	Work-orientied concepts of organisation, acquisition of skills and staff management	Planning of working conditions in craftmen's business in accordance with human needs	
	Concepts of the acquisition of skills by special groups of employees		
	Demographic trend and effects on working conditions and employees structures		

Annex on *AuT* focal areas

3. Sweden

3.1 AMFO - The Swedish Work Environment Fund

The Swedish Work Environment Fund (AMFO) was created in 1972 "to meet the needs of the entire Swedish labour market as to research and development (R&D), training and information". The main task of the Fund is to finance projects -- about 1,500 a year.

Decisions on projects are taken by the Board, consisting of trade union and employer representatives.

AMFO's annual budget (+MECU 120m - SEK 900m) is financed to 90% through the work environment levy and to 10% through interest on capital. Government has earmarked certain proportions of the levy for specified activities; the largest is for information and training on co-determination at work - MECU 3O.6, SEK 230m -followed by MECU 9.6 - SEK 72.6m - for activites of the regional safety delegates.

The activities of the Fund itself essentially concern grants for research and development projects (MECU 14.5 - SEK 109m -annually). These relate (in order of financing totals) to:

- psycho-social aspects, work organisation, co-determination;
- chemical hazards;
- work postures/loads;
- workplace design;
- physical hazards;
- research appointments;
- local safety work/occupational health; and
- work accidents.

The Fund also has a significant publications and information dissemination programme. A regular newsletter and final programme reports are published in English.

The Fund has detected the following as **trends** which it is promoting:

- Several traditional working life areas remain in the limelight: musculoskeletal injuries; accidents (including the risks connected with computerisation); chemical hazards, noise and vibration; the impact of the psycho-social working environment.
- Newer areas are continuing to grow in importance: the interrelationships between organisation, leadership and participation (the so-called LOM programme examining changing patterns in this field is being concluded in 1990); learning and the development of competence at the workplace ("It is important to tackle these

questions side by side with the more traditional aspects of the working environment, since the key to action is often to be found in what can be broadly termed the decision-making systems of companies and administrations" - Plan of Activities, 1990-1992 - p. 23).

- The need to design dissemination procedures according to the key characteristics of target groups. Growing importance in this respect is attached to ensuring that the various educational systems are aware of the results of experiments and research. Thus, the Fund has started to finance university chairs since the late 1980s (supporting notably the Gothenburg Centre for Work Sciences - cf below), trying to introduce more working life concepts into normal training and information channels.
- Influencing the design processes of the manufacturers of capital goods. The main argument here is that preventive solutions to working life issues are more effective and economical than the curative. Since Sweden is relatively small internationally, the Fund feels compelled to cooperate more with institutes and firms in other countries (a position epitomised by an agreement with the German AuT in the wood-working sector - cf EUROPE below).
- The Fund is taking more integrated and longer term stances, with greater focus on programmes, rather than individual projects (which are administratively heavy - even though the staff can decide on making grants of up to ECU 67,000). Programmes tend to be established for 5 years. Cooperation with other national organisations is sought - for instance STU, the national board for technical development.

In addition to those programmes constituting part of the project areas indicated above, programmes have been launched in the following domains:

- MDA - people-computers-work, a joint 4-year MSEK 64 programme with STU. The programme is aimed at promoting the development of computer technology at work that is adapted to human needs and requirements. It operates by means of inter-disciplinary work between researchers in technology/ computer sciences and social sciences/ humanities and cooperation between researchers and users.
- tackling musculoskeletal injuries;
- raising small business' awareness of the good economics of better working conditions;
- equal opportunities;
- local environment work and occupational health services; and
- learning and competence.

Another possible programme area concerns logistics, which plays an increasingly important role in societies consisting of networks of companies, particularly around large multi-national corporations.

A unique feature of AMFO activities was its 1987 three-year export development programme to explore the possibilities of increasing Sweden's exports of products and services which have a "work environment content".

3.2 Gothenburg Centre for Work Science

The Fund has taken the initiative to create and agreed to finance (MECU 3.4 - SEK 25m) for an initial three-year period the Gothenburg Centre for Work Science. The aim of the Centre is to

> *"find new and more efficient methods to recruit young, talented students into the special fields of work-orientated knowledge."*

The Centre runs two research programmes: "Management and Organisation of R&D Activities in the 1990s" and "The Future of Work Organisation" including alternatives to assembly systems, development of multi-skilled employees in relation to job security, productivity and quality of work and social values, incentive systems and work schedules.

3.3 Arbetslivscentrum (ALC)

The national institution that has focused on working life research is the Arbetslivscentrum. It has recently been reviewed (for the third time since 1987), as a part of a broad review of research on working life in Sweden (cf below). Established in 1977 to help implement the new co-determination law ("MBL"), ALC has had a staff of up to 60 researchers who have concentrated (at least to 1987) on five areas of research:

- public sector;
- labour law;
- management and new forms of work organisation (including contributions to major AMFO programmes);
- new technology and work;
- women and work.

3.4 Other Working Life Bodies

The range of organisations involved in Sweden in working life issues, including research, can be appreciated by the following list taken from Working Environment International 1990.

The Board of Occupational Safety and Health for Government Employees (SAN)
Information, guidelines and training in the civil service sector.

Statens arbetsmiljönämnd
Box 16386
S-103 27 Stockholm
+46 8 700 1300

The Construction Industry's Organsiation for Working Environment, Safety and Health
Founded in 1967, it is the first and largest of a growing number of industry-specific health services.

Bygghälsan
Box 94
S-182 11 Danderyd
+46 8 753 37 75

The Forest Operations Institute

"Skogsarbeten"
Box 1184
S-164 22 Kista
+46 8 750 72 20

The Foundation for Safety and Health in the Cooperation Union (KAN)
Training and information in the cooperative sector.

Kooperationes arbetsmiljönämnd
Box 20203
S-104 60 Stockholm
+46 8 44 98 20

The Joint Industrial Safety Council (ASN)
Assists private industry and trade in their local safety work - primarily through training and information. It is the leading producer of basic educational material in the field of working environment.

Arbetarskyddsnämnden
Box 3208
S-103 64 Stockholm
+46 8 22 94 20

The Laboratoy of Industrial Ergonomics (AML),
at the Department of Work Science, Royal Institute of Technology. Applied research and development in the field of working environment, technology and work organization.

Arbetsmiljölaboratoriet
S-100 44 Stockholm
+46 8 790 61 35

The Ministry of Labour's Division for Employment Conditions
Questions concerning safety and health, occupational health services, vocational rehabilitation, working time, security of employment and codetermination. Head of several authorities in the field.

Arbetsmarknadsdepartementets arbetslivsenhet
S-103 33 Stockholm
+46 8 763 13 49

The National Board of Occupational Safety and Health (The Board)
Central authority for matters concerning occupational safety and health. The Board is responsible at national level for the enforcement of the Work Environment Act. The Board issues regulations and recommendations under the Act.

The Board also conducts informational programs.

Under the direction and supervision of the Board the Labor inspectorate (yrkesinspectionen) is responsible for the enforcement of legislation. The Inspectorate is divided into 19 districts.

Arbetarskyddsstyrelsen
S-171 84 Solna
+46 8 730 90 00

The National Chemicals Inspectorate (KemI)
issues regulations and supervises the obervance of legislation on chemical products. Carries out health and safety assessments. Responsible for pesticide approvals.

Kemikalieinspektionen
Box 1384
S-171 27 Solna
+46 8 730 57 00

The Institute of Environmental Medicine (IMM)
at the Karolinska Institute focusses on the chemical and physical environment. Together with the Departments of Environmental Hygiene and Toxicology at the Karolinska Institute, IMM is Sweden's WHO collaboration centre for environmental health effects.

Institutet för miljömedicin
Box 60208
S-104 01 Stockholm
+46 8 728 64 00

The National Institute of Occupational Health (AI)
Sweden's leading research institution in the field of occupational health. The Institute is also responsible for the training od occupational health service personnel.

Arbetsmiljöinstitutet
S-171 84 Solna
+46 8 730 91 00

The National Rescue Services Board
The Board is directing the development of civil defense end rescue services. Responsible for land transportation of dangerous goods and for coordinating and developing actions taken in connection with discharges of oil and other chemicals.

Statens räddningsverk
Karolinen
S-651 80 Karlstad
+46 54 10 30 00

The Swedish Centre for Working Life (ALC)
Research and information emphasizing codetermination at work and the application of labor legislation.

Arbetslivscentrum
Box 5506
S-114 86 Stockholm
+46 8 790 95 00

The Swedish Confederation of Professional Associations (SACO)
Represents about 310.000 employees.

SACO
BOX 2206
S-103 15 Stockholm
+46 8 22 52 00

The Swedish Confederation of Professional Employees (TCO)
Represents about 1.300.000 salaried employees.

Tjänstemännens
Centralorganisation (TCO)
Box 5252
S-102 45 Stockholm
+46 8 782 91 00

The Swedish Council for Management and Work Life Issues
Research.

FA-rådet
Box 5042
S-102 41 Stockholm
+46 8 667 97 35

The Swedish Employer's Confederation (SAF)
Represents about 43.000 companies with 1,2 million employees.

Svenska Arbetsgivareföreningen
S-103 30 Stockholm
+46 8 762 60 00

The Swedish Fire Protection Association (SBF)
An independent non-profit organization engaged in fire prevention and protection

Svenska
Brandförsvarsföreningen
S-115 87 Stockholm
46 8 783 70 00

The Swedish Foundation for Occupational Health and Safety for State Employees
About 125 service centres all over the country.

Statshälsan
S-371 87 Karlskrona
+46 455 349 00

The Swedish Management Group
Training, education.

Box 1213
S-181 24 Lidingö
+46 8 762 74 00

The Swedish Trade Union Confederation (LO)
Represents 2,3 million bluecollar workers in the public and private sectors.

Landsorganisationen i Sverige
S-105 53 Stockholm
+46 8 795 25 00

The Swedish Vocational Training and Working Environment Council of the Transport Trades (TYA)
Safety and health at work in the transport sector.

Transportfackens Yrkes- och
Arbetsmiljönämnd
S-171 26 Solna
+46 8 734 52 00

The Swedish Work Environment Association (FFA)
A non profit organization with the aim of stimulating measures to create safe working conditions. Major activities:

- publication of "Arbetsmiljö", "Working Environment" and other literature.
- participation in exibitions and conferences.
- other information services in the work environment field.

Föreningen för Arbetarskydd
Box 5970
S-114 89 Stockholm
+46 8 16 67 40

The Swedish Work Environment Fund
The Fund finances research and development, training and information with the aim of improving the working environment in a wide sense. The Fund has extensive financial resources at its disposal - well over SEK 700 million anually - as a result of a special payroll tax paid by all amployers.

Arbetsmiljöfonden
Box 1122
S-111 81 Stockholm
+46 8 796 47 00

The Work Environment Association of West Sweden
Courses and conferences, sales of safety equipment.

Västra Sveriges
Arbetsmiljöförening
Pusterviksgatan 15
S-413 01 Göteborg
+46 31 13 64 43

3.5 Changing Dimensions

A government-appointed Commission is reporting on improving "the 400,000 worst jobs" in Sweden. In this connection, a new fund, Arbetslivsfonden (a temporary payroll tax paid by employers) has been created (+ECU 2bn -- SEK 15bn) to improve the working environment. The main reason for establishing the Fund was to siphon off inflationary demand in the economy. It will seek to return the money to the contributing companies particularly through projects for the disabled and longer term unemployed which the companies themselves jointly determine.

The report Arbetslivsforskning: Inrikning - Organisation - Finansiering, published in June 1990, examines the current state of working life research in Sweden and makes recommendations for the future. The key findings are as follows:

- resources devoted to working life research are sparse compared with R&D for production technology and management;
- there is "unreasonable concentration" on traditional disciplines; there is insufficient multi-disciplinarity;
- insufficiently impacts national policies;
- dissemination of results is insufficient; and
- evaluation procedures are weak.

Recommendations relate to the nature and activities of both AMFO and the ALC. They also include the need for research to be undertaken in five areas:

- international working life, focusing on Sweden's interaction with the rest of Europe;
- human, work and new technology (there has been too much focus in the past on "old" technology);
- work in relation to other roles in life;
- rapidly changing working life; and
- the growing service sector.

Addresses

AMFO, The Work Environment Fund,

Gothenburg Centre for Work Research, Chalmers Teknikpark, S-412 88, Gothenburg. 46 31 724 120

ALC, The Working Life Centre, Box 5606, S-114 86

Arbetslivsfonden, Box 4014, S-131 04 Nacka

4. France

4.1 ANACT - The National Quality of Working Life Agency

The main organisation in France concerned with work and technology is ANACT - the national agency for improving working conditions.

Founded in 1973 as a joint venture between government and employers and trade unions, and with an annual budget of MECU 9 (FFR 60m), its purpose is to encourage corporate socio-economic change/modernisation. The director of ANACT plays a coordinating role in carrying through the government's policy of fostering "negotiated modernisation" - ensuring that enterprises manage change through a dialogue process with all groups of employees. ANACT is thus concerned with action rather than research.

The 60-odd staff operate nationally and regionally in six main areas related to improving working conditions in processes of technological and organisational change:

- safety and health (covering: chemical and physical risks; work-stations for the handicapped; the impact on companies of an aging workforce);
- managing industrial projects (covering: work space; and worker involvement in the design stage);
- career development and the management of employment (ANACT seeks out firms which have received public grants and are representative of the industries in question);
- working time (where there is considerable interest in what is taking place elsewhere in Europe);
- production management, particularly changing physical and mental stress related to just-in-time approaches;
- computerisation in the tertiary sector, including the ergonomics of software development.

ANACT helps enterprises and administrations to enhance the relationships between work and technology both directly (through advice) and indirectly through a broad series of media activities: brochures, handbooks, check-lists, videos, etc. ...

ANACT focuses on particular branches: butchers, foundries, industrial cleaning, building, wood-working, hospitals, "collective restaurants", leisure parks, agriculture and industrial furniture.

4.2 FACT - Fund for Improving Working Conditions

ANACT staff also (together with the labour inspectorate) advise FACT - Le Fonds pour l'Amélioration des Conditions de Travail on making grants. The grants in question (Fact has an annual budget of some MECU 3 - FFR 20m) contribute to the costs of actions by enterprises and branch organisations in two main areas:

First, exemplary and innovative action to improve working conditions (both improved processes and organisation). Three types of enterprises are targeted:

- those individually seeking to improve conditions at the work place;
- those acting with other firms to improve the work place;
- those designing ergonomic equipment.

Second, studies of the social impacts of modernisation on the enterprise or branch levels: risk prevention, physiological conditions of work, the organisation and contents of work and competences to be drawn on, organising working time, training, the nature and numbers of jobs, ...

Projects have to go beyond legal requirements and be approved by a representative body of the employees: works councils, safety committee, etc. Results have to be made available to the general public. Grants can be made of up to 50% of study costs, 30% of the additional amount of the investment for improving working conditions and 30% of the staff time spent on the project.

4.3 Programme Mobilisateur Homme, Travail et Technologie

Research relating to work has constituted a part of the activities of the ministry of research and technology since 1982. More specifically, the Programme mobilisateur Technologie, Emploi, Travail was established in 1984, with a change of name, though virtually nothing else, in 1989 to Homme, Travail et Technologie - HTT.

The aim of the HTT is to develop scientific knowledge which is useful to the improved implemention of new technologies and new forms of "productive organisation". It seeks to better understand their economic, social and human dimensions. Equally important is the nature of the research which is joint in two senses: In the first place it is interdisciplinary in character involving both human/societal and engineering sciences. Secondly, it directly involves enterprises (including consultancy bureaux), public administration and professional bodies in formulating the issues, carrying out the research as such, funding it and using the results. Its medium term goals are

- to strengthen new disciplines, new focuses of research and lasting research/enterprises cooperation;
- to ensure that previous research results are used, particularly by smaller firms;
- to build up new knowledge for various Programme actions such as training, employment and local development, ...

4.4 PIRTTEM - Programme Interdisciplinaire de Recherche sur la Technologie, le Travail, l'Emploi et les Modes de Vie

The main body of scientific researchers in France is the Centre national de la Recherche scientifique - CNRS. It employs 25,000-odd researchers in its 1,400 research units, organised by disciplines. From 1982 research has been developed in interdisciplinary research programmes: PIRs.

One of the first such programmes to be created was PIRTTEM - Programme Interdisciplinaire de Recherche sur la Technologie, le Travail, l'Emploi et les Modes de Vie. This links human and societal sciences, life sciences and physical sciences for the engineer to develop fundamental research capacities. The aim of PIRTTEM is

> *to describe, analyse and evaluate the causes, the effects and what is at stake as regards recent technological change in the fields of production, social structures and life styles.*

PIRTTEM funds can only be drawn on by members of CNRS teams, enabling them to undertake additional research (rather than covering existing staff, etc costs). However, other research groups and enterprises can make use of the call for tender schemes if they link up with a CNRS team.

Both HTT and PIRTTEM put out a coordinated annual tender for research contracts (the amount available annually being of the order of MECU 0.7, with grants - of between ECU 4,000 and 70,000 - being given for 2-3 years). For 1989 and 1990 the topics have remained essentially the same:

- Modernising enterprises: the topics covered are the development of management, evaluation and remuneration tools, processes for increasing technological innovation, and drawing on various types of networks;
- Modernising services, including public services and administrations:
- Human systems/technical systems: the training of and learning by relatively unskilled adults
- Human systems/technical systems: ergonomics and productivity (the aim being to develop the analysis of work and design ergonomics to improve the quality, reliability

and the efficiency of work processes and reduce their unpleasantness, particularly in complex socio-technical systems);
- Human systems/technical systems: health and the working environment.

Addresses

ANACT, 7 Boulevard Romain Rolland, F-92128 Montrouge

FACT, Ministère du Travail, de l'Emploi et de la Formation Professionnelle, Direction des Relations du Travail CT 1.2, 1 Place Fontenoy, F-75007

Hommes, Travail et Technologie, Ministère de la Recherche et de la Technologie, 1, rue Descartes, F-75231 cedex 05

PIRTTEM, 15 quai Anatole France, F-75700

5. Great Britain

5.1 The Work Research Unit

The Work Research Unit was set up in 1974 as part of the Department of Employment. It was overseen at that time by the Tripartite Steering Group on Job Satisfaction (since defunct), whose meetings were chaired by a minister from the Department. The purpose of 0the WRU has remained broadly unchanged since its inception. It is

> *to encourage and stimulate industry and commerce to adopt measures that can lead to greater effectiveness and better economic performance through increasing the quality of working life (QWL).*

This means that balanced importance has been attached to ensuring that improving the quality of working life goes hand in hand with improving an organisation's economic efficiency and effectiveness.

Following a period of evaluation as regards both its existence and institutional attachment in the early 1980s, the WRU has since 1985 been incorporated into the advisory services of the Advisory, Conciliation and Arbitration Service (ACAS), a body financed by, but independent of, government.

The WRU, despite its name, has only been involved to a limited extent in undertaking research -- and this was mainly soon after its creation. Nor, as part of ACAS, has it had money to commission research from third parties. However, one of its four prime missions is to follow research developments. But with a small staff (10 professionals) and its other prime missions being given higher priority (cfr below), the results of its research monitoring activities are necessarily limited. On the other hand, it has built up an extensive information service on QWL matters, which is available to the public. The four prime missions of the WRU are:

- to create awareness of QWL through mounting and participating in conferences and courses;
- to provide information in response to enquiries and to publish papers;
- to monitor QWL activity and research; and
- to provide advice and assistance to organisations through project work and visits.

All the Unit's staff are involved in company projects which, increasingly, are referred to it by ACAS field advisers who are located in offices throughout England, Scotland and Wales. The Unit's quarterly publication WRU News and Abstracts is mailed to some 2,000 individuals and organisations.

Some of the main themes relating to work and technology being handled by staff members at present relate to the management of technological and organisational change, payment and job evaluation systems in the context of change, employee involvement, Total Quality Management, teamwork and organisation culture.

5.2 Ministries

The main governmental funding body for technological research and development is the Department of Trade and Industry (DTI). However, DTI does not appear to be interested in the inter-relationships between work, organisation, human resources and technology. Thus, its main programmes (ALVEY in information technology, ...) are "purely technological".

The Employment Department - ED, on the other hand, has a research programme aimed at filtering into the employment policy division relevant research findings. The research division thus commissions research of third parties relating to emerging employment issues. At present, surveys and research projects connected with technology concern:

- The 1987 British Social Attitudes Survey. This is based on a random sample of 3,000 households which were asked questions about the impact of new technology on work. The same questions were repeated in the 1990 Social Attitudes Survey. Overall, there was a positive feeling towards new technology.
- The workplace industrial relations survey (WIRS), 1984, which demonstrated an positive acceptance of new technology by the bulk of the British workforce;
- The impact of new technologies on the quantity of employment, on employment flexibilities, on skills and on the impact of future training policies. Complementing

WIRS3, DE is carrying out (with the Training Agency) a major national survey of employers' recruitment and skills training practices;

- Teleworking. Various forecasts have been made of the future extent of teleworking (up to 4 million persons by 1995). These appear not to have paid sufficient attention to the associated barriers and obstacles. A set of research projects is being commissioned by ED to understand these barriers with a view to helping towards the development of government policies which can reduce them;
- The impact of technology and changing legislation on work (the abolition of the Dock Labour Scheme);
- Labour market flexibility. Research is focused on the extent of part-time working, why it is used, whether it really is second-class labour, temporary employees and peripherals.

5.3 Economic and Social Research Council

The main funding body for the area "work and technology" is the **Economic and Social Research Council - ESRC**. The Council has four committees charged with developing new ideas for research programmes, the largest of which is Industry, Economy and Environment Research Development Group - IEERDG. (The other research development groups cover society and politics, human behaviour and development and the research resources advisory group.) IEERDG spent some 45% of the MECU 18.6 (£13m) available for initiative research in 1989-90 in 15 areas including competitiveness, the history of innovation, new technologies and the firm (NTF) and the Programme on Information and Communication Technologies (PICT). NTF and PICT are outlined below.

NTF is part of a matrix of work currently proceeding on technology-related issues. It has been formulated after discussions with industry, government and academics. The object is to advance understanding of the factors that relate to the initiation of, and response to, new technologies in firms. The emphasis is on the processes and problems involved in technological change within organisations. The focus is the firm in its institutional and market context.

There are five interrelated elements to the initiative. The first is a core on which others feed and to which they contribute: the technological change process and outcomes within the firm (how firms select and implement change, what the outcomes are and what this teaches about technological change). The other elements are:

- the external sources of finance for new technology projects;
- skill shortages and the methods firms employ to overcome them;
- the relationships between technological change and changes in product markets; and

- international collaborative arrangements between firms in order to access, develop, market or exploit new technologies.

The project costs MECU 1.72 (£1.2m) over 5 years, earmarked as follows:

89/90	**90/91**	**91/92**	**92/93**
0.67m	0.63m	0.28m	0.18m

The **PICT** initiative examines some of the fundamental long-term economic, social, managerial and policy implications of diffusion and development of new information and communication technologies. Its starting tenet is that the successful development, adoption and diffusion of new technologies is closely related to appropriate economic, institutional, organisational and policy changes. It was established to complement the large-scale national and European Community science and engineering research programmes, such as ALVEY and ESPRIT. PICT has adopted a strong interdisciplinary orientation particularly in the selection of research centres and their networking: a central coordinating unit and six interdisciplinary research units.

The annual MECU budgets for PICT for the academic years are:

89/90	**90/91**	**91/92**	**92/93**
1.2m	1.05m	0.48m	--

Addresses

ACAS, Advisory, Conciliation and Arbitration Service, 27 Wilton Street, London, GB-SW1X 7AZ

Employment Department, Caxton House, Tothill Street, London, GB-SW1H 9NF

ESRC, Economic and Social Research Council, Cherry Orchard East, Kembrey Park, Swindon, GB-SN2 6UQ

6. The Netherlands

The Ministry of Social Affairs and Employment provides a little more than MECU 9 per annum for research related to social and technological innovation. This is used primarily for funding institutions dealing with physical working conditions and medical factors relating to work. An additional MECU 5 is spent on commissioned research into current policy issues related to working conditions.

Technology policy is the responsibility of the Ministry of Economic Affairs. Its objective is "to contribute towards sustained economic growth and thriving social development by strengthening the technological base." This policy is implemented through 3-5 year "incentive programmes". A major problem experienced in this respect is that of compartmentalisation: programmes for connected areas such as quality control, technology assessment, computerisation and automation, etc remain ... unconnected.

However, in its Technology Policy Survey 1989-1990 the Ministry points to the importance not only of the educational system (both for basic and recurrent education for using technology and for the diffusion of knowledge of the practical applications of technological innovations) but also to the fact that the "close correlation of technological change and the social environment is extremely important". ("Another essential condition for the use of the growth potential is social acceptance of new technologies".)

6.1 TAO and MERIT

One broadly inter-disciplinary 5-year programme is TAO: "Technology, Work and Organisation", launched in 1986 (but only starting in 1988). It is jointly run by two university centres, with MERIT (Maastricht Economic Research Institute on Innovation and Technology) entrusted with industry and Erasmus University with services. It has a total budget of MECU 2.74 (HFL 6.3m).

The roots of TAO lie in the government's concern with the social and socio-economic consequences of new information technologies. The aim of the programme is to contribute to a balanced development of new product and process technologies. This means embedding social criteria into projects from their very beginning: from the design process, through undertaking technology assessments, and into the whole process of managing the introduction and use stages.

TAO provides for research being undertaken in a variety of dimensions, all projects being decided on and followed on a cooperative basis between labour and management. First, the following areas are covered:

Industry Services

- Design strategies
- Management accounting
- Personnel policies
- Industrial relations
- ...

Secondly, projects in both areas distinguish two main flows: the one, the renewal and improvement of what exists; the other, innovation.

And, thirdly, in addition to straight-forward case studies, two types of research process are used: on the one hand, long-term cooperation with a single company involved in action research. For instance, a project at the NMB Bank was directed at the integral design of a new information system in stockbroking. On the other, collective research over a period of five years with a reference group of companies experiencing a common issue. For instance, a project is examining the process of factory design and the use of facilitators in the innovation process.

Methodologically, research has to conform to four conditions:

- research has to lead to normative results: outcomes should enable organisations to design future technological innovation in ways which balance with their human resources; research should be preventive, rather than curative;
- research must comply with accepted scientific standards;
- research must be directed towards the organisation as a whole; and
- research must be multi-disciplinary.

Research being carried out in the manufacturing and service areas covers respectively:

Manufacturing	**Services**
- The management of innovation	- Innovation processes
- Overall production restructuring	- Organisation of primary processes
- personnel policy	- Overall design of information systems
- Managerial accounting	- Strategic planning
- Industrial relations	- Redesigning personnel policy and industrial relations

The MECU 2.74 programme budget over five years to 1993 should be doubled through corporate resources.

6.2 COB/SER

A second on-going research programme relating to a large extent to work and technology is run by COB/SER, the Social and Economic Council's joint committee for Business Development. As a successor organisation of the national productivity centre, COB/SER is a jointly-governed (trade unions and employers) committee commissioning research and projects aimed at improving the functioning of enterprises as production and work organisations. Its three main areas of concern are:

- general management issues;
- social aspects of work organisation; and
- technical, organisation and economic management issues.

Several COB/SER projects concern quality of working life issues, with particular concern for helping in the development of various education and training systems.

COB/SER, in addition to its operating budget (covered directly by the SER, the multi-partite Social-Economic Council, receives annually MECU 0.43 (HFL 1m) from each of the ministries of economics and social affairs and employment for its research and projects.

6.3 NIA - The Dutch Institute for the Working Environment

The purpose of NIA, the Dutch Institute for the Working Environment, is to improve the quality of the working environment and, as a possible consequence, of the effective performance of work organisations. The concept of "working environment" embraces the physical, chemical, biological, psychological and organisational conditions in which a job is performed, as well as the job itself.

NIA employs 155 persons (including some 110 professionals) and has a budget of some MECU 7.4 (Hfl 17m). Its four divisions cover information, research, open training and in-house training and advice. The prime focuses are training safety experts, consultancy on working conditions, and research and consultancy on sickness absenteeism.

Research related to work and technology concerns organisational change (action research) and the use of (new) toxic substances.

Addresses

COB/SER, Postbus 90405 NL-2509 LK The Hague

MERIT, Maastricht Economic Research Institute on Innovation and Technology, University of Limburg, P.O. Box 616, NL-6200 MD, Maastricht

NIA, Postbus 5665, NL-1007 AR Amsterdam

7. Belgium

7.1 IACT/IVA - The Institute for Improving Working Conditions

Since 1980, Belgium has been developing a federal structure. However, there remain various institutions and programmes operating nationally, one being IACT-IVA, the institute for improving working conditions. It was founded in 1978 at the same time as a national Work Humanisation Fund (whose mandate was not renewed in 1982).

Like ANACT in France, IACT/IVA is a public body under the tutelage of the ministry of labour and employment with a joint advisory council. Its purpose is to stimulate improvements in working conditions through information, advice and assistance to enterprises, mainly in the private sector but more recently also in the public sector. It does not finance research and only carries out studies which prepare for the activities it undertakes on its own initiative. By "working conditions", it understands:

- job contents and stress;
- physical working conditions and prevention of noise, temperature, toxic, etc pollution;
- behavioural conditions of work: concertation, participation, communication, bureaucracy, etc.; and
- work organisation: work systems (such as shifts, assembly line, part-time), flexibility, new technologies and their impacts.

At present, two thrusts are particularly important:

- working conditions in the health care sector, based on a guide developed by the Institute;
- developing the methodology for tackling problems of working at visual display units, with the help of external experts and the financial support of the European Commission.

It currently has a staff of 20 (below its functional capacity) and an annual budget MECU +1.1 - BEF 50m (exclusively from the ministry of labour and employment - 90% - and sales of publications).

7.2 Research Bodies

In the Flemish Region, STV (the Flemish Foundation for Technology Assessment) was established in 1984 as the social dimension of the Flemish Executive's policy of carrying out the "third industrial revolution in Flanders". STV has an annual budget of MECU 1.1 (BEF 50m) and employs 15 researchers. Its main activities are:

- information on the social aspects of technological innovation;
- research on the social consequences of technological innovation;
- accompanying and advising the social partners (including the Flemish Executive) on innovation processes.

Areas tackled by STV have included involvement from the earliest design stage of the Regional Executive's RITA project of reorganising and introducing information technology in public administration. The STV researchers elucidated questions linked with public administration decision-taking, the relationships between designers and users, likely skill changes and the needs for training and upgrading.

In 1989 the national ministry of scientific policy launched a research programme in the social sciences for the period 1989-93 (MECU 17.5 - BEF 750m), linked to the European MONITOR programme (with its three sub-programmes FAST, SAST and SPEAR-cf "EUROPE" below). This was the continuation of a series of programmes launched in 1975 and subsequently continued through the national FAST programme aimed at: "studying the development of a series of scientific and technological phenomena and evaluating their social impact with the aim of inputing into policies".

Some of the research already carried out has assessed the implications of technological developments in such fields as law, work and employment, teaching and training, and the services sector.

The 1989-93 programme has four aims:

- to develop research tools, notably through setting up supporting centres, "points d'appui" (annual budget: MECU 1.75). These are inter-linked, specialised, inter-university, inter-disciplinary research groups with longer term perspectives which closely collaborate with official data-collection bodies to improve data collection and availability. Areas covered are: employment, work and training; demographic issues; society and health; incomes, poverty and social security; immigration; gender studies; political and administrative system;
- to inform political, economic and social circles about new societal challenges. To this end studies (cf below) are grouped in sub-programmes (total annual budget of MECU 1.95) and target groups (users) are being directly involved in the studies;
- to train researchers, including the provision for international collaboration;
- to make better use of research, ie. improving the process of diffusing knowledge.

The annual administration cost of the programme is MECU 0.25. The topics being examined in the sub-programmes are:

- Corporate socio-economic restructuring (this being a follow-on of the Belgian FAST programme) - MECU 1.1;

- The issue of migrants - MECU 0.45;
- Technology assessment. This programme is looking both back to the consequences of certain decisions made in the past and looking forward to the possible impact of new technologies, as laid down in a national collective agreement. It is closely linked with other European (STOA, MONITOR) and Belgian (STV) programmes - MECU 1.35;
- Social services in the informal sector - MECU 0.95;
- The impact on society of the ageing of the population and of the changing structure of the family - MECU 0.95;
- Social inequality and dualisation in society - MECU 0.7; and
- The legitimacy and effectiveness of public services - MECU 0.95.
- Services to enterprises - MECU 0.95

Addresses

IACT/IVA, 60 rue de la Concorde, B-1050, Brussels

Stichting Technologie Vlanderen, Jozef II Straat 30, B-1040, Brussels

Programme de Recherche en Sciences Sociales Point d'Appui "Emploi, Travail et Formation":

- Prof. M. Alaluf, Institut de Sociologie, ULB, Avenue Jeanne 44, B-1050, Brussels
- Prof. J. Bundervoet, Centrum Arbeids- en Bedrijfssociologie, KUL, E. Van Evenstraat 2c, B-3000, Leuven

8. Denmark

The Danish labour market organisations (trade unions and employers) have always shunned imposed legislation, preferring direct cooperation for improving the quality of working life. Thus, the 1977 Act on the Working Environment aims to create a basis to assist enterprises to resolve their safety and health problems themselves, with guidance being provided by labour market organisations.

The bodies assisting enterprises at the national level are the Working Environment Council (the national overseer of all relevant actions) and the Working Environment Fund, whereas at the branch level there are trade safety councils.

8.1 AMF - the Working Environment Fund

AMF seeks to promote a better working environment by:

- the provision of information: AMF produces a wide selection of books, pamphlets, films and tapes. These strive to be neutral, objective - a recent publication has presented the "psychical work environment". There is also a travelling exhibition, Informobil. An annual prize is given for the best work environment company. And an English language newsletter was first produced in 1990, summarising the significant findings of research financed by AMF (cf below);
- the provision of training: 15,000 participants take part annually in AMF programmes, especially company security representatives; a recent innovation has been the organisation of a seminar on "the working environment and corporate strategy";
- supporting research. The external research in question (AMI projects in particular are financed) is concerned with practical applications to the working environment - the dangers of organic solvents, the design of slaughterhouse knives, etc. Research is reactive (rather than proactive) in character, and specific areas are not selected. Total annual funds amount to MECU 5.1 (DKK 40m), paid for by insurance contributions from all employers (including those in the public sector).

8.2 Trade Safety Councils

The 12 trade safety councils (which cover education and social and health services as well as 10 more traditional industries) have four main functions:

- to survey the problems pertaining to special working conditions of the industry/trade concerned;
- to assist individual enterprises in solving problems relating to working conditions;
- to help to draft industry-specific regulations; and
- to help to draft guidelines for improving safety and health activities within the industry.

8.3 AMI - The National Institute for Occupational Health

AMI, literally "the working environment institute" but termed in English "National Institute for Occupational Health", is a part of the Danish inspection service with university-equivalence. It has five functions:

- information and advice;
- workplace investigations and related activities including toxicological evaluations;
- registrations of chemical substances and products;
- post-graduate education; and
- research.

Its annual budget is of the order of MECU 5.7 (DKK 45m).

8.4 Industri- og Handelsstyrelsen - IHS

The Danish government's main instrument for implementing its policy to promote industry and trade is the IHS, the ministry of industry's "agency for industry and trade". It has a variety of research and support programmes relating to technology and work.

In 1984, the government launched a MECU 180 (DKK 1.4bn) R&D programme, administered by IHS to enhance the use of information technology in Danish working life. In a smaller way, the programme is being extended to 1992 with a MECU 6.4 (DKK 50m) budget for promoting the best possible use of advanced production systems through ensuring that their development takes place within the broader socio-economic context of the company. The programme is supporting action research aimed at:

- improving the connection between advanced production systems and work organisation within companies;
- designing such systems into total corporate strategy; and
- improving the interrelationships between all functional areas involved in the enterprise: development, building, pre-production, production, inventories, service, etc).

In 1988 IHS launched a 3-year, MECU 3.8 (DKK 30m) "Management and cooperation for technological innovation" programme. The programme seeks to enhance the competitiveness of Danish industry by providing grants to companies which directly involve their total workforces in all phases of the introduction of new technologies. The three prongs of the programme concern:

- the development of know-how, analyses and experiments (including demonstration projects);
- the communication of available know-how (including educational material and study groups as well as written and audio-visual material); and
- training and upgrading consultants who help with the concerted introduction of new technology in smaller firms.

As regards sectors, another IHS project (MECU 0.64 - DKK 5m in 1990) is striving to improve the total efficiency of the building sector by focusing on enhanced communications, particularly the interactions between computer systems and smaller firms' building specifications. One significant aspect is the attempt to ensure, by means of a common language, that smaller firms can work efficiently using the most appropriate techniques, technology and organisational systems and yet avoid subsequently becoming embroiled in legal wrangles.

Addresses

AMF - the Working Environment Fund,Vermundsgade 38, DK-2100 Copenhagen 0

AMI - The National Institute for Occupational Health

Industri- og Handelsstyrelsen - IHS The National Agency for Industry and trade, Tagensvej 135, DK-2200 Copenhagen N

9. Norway

Of the variety of bodies and programmes focusing on work and technology, two are outlined here: HABUT and SBA, with indications provided on two other programmes.

9.1 Habut

HABUT represents the Norwegian initials for the "basic agreement on corporate development". It was established under the 1982 Basic Agreement between the confederation of trade unions (LO) and that of employers (currently NHO). It was renewed in 1986 and again in 1990.

The aim of HABUT is to "increase value creation by the broad participation of employees" in their companies. In the 1990 renewal, particular importance is attached to the need to relate the development work more closely in the future to corporate strategy. And more attention is focused on the branch level (good results have been achieved from branch projects in the fishing, textiles and car dealers' branches).

As an organisation HABUT provides funds for jointly designed change programmes within private enterprises. Some 300 projects have been launched since 1982. Most start at a "retreat" where a broad cross-section of the company examines issues of the future (though the greatest achievement is usually that people come together for the first time to review, dispassionately and under little time pressure, their problems and opportunities). When one or a very small number of problems is focused on at these "mapping" or "search" conferences, HABUT can finance the work, part-time, of project personnel. Areas selected for their particular importance in this respect are:

- Quality
- Capital investment and the use of equipment
- Competences
- Organisation development, structures and human relations
- The development and appropriate use of technology for organisation change
- Long-term strategies, including technological policies towards the local community in question
- Investment policies
- Product and market development
- Improving working conditions.

9.2 SBA - the Norwegian Work Life Centre

SBA is the most recent (started in 1988) and all-embracing national institution established to foster corporate economic development through more human organisations. It is a national five year programme jointly sponsored by trade unions (blue- and white-collar), the confederation of Norwegian Business and Industry (NHO) and the government. It has an annual budget of MECU 1.6 (NOK 12.5,) in the form of membership contributions.

Its main purpose is "to promote participative strategies in work life to improve productivity and change, enhance quality performance, to raise competitiveness and thereby secure employment in Norway." Its further aim is "to empower employees on all levels to take a larger role in corporate strategy and organisational development and change". And it constitutes a forum for evaluating the experiences and value of participative forms of work which can be used by membership.

It sees two major shortages to further development in this area: the lack of competent consultants and insufficient training and further education capacities.

As regards project selection, SBA finances, follows and fosters projects in the private and public sectors which fit the following set of criteria:

- where the conditions favour participative strategies;
- where there are channels for diffusion of experiences to other workplaces and work life institutions;
- where the whole organisation is mobilised, with all the professions and parties involved participating in the practical development efforts;
- where results are achieved in the form of increased productivity and change;

- where local research or education institutions work together with the parties in the individual enterprise.

Each project is evaluated by its local results, with the overall programme being evaluated by the sum of the results within the projects, as well as the dissemination achieved.

Twenty-one projects are currently in operation in the private, health care, municipal, the state, and the educational sectors. Importance is also being attached to the development of various networks: of internal change agents, of change managers, of local trade unions leaders and at the international level.

9.3 Other National Programmes

NORAS, the Norwegian Research Council for Applied Social Research, finances research especially directed towards public policy making in the public sector. One programme is LOS to examine and improve management and organisation in the public sector. It is similar in conception to the Swedish LOM programme.

The MECU 13 (NOK 104m) BUNT programme of the Royal Norwegian Council for Scientific and Industrial Research (NTNF) promotes business growth in smaller firms through new technology, having in mind human dimensions.

Addresses

HABUT, Postboks 5250 - Majorstua, N-0303 Oslo 3.

SBA, B.p. 551 Sentrum, N-0105, Oslo 1

NORAS

NTNF, P.O. Box 70 Tàse, N-0801 Oslo 8

10. Switzerland

Between 1983 and 1989 the Swiss goverment financed a national research programme, PNR 15, focused on "The Quality of Working Life".

The aim of launching the programme was to motivate research workers and practitioners to plan and evaluate observations and measures aimed at "humanising work". Work is considered to be "human" if it has the following characteristics: it prejudices neither the health nor the well-being of the workers (or at least only temporarily), it enables individual or collective influence on working conditions, it meets the needs of the worker and it enables him/her to develop personally. In introducing the concluding publication of the project (February 1990), the chairman of the expert group, François Stoll, writes that

> *"the link between the lack of quality of the workplace and the increase in alcohol and drug consumption is too rarely taken into account. If the available means for improving the quality of working life are broadly known, work stations are rarely designed accordingly. Particularly in sensitive sectors, employers are little inclined to give the same importance to quality of working life as to productivity."*

The programme of research covered the following areas:

- work as a value;
- work organisation and new technologies (office computers, distance working, CAD, industrial robots and scanners in retail outlets);
- organising working time;
- health protection in the building industry; and
- management and types of work.

PNR 15 had a total budget of MECU +3 (FS 5m) over a 4 year period.

Addresses

PNR 15, Secrétariat de la Division IV, Fonds national suisse de la Recherche Scientifique, Case postale, CH-3001, Berne.

11. Greece

The Greek Productivity Centre established an Institute of Labour Economics and Organization in 1990.

The activities of the Institute concern research, the transfer of know-how and training. Research will be examining the following areas:

- how labour as a factor of production can be optimally combined with other factors of production, particularly capital and technology, with a view to raising the overall productivity of the Greek economy;
- occupational safety and hygiene and working conditions more generally;
- issues arising out of labour law and labour relations; and
- the impacts of changing demographics.

Advice will be available to the Greek State on subjects relating to employment and to enterprises and organisations.

Addresses

Greek Productivity Centre, 28 Kapodistriou Street, GR-10682

12. Ireland

There is no national "Work and Technology" research programe in Ireland. Whatever activities which are conducted in this field are carried out by EOLAS, the national council for science and technology (which operates in conjunction with the Industrial Development Authority to develop Irish enterprise and is the FAST correspondent in Ireland), and the Irish Productivity Centre.

The Irish Productivity Centre is a joint body governed by representatives of employers' associations and trade unions. It has a series of activities for improving labour-management cooperation, including research projects. Several current projects are based on the concept of "shared learning": twinning non-competing organisations to examine each others' problems in order to find solutions. Projects are currently being undertaken in the following domains:

- Employee involvement in state and private enterprises;
- Integrated rural development to try to ensure that rural settlements are maintained on a viable basis using modern technology and improving the management of workers' cooperatives;
- Breweries and adaptation to technological change; and
- Quality of working life in smaller firms.

Addresses

Productivity Centre, 35 Shelbourne Road, Dublin 4

13. Finland

It is less technology than the management of change more generally (particularly strikes and negotiations) which has led to the Finnish JOY project (a Finnish acronym for "management, organisation, cooperation"). Large factories, particularly in the paper and pulp and engineering sectors, have experienced strikes following from major investments, new equipment and changing working conditions.

Like other Nordic countries, Finland has a law on cooperation within the company. It aims to develop the working environment and increase productivity. For this, research is considered important. JOY is also inspired by the Swedish LOM and Norwegian HABUT projects. It is being developed into a nation-wide framework project, growing outward from the two manufacturing sectors concerned to the municipalities. It seeks to create at the workplace level joint (labour-management) development projects.

The major problem is seen by the JOY administrators to be that of diffusion of available knowledge. There is, they say, an evolution in leadership styles and participation (figure 1) and a need to develop greater vision (figure 2). Research in these respects has a significant normative role.

Addresses

Committee for Labour Relations, Ministry of Labour, PL 524, SF-00101 Helsinki

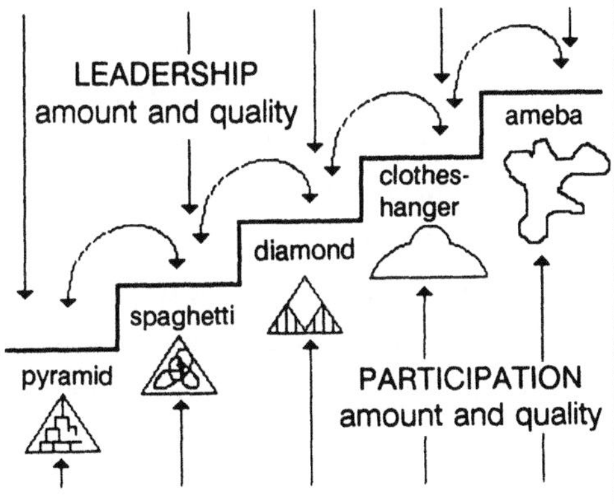

Fig. 1

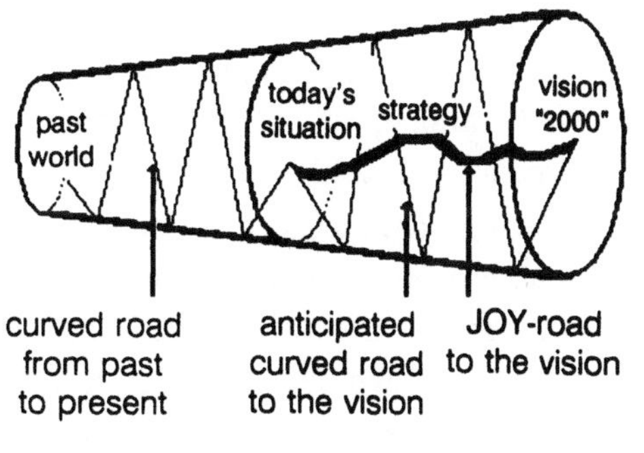

Fig. 2

14. Europe

"Europe" can be seen in two ways: first, pan-European projects, particularly those financed by or closely connected with, the European Community; secondly, trans-national cooperation between institutions or individuals from two countries.

14.1 European Community

Approaches specifically connecting work and technology are apparently of infinitessimal importance in the large Community research projects (which in fact constitute Europe's "industrial development policy"), but constitute the core of the FAST programme and the work of the European Foundation for Living and Working Conditions.

1. **EUREKA** was launched in 1985 to "raise the productivity and competitiveness of Europe in the fields of advanced technology through collaboration among enterprises and research institutes."

   It currently has some 300 projects with an estimated cost of ECU 6.4bn in the following areas of technology: medical and bio-technology; communications; energy; environment; information technology; lasers; new materials, robotics and production automation; and transport. Some individual projects are specifically based on the latest concepts of new production technologies and organisation: 100 new machines produced using the concepts developed in the "Adaptive Garment Manufacturing Unit", aiming to help the industry meet intense international competition, had already been sold after 3 years. HALIOS promotes and improves the development of equipment and new technology for fishing vessels to make them more profitable, more efficient and safer; individual projects include "safety on board and ergonomics" and "safety net" (for overboard survival) sub-projects. Moreover, "supportive measures" have been taken in some of these areas: in the field of lasers, preventive measures are being taken to identify and solve safety and health problems before they actually arise.
2. **ESPRIT** (European Strategic Programme for Research and Development in Information Technology) appears to be almost purely technological. Of projects in operation in 1988 (cf MERIT-TAO, 1989) at least three brandished "human factor" in their titles: the Human-centered CIM project examined the socio-organisational aspects of IT; a second, though discontinued prematurely, produced a set of criteria for the inclusion of human factors in the design of CIM systems and methods of enhancing their usability; and the third, HUFIT - Human Factors in IT, sought to help manufacturers of IT hardware design human factors into office systems.

3. **FAST** (Future Assessment of Science and Technology) was set up in 1978 for a five year period. This was extended for a further five years in 1983 (with a total budget of MECU 8.5), and again in 1988 (FAST-MONITOR). The aim of the programme is to identify new priorities for research and development. The programme has two wings: the "research task" assesses the implications and consequences of scientific and technological developments for the economic and social developments of the European Community. This task was performed in ways characterized as "multi-dimensional" (to avoid technological determinism) and "trans-disciplinary" (the collaboration of "soft" and "hard" sciences). The "organisational task" involves the cooperation between research institutions in Community countries involved in technology assessment. Activities focused on five areas (the first of which is most relevant in the present context):

   1. the relationships between technology, work and employment consisting of three subject areas: critical social groups (farmers, teachers, brain-workers and the over-50s); families of technologies (robots, light technologies, new materials, key technologies); and employment;
   2. the transformation of the services and new technologies;
   3. the function of communication;
   4. the future of the European food system, and
   5. the integral development of naturally renewable resources.

   The **MONITOR** programme (MECU 22) is the continuation of the previous FAST programme. Its aim is "to define more clearly the orientations and new priorities to be assigned to Community Research and Technological Development (RTD) policy while taking into consideration other Community policies". Its three-pronged programme consists of: strategic analyses to formulate proposals for Community action in RTD - SAST; forecasting the development of science, technology, the economy and society and their interrelationships - FAST; and evaluating the efficiency and impact of the Community's programmes on and resources for RTD - SPEAR. Thus, a series of national studies is being carried out on "Anthropocentric production systems", with MERIT making, within the framework of the European MODEM network, "a review of the relevance placed on anthropocentric technologies in national R&D programmes in all of Europe". And FAST has prepared a MECU 250 R&D proposal on Human work in advanced technical environments: A contribution to European competitiveness in the beginning of the Twenty-First Century.
4. The European Foundation for Living and Working Conditions (Dublin) has a governing structure consisting of representatives of national governments, trade unions and employers' associations together with Commission representation as such. Its rolling four year programme has laid down six areas of activity:

1. Developing social dialogue and industrial relations
2. Restructuring working life
3. Promoting health and safety
4. Protecting the environment, the worker and the public
5. Raising the standard and quality of life for all
6. Assessing technologies of the future

A review of the projects undertaken in these fields indicates a monitoring and reactive approach, with no projects seeking, on a multi-national and multi-partite basis, to bring about the simultaneous development of technologies and work structures and organisation.

An overview of the 1989-92 programme was published with the 1989 Annual Report.

14.2 Trans-border Cooperation

In the course of this review only one trans-national venture came to light which is jointly financed by two national research bodies:

A German-Swedish research project was started in 1987 to develop machines optimising working environment conditions in general and minimising the emission of wood dust in particular (for its carcenogenic properties - especially that of beech and oak). The project has been divided into a dozen parts, each trying to improve one type of machine.

The project was based on complementarity: there is virtually no manufacturer of such machinery in Sweden, but considerable use of it, whereas manufacturers are found in Germany. The national programmes (AMFO and the then HdA) were able to provide the direct contacts and part of the funds for the R&D.

It was found that approaches to project management differ between the countries: the Swedish researchers initiated the project and received AMFO finance, whereas the German researchers were designated to work on an already existing project. The results have not only been incorporated into the design of new machinery, but efforts are also being undertaken to renovate existing machines. CNC and FMS approaches still generate dust problems.

Addresses

EUREKA, 19H avenue des Arts, Bte 3, B-1040

MONITOR, Directorate-General Science Research and Development, 200 rue de la Loi, B-1049

European Foundation for Living and Working Conditions, Loughlinstown House, Shankill, Co. Dublin, Ireland.

List of Authors

List of Authors

Gerd ALBRACHT, Ministerialdirigent, Dipl.-Chemiker, Social Affairs Ministry of Hessen, Dostojewski-Str. 4, D-W 6200 Wiesbaden, Germany

J.F. ALLEGRO, Dr., Prof., Dir. Nederlandsinstituut voor Arbeitsomstandigheden, De Boelelaan 30/32, NL-1007 AR Amsterdam, Netherlands

Carl A ASKLÖF, Dir. Arbetsmiljöfonden, The Swedish Work Environment Fund, Box 1122, S-111 81 Stockholm, Sweden

Hans-Jürgen BIENECK, Regierungsdirektor, Bundesministerium für Arbeit und Sozialordnung, Postfach 14 02 80, D-W 5300 Bonn 1, Germany

Lars BJÖRK, Programme Manager, The Construction Industries Organization for Working Environment, Safety and Health, Box 94, S-182 11 Danderyd, Sweden

Siegfried BLEICHER, Mitglied des Vorstandes der IG-Metall, Postfach 11 10 31, D-W 6000 Frankfurt 11, Germany

Karl BOTT, Social Affairs Ministry of Hessen, Dostojewskistr. 4, D-W 6200 Wiesbaden, Germany

Jaap BOONSTRA, Drs., Nederlandsinstituut voor Arbeitsomstandigheden, De Boelelaan 30/32, NL-1007 AR Amsterdam, Netherlands

Monica BREIDENSJÖ, M.A., Head of Section of Working Environment, The Swedish Confederation of Professional Employees, Box 5252, S-102 45 Stockholm, Sweden

David W BUDWORTH, Dr., Co-ordinator, Joint ESRC-SERC Committee, 10 Sydney House, Woodstock Road, Bedford Park, London W 4 1DP, United Kingdom

Ulrich DÄUNERT, Dr., Leiter der Projektträgerschaften des BMFT für Arbeit, Umwelt und Gesundheit, Südstr. 125, D-W 5300 Bonn 2, Germany

Ludger DEITMER, Dr., Dipl.-Ing., Projektträger Arbeit & Technik, Institut Technik & Bildung, Universität Bremen, FB 11, Grazer Str. 2, D-W 2800 Bremen 33, Germany

Peter DOCHERTY, Dr., Institute for Management of Innovation and Technology, Stockholm and Statskontoret, Stockholm, Box 3 41 07, S-100 26 Stockholm, Sweden

Andreas DRINKUTH, Leiter der Abt. Autonomie/Technologie/HdA, IG Metall, Postfach 11 10 31, D-W 6000 Frankfurt 11, Germany

Per H. ENGELSTAD, Prof., HABUT-Programme, Arbeidsforskiningsinstituttet, The Work Research Institute, P.b. 81 71 Dep., N-0034 Oslo 1, Norway

Tomas ENGSTRÖM, Assistant Professor, Ph.D., Chalmers University of Technology, Department of Transportations and Logistics, S-412 96 Gothenburg, Sweden

Ingra FREIGANG-BAUER, Dipl.-Soz., Rationalisierungs-Kuratorium der Deutschen Wirtschaft, RKW, Arbeits- und Sozialwirtschaft, Postfach 58 67, D-W 6236 Eschborn, Germany

Ekkehart FRIELING, Dr., Prof. Gesamthochschule Kassel Universität, Institut für Arbeitswissenschaft, Heinrich Plett-Str. 40, D-W 3500 Kassel, Germany

Wendell FLETCHER, Senior Analyst, Office of Technology Assessment Congress of the United States, Washington, D.C., 20510, U.S.A

David GRAYSON, Dir. Work Research Unit, Advisory Conciliation & Arbitration Service, ACAS, 27 Wilton Street, London SW1X 7AZ, United Kingdom

J. Friso den HERTOG, Dr., Prof. MERIT - University of Limburg, P.O.Box 616, NL-6200 MD Maastricht, Netherlands

Jean-Yves HOCQUET, Deputy-Director of the ANACT, agence nationale pour l'amélioration des conditions de travail, 7, bd romain, rolland, F-92128 Montrouge, France

Antony C HUBERT, Secretary General European Association of National Productivity Centres, Rue de la Concorde 60, B-1050 Bruxelles, Belgium

Christian JACOBSSON, Msc. Occupational Psychology, Department of Psychology, University of Gothenburg, P.O.Box 1 41 58, S-400 20 Gothenburg, Sweden

Fritz-Jürgen KADOR, Dr., Bundesvereinigung der Deutschen Arbeitgeberverbände, Postfach 51 05 08, D-W 5000 Köln 51, Germany

Gisela KIESAU, Dr., Prof., Dir. Bundesanstalt für Arbeitsschutz, Vogelpothsweg 50-52, D-W 4600 Dortmund 1, Germany

Jost-H. von dem KNESEBECK, Dr., Bundesministerium für Forschung und Technologie, Postfach 20 02 40, D-W 5300 Bonn 2, Germany

Hubert KRIEGER, Dr., European Foundation for the Improvement of Living and Working Conditions, Loughlinstown House, Shankill, Co. Dublin, Ireland

Karl KUHN, Dr., Bundesanstalt für Arbeitsschutz, Vogelpothsweg 50-52, D-W 4600 Dortmund 1, Germany

Peter LIDEHÄLL, Dr., Arbetsmiljöfonden, The Swedish Work Environment Fund, P.b. 1122, S-111 81 Stockholm, Sweden

Rolf LINDHOLM, Senior Vice President, Swedish Employers' Confederation, S-103 30 Stockholm, Sweden

Eckhard LÜBBERT, Dr., Ministerialrat, Bundesministerium für Forschung und Technologie, Postfach 20 02 40, D-W 5300 Bonn 2, Germany

Christer MARKING, Arbetsmarknadsdepartementetn, I Arbetslivet, Slottsbacken 6, S-103 33 Stockholm, Sweden

Lars MEDBO, M.Sc., Chalmers University of Technology, Dept. of Transportation and Logistics, S-412 96 Gothenburg, Sweden

Flemming NORRGREN, Gothenburg Center for Work Science, Chalmers Teknikpark, S-412 88 Gothenburg, Sweden

Klaus NORTH, Dr. International Labour Office F/MAN, C.P. 500, CH-1211 Genève 22, Switzerland

Paul OEHLKE, Dr., Projektträger Arbeit & Technik, Institut Technik & Bildung, Universität Bremen, FB 11, Grazer Str. 2, D-W 2800 Bremen 33, Germany

Hans PORNSCHLEGEL, Prof. Projektbegleitung Verbundvorhaben "Logistik" c/o Sozialakademie Dortmund, Hohe Str. 141, D-W 4600 Dortmund 1, Germany

Klaus SCHNEEWEISS, Zentralstelle für Arbeitsschutz bei der Hessischen Landesanstalt für Umwelt, Unter den Eichen 7, D-W 6200 Wiesbaden, Germany

Bo TENGBERG, Officer, Swedish Trade Union Confederation Miljöenheten, S-105 53 Stockholm, Sweden

Rainer THIEHOFF, Dr. Bundesanstalt für Arbeitsschutz, Vogelpothsweg 50-52, D-W 4600 Dortmund 1, Germany

Joseph E. THURMAN, International Labour Office, Conditions of Work and Welfare Facilities Branch, Working Conditions and Environment Department, C.P. 500, CH-1211 Genève 22, Switzerland

Axel WALTER, Dr., Social Affairs Ministry of Hessen, Dostojewskistr. 4, D-W 6200 Wiesbaden, Germany

Werner WOBBE, Dr., Commission of the European Communities, Rue de la Loi 200, B-1049 Bruxelles, Belgium

Klaus WEST, Dr. Wissenschaftlicher Mitarbeiter, Lehrstuhl für Soziologie, Sozialakademie Dortmund, Hohe Str. 141, D-W 4600 Dortmund 1, Germany

GPSR Compliance

The European Union's (EU) General Product Safety Regulation (GPSR) is a set of rules that requires consumer products to be safe and our obligations to ensure this.

If you have any concerns about our products, you can contact us on ProductSafety@springernature.com

In case Publisher is established outside the EU, the EU authorized representative is:

Springer Nature Customer Service Center GmbH
Europaplatz 3
69115 Heidelberg, Germany

Batch number: 10726298

Printed by Printforce, the Netherlands